CATHODE PROCESSES IN THE MERCURY ARC

KATODNYE PROTSESSY RTUTNOI DUGI
I VOPROSY EE USTOICHIVOSTI

КАТОДНЫЕ ПРОЦЕССЫ РТУТНОЙ ДУГИ
И ВОПРОСЫ ЕЕ УСТОЙЧИВОСТИ

CATHODE PROCESSES in the MERCURY ARC

by

I. G. Kesaev

with a foreword by
JEROME ROTHSTEIN

Authorized translation from the Russian

c/b

CONSULTANTS BUREAU
NEW YORK
1964

ISBN-13: 978-1-4684-1553-7 e-ISBN-13: 978-1-4684-1551-3
DOI: 10.1007/978-1-4684-1551-3

The Russian text was published by the State Power Engineering Press in Moscow in 1961, as Trudy No. 67 of the VEI (V. I. Lenin All-Union Order of Lenin Institute of Electric Engineering).

The translation contains revisions made by the author subsequent to the publication of the Russian original.

Кесаев Игорь Георгиевич

Катодные процессы ртутной дуги
и вопросы ее устойчивости

Library of Congress Catalog Card Number 63-17639

Softcover reprint of the hardcover 1st edition 1964
227 West 17th Street, New York, N.Y. 10011

FOREWORD

Although the electric arc is perhaps the oldest field in electronics, it is still one of the least well understood. The beautiful and complex phenomena accompanying the conduction of electricity in gases have led to the development of spectroscopy (and thus of atomic structure and quantum theory), discovery of isotopes, of the electron itself, of X-rays and thus to Moseley's discovery of atomic number, to the discovery of electron emission, and thus to modern electronic technology as well as to much of modern physics. The properties of the plasma state of matter, which until relatively recently could be produced only by the passage of electricity through a gas, have helped explain many natural phenomena ranging from lightning, the aurora, and the ionosphere to solar, stellar, and mushroom-cloud physics. Many plasma devices, such as neon signs, sodium and mercury lamps, thyratrons, ignitrons, and mercury pool rectifiers, have been with us for decades, and more are still being invented. To name but a few recent ones, we have plasma torches, plasma propulsion devices for space flight, and magnetohydrodynamic generators. Solving the problem of containing a high-energy plasma would lead to controlled energy generation by nuclear fusion. Plasma physics is thus more vigorous than ever in spite of its venerable old age.

Kesaev's book is concerned in its entirety with one of the least satisfactorily understood aspects of the relation of a plasma to the electrodes bounding and maintaining it. Without doubt it is a significant contribution to the field. It has both the virtues and shortcomings of a treatment based on classical macroscopic physics. Among the virtues, in addition to detailed presentation of many experimental results, is the fact that future theoretical work would seem more likely to extend, refine, and provide a microscopic or fundamental justification for the relationships he deduces than to overthrow them. The experiments utilize a variety of techniques, including spectroscopy, high-speed photography, oscillographic studies, shaped magnetic fields, and the laborious collection of arc spot lifetime statistics as a function of the many significant parameters. While these techniques are not new, or even pushed near their limits, their combination in pursuit of a single objective is impressive. One can raise an occasional eyebrow over the neglect of some references or about a few impressions given by the text with respect to historical priorities, and it is possible to disagree with the author occasionally in his assessment of previous work, but these are unimportant points in comparison to the real contribution he has made to the field.

The book is written at a level easily understood by physicists and most engineers. Because it deals with a subject which, though old, seems to be continually renascent, and which has great scientific and technical importance, it should be of interest and value to many members of the scientific and technical community.

Jerome Rothstein
Senior Staff Scientist
Laboratory for Electronics, Inc.
Boston, Massachusetts

PREFACE

The investigation of the electric arc and its diverse modifications during the many years since V. V. Petrov's first experiments has produced much outstanding theoretical and experimental work, opening the way to extensive technological application of the arc discharge. Nevertheless, our knowledge of the physical processes taking place in the arc is still extremely limited and inaccurate. This is particularly true of the cold arc, a technically important type of discharge including a wide range of metallic arcs which develop in gas or in vacuum with the help of the vapor from the metal forming the cathode. Lack of exact information on the cold arc is a serious drawback to the development of work on new, more advanced gas-discharge apparatus, making the search for the answers to the questions involved a haphazard process, employing purely empirical methods. By no means least responsible for the backward state of the physics of the phenomena are the obvious lack of system in studying the arc and the absence of any clear ideas as to what should be studied next or what the fundamental problems are. At the present time, in view of the extensive work being done to provide the national economy with new technology, it has become necessary to seek out new approaches and organized methods of investigating the arc discharge. These studies must be of the nature of a planned operation, conducted according to a previously established broad program, and there must be a single guiding idea at the basis of the program to unite the individual aspects of the work. It is clear that before any such program can be set up the existing information on the arc discharge must be systematized and subjected to a critical review, and the investigational work itself must be conducted in intimate relation with practice. An attempt at just such a planned investigation of the cold arc, made by the author in the period from 1957 to 1959 in the Gas Discharge Apparatus Laboratory of the V. I. Lenin All-Union Electrical Engineering Institute, forms the basis of the present book.

In setting up the work, the broad scope of the problem made it necessary first of all to limit the questions most closely associated with it. The investigation was limited to the low-pressure arc with a mercury cathode, one of the most typical forms of cold-cathode arc widely used in practice. The stability of the arc, a problem which in its essential features has so far received no attention, is of central interest in this work. The research program itself and its constituent parts are clear from the plan of presentation and the contents of the individual parts of this book.

The first chapter gives a systematic review of the literature on the cold arc. Including as it does only the most important information on this type of discharge, it is not intended to be complete. The review may be divided into three fundamental parts. The first part gives a discussion, in roughly chronological order, of the results of experimental work on the arc. The second part is devoted to a description of the theories that have been advanced at various times. The third part gives a résumé of the state of the problem of the mechanism of the cold arc, an analysis of the difficulties encountered in solving this problem, and an evaluation of the various possibilities for selecting the most satisfactory of the existing theories.

One source of new information on the arc, completely unused up to the present time, is the investigation of the stability of the arc. Chapter 2 is devoted in its entirety to this problem and begins with a presentation of the work done by the author. It contains information on methods and results of studying spontaneous quenching of the mercury cathode arc, of the oscillational processes occurring at the cathode, and of a number of other transient phenomena in the cathode region of the discharge; this leads to some conclusions concerning the internal instability of the mercury cathode arc. This idea unites the various parts of the work and forms the point of departure for subsequent chapters.

Chapter 3 gives the results of investigations of the structure of the cathode spot on mercury, made by various continuous scanning methods on an enlarged image of the spot. It was possible in this way not only to discover the fine structure of the cathode spot but to follow in detail the changes which it undergoes. It was found that the cathode spot consists of individual unstable cells carrying, on the average, the same current and having a definite lifetime. A consideration of the conditions under which the cells decay and are replaced by new cells has made it possible to establish the existence of a coordination mechanism which controls the decay and formation of the cells. According to these data the steady state of the arc is actually a state of dynamic equilibrium, the processes of decay and formation of cells running in parallel. This point of view leads to a new conception of discharge control; it is developed in the second part of the chapter through an analysis of the transition of the arc from one steady-state condition to another.

The results of the studies made on the structure of the cathode spot lead to the idea that the spot on mercury is in a state of continuous reconstruction, consisting not only in the replacement of decaying cells by new ones, but also the reconstruction of the cells themselves. In view of this the question arises as to whether or not it is possible, on the basis of this idea of the instability of the spot and its continuous reconstruction, to explain its variability on metals. A detailed analysis of the reconstruction of the spot on mercury including the essential role played by the inherent magnetic field of the arc makes it possible to formulate a general rule giving the direction of the reconstruction in a low-pressure arc with a liquid mercury cathode. The applicability of this rule, called the maximum field principle, is shown by a quantitative study of the three principal forms of motion of the cathode spot, namely: ordered motion in a magnetic field, division, and chaotic movement over the cathode (Chapter 4). Thus it is possible to reduce all the various forms of motion of the cathode spot to a general cause, namely, the instability of the spot on cold cathodes.

In presenting this work for publication, the author considers it necessary to add some comments. First of all, the author does not consider all the results of the work, such as the maximum field principle mentioned above, or the ideas used to form a model of the cathode spot which lie at the basis of the principle, as immutable truth. They must be regarded as conclusions arrived at on the basis of a definite group of observations, serving the purpose of systematization. They are also useful in that they concentrate attention on some of the questions which may be very essential in solving the problems of the cold arc. It is quite natural, however, that in the course of further work even these conclusions may be corrected or discarded and replaced by new ones depending on the results obtained when they are confronted by experiment. The investigational process consists in essence in this sort of change and gradual refinement of ideas. In view of this the author will be grateful for critical comments on this work, which can only aid in gaining precise knowledge.

In completing the work and preparing the illustrations, great aid was rendered by technicians E. I. Tsopina, Yu. K. Tsopin, and L. A. Levshenkova, to whom the author expresses his sincere gratitude.

The Author

CONTENTS

Chapter 1

THE PRESENT STATE OF KNOWLEDGE OF THE COLD ARC

1. General Description of the Arc Discharge

Among the various forms of electrical discharge in gases and metallic vapors, the arc discharge has, at the present time, achieved especially great importance from the point of view of engineering application. This is to a considerable degree due to the fact that, unlike silent or glow discharges, the arc is a high-current form of discharge in which the current can be increased practically without limit. However, from the physical point of view, the arc is more accurately described by its other properties [1,2], namely:

1. The discharge is ignited at a relatively low voltage, of an order of magnitude approximating the ionization potentials of gases.

2. The current density is extremely large at the cathode surface, reaching 10^6 A/cm^2 or more.

3. There is a high concentration of neutral and charged particles in the cathode region of the discharge.

The first of these properties led Stark to define an arc as a discharge in which the gas is ionized exclusively by electrons [3]. It follows directly from Stark's definition that a definite fraction of the electrons taking part in the discharge must be created at the cathode of the arc or emitted by the cathode itself. Thus, the processes taking place at the cathode became very important for an understanding of the mechanism of the arc. All subsequent study of the arc has served to confirm the idea that the fundamental mechanism of the discharge responsible for regenerating the charges is concentrated in a thin layer of gas at the cathode surface. Highly suggestive in this respect is the nonuniform electric field distribution between the electrodes of the arc. The fields around the electrodes of the arc are greatly distorted as a result of the formation of large space charges. As in glow discharge, the field distortion in the arc is always of such a nature as to form an especially strong field in the thin layer of gas at the cathode surface, or the so-called cathode-drop region. However, in contrast to glow discharge, where the cathode drop reaches several hundred volts, the cathode drop in the arc is approximately the same as the ionization potential of the gases (or vapors) in which the arc is maintained. The very marked reduction in cathode drop observed on passing from the glow discharge to the arc leads one to the conclusion that substantial changes occur in the mechanism by which the charges are regenerated.

From a quantitative standpoint, the transition to an arc is marked by a considerable increase in efficiency of the charge-regeneration mechanism, as can be seen from the remarkably high current densities existing at the cathode of the arc. In the most general way, this increase in intensity of the regeneration processes in the arc is related to the fact that the arc is a form of discharge which is characterized by high particle concentrations in the cathode region. This is obviously primarily true of neutral particles, because the exceedingly high charge concentrations required for the arc can only be produced if there is a quite high density of neutral particles at the cathode. In the high-pressure arc this condition is fulfilled in a natural way by the atoms in the gaseous medium filling the space between the cathode and the anode. In the low-pressure arc, the peculiar thing about the discharge is that it automatically creates its own medium from local heating and violent evaporation of the metal in the cathode. This latter fact makes it possible to assert that low-pressure arcs do not exist in nature with a cold cathode in the literal sense of the word.[1] In accordance with generally accepted terminology, however, we will use in this book the term "cold-cathode arc" for the type of arc discharge in which no point on the solid or liquid cathode is heated to the temperatures required for appreciable thermionic emission of electrons to occur. This type of discharge necessarily includes all metallic arcs in which the cathode is made of a metal belonging to the low boiling point group. However, this type of arc is also possible with metals having a high boiling point such as tantalum and tungsten. The difference between the two types of cathode is that when using cathodes with a high boiling point the cold arc can, under favorable conditions, change to a hot arc with an electron-emitting cathode, while a change of this sort is impossible with cathodes of the first type.

[1]This statement refers to the normal arc with high current density at the cathode.

In this book we shall be interested principally in data on the low-pressure, cold-cathode arc, for the following reasons. This type of arc has found exceedingly wide application in the form of the mercury arc used in rectifier work. From the point of view of the physical processes occurring at the cathode, this type of arc presents an independent, well-defined group of phenomena, which have so far presented a continuous chain of riddles. In contrast to the high-pressure arc, the conditions under which an experiment is made on low-pressure arcs are of a comparatively simple sort, since the arc is formed simply in the vapor of the metal forming the cathode without any extraneous medium being involved. Nevertheless, the arc is complicated enough even under these conditions. In its structure, the low-pressure arc is found to consist of two regions, greatly different both in external appearance and in the role which they play in the discharge. One of the regions is pressed close to the cathode and appears as a brightly luminous spot, under ordinary conditions performing rapid random movements over the cathode. This form of cathode region in the arc is responsible for the fact that the name cathode spot has become firmly attached to it. The other part of the discharge occupies the greater part of the space between the cathode and the anode and looks like the not very bright diffuse luminescence usually spread over the entire cross section of a tube containing a rarefied gas or metallic vapor. This is the so-called positive column; it plays the role of a simple gaseous conductor connecting the cathode spot with the anode. The purely auxiliary role played by the positive column is apparent from the fact that when the distance between the cathode and the anode is reduced to about 3 cm or less this part of the discharge disappears, while the cathode spot remains on the cold cathode under all conditions as long as the arc discharge continues to exist. On closer examination it is found that the cathode spot can break apart into a number of separate spots, the number of which increases as the current increases. A remarkable property of the arc is that practically the whole current of charged particles passing through the cathode surface of the arc is confined within the limits of these spots. This concentration of the energy field and the particles in exceedingly small portions of the cathode surface must lead to instantaneous boiling of the metal at each of the cathode spots, which is one of the fundamental processes required to maintain the low-pressure arc.

It obviously follows from what has been said that all the peculiar phenomena associated with the arc have to do with processes taking place in the small critical region of the discharge near the cathode, which is ordinarily called the cathode spot. Actually, it is precisely the processes occurring at the cathode spot that make the difference between the arc and other forms of discharge, while the positive column of the arc may in principle be no different from the corresponding part of a glow discharge. Thus, there are reasons to assume that the problem of making a study of the arc, including the mechanism by which it operates and its most characteristic properties, reduces to making an investigation of the cathode spot. This is why exclusive attention is given below to the cathode phenomena in the arc.

Before we pass to a review of the work that has been done in this field, it may be well to define some of our ideas on the cathode spot, and draw a dividing line between arc and spark discharges. Originally, the term "cathode spot" was used for the brightly luminous part of the cathode of a carbon arc, which operates at a very high temperature. Subsequently the term came to be applied to mercury and other metallic arcs. Since in this case the brightest part of the discharge is the negative luminescence, when applied to metallic arcs this term would obviously refer to the part of the discharge that we have been talking about. In contrast to this, some authors understand the cathode spot to be that part of the surface of the cathode occupied by the discharge. In this book the term "cathode spot" will mean the whole cathode part of the discharge, including the active surface portion of the cathode itself as well as the cathode-drop region and the bright negative luminescence adjacent to it. Including the active part of the surface layer of the cathode in the definition is justified by the fact that the discharge phenomena obviously begin somewhere in this part of the cathode. As Loeb [2] has pointed out, as a result of the instantaneous liberation of large amounts of energy in small areas of the cathode, the surface layer of the metal in these areas will be in a vapor state with a large particle density of a sort unusual under terrestrial conditions. It is possible that a definite fraction of these particles are already in the ionized state. In this case it is natural to consider the surface region of the cathode as an essential part of the discharge.

Approximately the same conditions will occur at the cathode of a spark discharge, which is in many respects suggestive of the arc. In view of this the question arises as to where the line should be drawn between an arc and a spark. In this book, all those forms of discharge will be considered to be arc discharges which have low cathode drop and high current density at the cathode, but with the additional condition that the rate of growth of current in the discharge is not greater in order of magnitude than 10^7 A/sec. As is shown by experiment, at current growth rates of the order of 10^8 A/sec and above, the nature of the phenomena at the cathode and in the volume of the gas changes somewhat. Here, in particular, an intense spark spectrum is observed in the luminescence from the discharge, along with spontaneous formation of new cathode spots on the cathode, which show a number of peculiarities of behavior. From all indications, this type of discharge may be classed with spark discharges.

A. PROPERTIES OF THE CATHODE SPOT ON MERCURY AND OTHER METALS

As a result of many years of practical experience and special investigations in the arc-discharge field, a large number of isolated observations dealing with cathode processes in the arc have accumulated. The greater part of these observations have been made in experiments on carbon and mercury arcs, and only comparatively few of them with other metallic arcs. We are justified in concluding from these observations that the cathode region of the arc is the most characteristic part of the discharge and that it behaves in an exceedingly complicated manner and is not easily accessible to investigation. Our immediate problem is to review and systematize the experimental data on the dimensions, the electrical characteristics, and a number of the more important properties of the cathode spot itself.

2. Extent of the Cathode Parts of the Arc

From the point of view of interpreting the physical processes at the cathode it is of interest first of all to inquire into the extent of the cathode regions in an arc discharge. In direct visual observation, the cathode spot appears as a luminous part of the metal surface of the cathode itself, which shows that this region of the discharge is of extremely small thickness. It is very difficult to define it any more accurately since the spot moves rapidly over the cathode in a random way, and the unevenness in the cathode surface is usually many times greater than the quantity to be determined. One of the most refined and successful attempts at making measurements of this sort is described by Smith [4], the results of which are still used in many calculations. The principle involved in the observations Smith made is shown schematically in Fig. 1. A magnetic field tangential to the mercury surface was used to move the cathode spot in a straight line along the axis of symmetry of the cathode, which in Fig. 1 is normal to the plane of the figure, which shows a cross section of the tube. An enlarged image of the negative luminescence g and its reflection in the mercury g' was photographed by means of a microobjective, mounted to one side and making an angle of about 3° with the cathode surface. As a result of the rapid motion of the cathode spot, the picture showed two parallel bands, corresponding to the negative luminescence and its reflection, close to one another. Allowing for the magnification by a factor of eight, in Smith's opinion the distance between the bands represented twice the width of the cathode dark space d, where the cathode drop is concentrated, while the width of each of the bands corresponded to the distance d_g that the negative luminescence extended along the axis of the discharge. From his experiments, Smith arrived at the following values for the two quantities desired: $d = 0.001$ cm, $d_g = 0.004$ cm.

The measurement technique just described, for all its external elegance, suffers from the fundamental defect that it is based on the doubtful assumption that the mercury surface is not perturbed. Actually, as may be seen from the optical system of the direct and reflected rays shown in Fig. 1, what is measured is the location of the luminescence referred to the level of the part of the mercury that is far from the cathode spot. On the part of the mercury where the spot is located there is usually a depression, surrounded by an irregular wall. The assumption made by the author, that at the rate of motion of the spot of the order of 70 m/sec found in the experi-

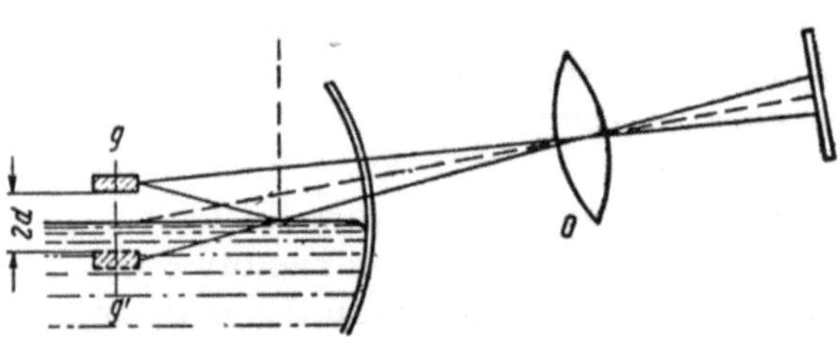

Fig. 1. Method of finding the length d of the cathode dark space in a mercury arc.

ments the distortions in question are completely absent, appears unlikely in view of how small the measured quantity is and in view of the fact that the image of the double line was observed by the author to disappear when the optical system was inclined by more than 4 or 5°. All this leads to the conclusion that the measured values of d and d_g are far from being right. These quantities seem to be so small that in general it would be difficult to set up the experiment so that the irregularities in the cathode surface could be neglected. In view of this, the direct method of measuring the size of the cathode parts of the discharge does not inspire a great deal of hope.

3. Dimensions of the Emitting Surface and the Current Density at the Cathode

The current density at the cathode is one of the most important characteristics of the arc. Knowing the current density in the cathode spot would make it possible to find the length of the cathode drop d and the field gradient at the cathode from the space-charge equation without resorting to a direct measurement of d, which is difficult to obtain. In view of this it is easy to understand the interest which has long been shown in measurements of the surface of the cathode spot. The problem has been comparatively simple when dealing with the carbon arc with its bright fixed cathode spot. In this case an enlarged image of the spot only had to be projected onto a screen with a millimeter cross-section scale on it in order to see what the apparent boundaries were. This is the method that was used by Güntherschulze [5] in 1922; he found that the surface of the spot in a carbon arc operating in the atmosphere was 0.00213 cm^2/A, which corresponded to a current density of about 470 A/cm^2. This figure was not greatly different from earlier values. The new feature in Güntherschulze's work was that he had a correct understanding of how the anode distorts the picture and so eliminated the distortion by moving the anode far enough away. This enabled him to establish the fact that the current is proportional to the spot surface; this led to the conclusion that the current density is constant. At that time the existence of a completely definite current density for any given cathode material was explained by saying that a definite concentration of current is required to get the cathode hot enough for thermionic emission of electrons to occur, the current concentration depending on the thermal conductivity of the material.

Hence it was perfectly logical to assume that the current density in an arc with metallic electrodes would be many times larger.

Finding the dimensions of the emitting surface on mercury and other metals is a considerably more complicated problem, because on cathodes of this type the spot is always moving rapidly and dividing. One of the first rough estimates of the surface of the spot on mercury comes from Stark [3,6], who gave values of the order of 0.02-0.1 cm^2. Almost two decades later Güntherschulze [7] made the first attempt to find the dimensions of the cathode spot on mercury by a photographic method. After some unsuccessful attempts at taking direct photographs of the spot as it was moving at random over the mercury, Güntherschulze was obliged to resort to mirror scanning of the image of the spot. This enabled him to get an enlarged image of the spot on a photographic plate in the form of a greatly elongated band. Assuming the spot to be round, he was able to get values for the spot surface at different currents. As in the case of the carbon arc, it proved to be proportional to the current, from which it was concluded that the current density is constant for the mercury cathode. The value found was 4000 A/cm^2. For a long time this figure was used in all the calculations as being the most reliable one.

The measurements made from 1922 to 1948 made no substantial changes in the value for the current density in the cathode spot. The only measurements in this group that are of any interest are those made on the current density in a fixed spot, described by Tonks [8]. Here, an enlarged image of a spot held in place on the boundaries of various metals wetted with mercury was photographed on a film moving at constant speed. This enabled Tonks to observe several interesting details in the behavior of the spot, which unfortunately did not receive any subsequent attention. In this paper in particular it was convincingly demonstrated that there is a focal structure to the cathode spot, which in the pictures seemed to consist of a number of smaller spots moving around at high speed with respect to one another. Nevertheless, these observations gave a value for the current density of the order of 9000 A/cm^2, only a little greater than the value given earlier by Güntherschulze, which is probably the result of Tonk's use of optical equipment with insufficient resolution.

Some considerable changes occurred in the value for the current density at the cold-arc cathode in 1948-1949 as a result of using new, more highly developed methods of observation. In the experiments made by Cobine and Gallagher [9] the cathode spot on mercury and other metals was given a rapid rectilinear motion by applying a magnetic field tangential to the cathode. In an arc with a mercury cathode, a value was found for the emitting surface using the data from measuring the transverse dimensions of the trace left on a photographic emulsion by a moving image of the cathode spot. Division of the cathode spot, which complicates the measurements under ordinary conditions, was practically eliminated in these experiments by using a strong magnetic field and small discharge currents. Here the width of the band was supposed to correspond more or less accurately to the transverse dimensions of the spot. At a current of 2.6 A the cross-sectional area of the spot proved to be equal to $3.9 \cdot 10^{-3}$ cm. Assuming the emitting surface to be a perfect circle, the authors arrived at a value for the current density on the mercury of the order of $2 \cdot 10^5$ A/cm^2. Using an ingenious perforation device, Cobine and Gallagher also found the longitudinal dimensions of the spot, which proved to be equal to approximately $8 \cdot 10^{-3}$ cm. Taking account of the difference between the transverse and longitudinal dimensions of the spot gave a somewhat smaller value for the current density. Some data on the current density for a number of solid cathodes were derived from measuring the width of the tracks left by the cathode spot directly on the oxidized surface of the metals. Thus, for cathodes made of Pb, Al, W, and Cu, values were found for the current density varying from $0.26 \cdot 10^5$ to $1.2 \cdot 10^3$ A/cm^2.

Somewhat higher current densities for various metallic cathodes were found in studies on high-current pulse discharges in high-pressure media. Thus, Somerville and Blevin [51] give values of $4 \cdot 10^4$ to $1.6 \cdot 10^5$ A/cm^2 for an aluminum cathode. Values of the same order of magnitude for various metals including Sn, Zn, Cd, Pb, Cu, and Ni were found by Emil Žižka [52].

Further progress in finding the current density in the cathode spot is intimately associated with the name of Froome [10-14], who, during the same period of time, carried out a series of excellent studies, in the course of which he used a high-speed Kerr light shutter to take instantaneous photographs of an enlarged image of the spot on mercury and on the liquid alloy of Na and Ca. The apparatus used in this work made it possible to take single or multiple pictures of the spot at 0.4-70 μsec intervals, while the exposure time could be varied from 0.1 to 6 μsec. Since he had enough magnification of the image and such short exposure times, Froome was able to reproduce the cathode spot in all its details in the pictures. Froome came to the conclusion from studying the pictures that the cathode spot on mercury and the Na-Ca alloy is a peculiar sort of bundle, consisting of smaller spots, each of which carries a current of about 1A and moves rapidly over the cathode. The current density which he found for these cathode types lies between $2 \cdot 10^6$ and 10^7 A/cm^2. In addition to this, Froome found that when a tangential magnetic field is applied to the cathode the small spots are drawn out into a chain approximately along the lines of force and move in close order in the direction of the normal to the chain. The thickness of the chain formed is small compared with its length. This explains the difference between the values for the current density given by Froome and those given by previous authors, who assumed that the spot was round.

In view of the great importance which the data on the current density at the cathode have for the theory of the arc discharge, the results of all these experiments are shown in Table I. In comparing these figures, the first thing that strikes the eye is the steady increase, by whole orders of magnitude, in the values found for the current density since the time the first attempts were made to get a value. This change in the values caused Bertele [15] to come to the far-reaching but completely unjustified conclusion that there is a gradual decrease in current density in the cathode spot as it ages. Bertele came to this conclusion as a result of making a formal comparison of the data of Güntherschulze, Cobine-Gallagher, and Froome. He ascribed the difference in the current densities given by these authors to their having observed the spot in different stages of its development. This interpretation of the results of the measurements was so arbitrary that it evoked righteous objections from one of the authors mentioned [16]. It is clear from the above history of the question that the change in our ideas as to the dimensions

TABLE I. Comparison of Experimental Values for the Current Density in the Cathode Spot on Mercury

Year of publication	Sources	Current density, A/cm^2
1922	[7]	$4 \cdot 10^3$
1935	[8]	$9 \cdot 10^3$
1948	[9]	$2 \cdot 10^5$
1948-1949	[10-12]	$2 \cdot 10^6 - 1 \cdot 10^7$

of the cathode spot has been a direct result of improvements in the methods of observation, which have made it possible to discover the fine structure of the spot and get something approaching the true value for the emission surface of the cathode.

It must be emphasized that all the above work uses essentially indirect methods of finding the size of the emission surface. In some cases (on a mercury cathode) it was identical with the surface S_L of the bright luminescence in the cathode region of the arc, while in other cases (on solid cathodes) the dimensions found were based on the eroded surface S_k of the cathode itself after subjecting it to the action of the discharge. But neither value can give a correct idea of the true dimensions, S_j, of the high current density region at the cathode or of the field gradients existing there. As a matter of fact, as we shall see below, S_k is in reality a relatively large region of the metal surface which has been melted as a result of the thermal energy given to the cathode by positive ions. As a result of the energy dissipation in the metal, the dimensions of this region may be tens or hundreds of times greater than those of the surface S_j, which receives the thermal energy. It is here fundamentally impossible to find the relation between the quantities S_k and S_j, since S_k depends principally not on the size of the surface S_j but on the energy supplied to the surface. Matters are no better when it comes to finding the current density from the size of the visible luminescent surface S_L of the cathode spot. First, the cathode spot has a complicated structure and is subject to continuous and exceedingly rapid changes, the rate of which we do not know accurately. As a result of this, the dimensions of the cathode spot found from the size of a photographic image may be greatly exaggerated even with the shortest exposures available in practice. The greater the resolving power of the apparatus used, the smaller will be the surface of the cathode spot derived from the observations. The increase in resolving power of the apparatus used is precisely what is responsible for the steady increase in the values found for the current density in the arc during the last few decades. In itself the progress made shows that this method of finding the current density is unreliable. Second, even if it were possible to get a true instantaneous image of the cathode spot, which Froome apparently came close to getting in his experiments, one would still have to reckon with the possibility of a considerable error in assuming that the luminous region is the same thing as the actual path of the current, which is what determines the maximum value of the electric field gradient at the cathode surface. Thus, an analysis of the methods used to find the current density at the cathode of the arc leads to the conclusion that by their very nature these methods will lead to obviously low values for the current density which do not give any clear idea of the magnitude of the electric field at the cathode of the arc, and this is something that has to be kept in mind when using them to calculate the cold emission of electrons from the cathode.

An additional large uncertainty has been injected into the problem recently by the indications obtained not long ago that there is a form of arc having unexpectedly low current densities at the cathode [17, 18]. In this form of discharge, called a type D arc, the cathode spot is formed on the end of a molybdenum rod surrounded by mercury and located in a magnetic field. From measurements made by authors who have described this form of discharge the current in the cathode spot is only about 10 A/cm^2, and in this respect it is more suggestive of a glow discharge, while the low operating voltage means that it should be included among the arc discharges. It is not at all impossible that in this form of discharge the current is always being rapidly redistributed between isolated centers where the current density is larger, and that these have remained unnoticed because the methods of observation used do not have enough resolving power to reveal them. On the other hand, rapid voltage fluctuations can occur in the discharge, and this complicates the whole picture. These and other possibilities require us to proceed with caution when we hear reports asserting that there is such a thing as a low current density arc, at least until some more extensive proofs are obtained.

4. Magnitude of the Cathode Drop U_c

For quite a long time the most reliable value of U_c for the mercury arc that appeared in all papers on the arc discharge was the figure 5.27 V, which was found about 1905 by Stark and coworkers [3] by using probes. Only after Langmuir and Mott-Smith came up with the exact theory of probes did it become clear that this measurement was in error. One of the first extensive investigations of the potential drop under ionizing conditions in the vicinity of the mercury arc cathode, made using the up-to-date theory of probes, is due to Killian [19]. The value of U_c found by Killian was very close to the ionization potential of mercury, $U_i = 10.4$ V. Similar measurements made by Nottingham [20] in an arc with copper electrodes gave the value $U_i = 20.5$ V, which is equal to the second ionization potential of copper. On the basis of these and a number of other data, the idea evolved that the cathode drop must be at least equal to the ionization potential of the cathode-metal vapor, which plays the role of the gaseous medium in which the discharge occurs. Hence the conclusion was drawn that the fundamental ionization process in the negative luminescence region is direct ionization of the gas by electrons accelerated in the cathode-drop region. Some doubts as to the validity of this conclusion were raised by Lamar and Compton [21], who several years later repeated Killian's measurements using the improvements in technique which he had recommended. The fundamental improvement in these experiments was the confinement of the cathode spot to a small droplike mercury surface, so that some degree of definiteness was introduced into the measurements of the distance between the probe and the cathode spot. At a distance of about 0.4 cm from the spot, the values of the potential in space measured with respect to the cathode were 10-10.2 V, depending on the current, which varied in the experiments from 4 to 11 A. Further from the cathode, the potential in space became somewhat less, which indicated that there was a region of negative field gradient directly in front of the cathode spot. The concentration and energy of electrons was also found to be decreasing in the direction from the cathode spot to the anode, but the decrease was considerably less sharp. The values measured by Lamar and Compton for the cathode drop in a mercury arc turned out to be less than the ionization potential of mercury even when the potential in space was extrapolated onto the cathode surface. All later measurements led to similar results. Thus, the data of Klyarfel'd and Sobolev [159] give a cathode drop of about 9 V for a low-pressure arc with a mercury cathode. Values this low for U_c eliminate the possibility of direct electron ionization of the mercury vapor. As Compton pointed out [22], the probability of the vapor being ionized by electrons with an energy of about 10.4 V or even several volts higher is too small for any such mechanism to produce the high ion concentration observed in the negative luminescence. As a matter of fact, if the energy of the field is expended in drawing electrons out of the metal, the residual energy of the electrons cannot be more than 5.4 V. This is only enough for stepwise ionization of the mercury vapor in the negative luminscence region. In contrast with direct ionization, stepwise processes have a relatively high ionization probability at energies only slightly in excess of the threshold value for the process in question. All this forces us to the conclusion that an important role is played by stepwise processes in ionizing the mercury vapor in the negative luminescence region, in view of which it is more nearly correct to associate the value of the cathode drop in a mercury discharge with the resonance potential than with the ionization potential of the vapor from the cathode metal. This conclusion is supported by the results of studies made on low-voltage arcs.

In spite of the apparently obvious character of the interpretation just given for the value of U_c, it must not be assumed that this is the only possibility or that it is based on well-established facts. In a recent paper [160] on new investigations of vacuum arcs, the author points out that there is a definite relation between the arcing voltage U_g and the thermal properties of the metal in the cathode. According to the data in this paper, U_g is a function of the product of the boiling point and the heat conductivity of the metal in the cathode. The values of U_g found for a number of metals all give a good fit to the same curve. An interesting fact in itself is that deviations from this rule are shown by ferromagnetic metals.

In making a final estimate of the degree of reliability of the existing data on the cathode drop U_c in various metallic arcs, we must take account of the following two facts. First, because of the exceedingly small extent d of the space-charge region at the cathode of these arcs, trying to find the true values of U_c with probes does not really solve the problem. This method only gives the potential in space at some comparatively large distance x from the cathode. Since the desired potential at the point $x=d$ may be appreciably different from the values found, an unavoidable error is made in the measurements. The problem is to reduce this error as far as possible and find out what it amounts to. To reduce the error, the measured values are usually extrapolated to the point $x = d$ or $x=0$. If we consider the properties of the usual gas-discharge plasma we see that in the negative luminescence region, where the charge concentrations are enormous, the potential in space will reach maximum values with a quite high field gradient near this region. If the nearest points in space where the potential is known are fairly far away from this region, extrapolation does not work and probably gives values of U_c that are just that much lower. Second, it must be kept in mind that the potentials found in the space near the cathode are in essence only values averaged over time. As was found by Granovskii and Bykhovskaya [23] the low-pressure mercury arc is a source of various kinds of oscillations. The existence of these oscillations can lead to complications in probe measurements in the cathode region and produces an obvious error in finding the value of U_C, an error which was not considered in the experiments just described.

5. Radiation from the Cathode Region of the Arc

From what has been said above it may be seen that the extremely small dimensions of the cathode spot eliminate the possibility of using probes to make a direct study of this region of the discharge. Under these circumstances, optical, and above all spectroscopic, methods become of special importance in investigating the spot. The spectrum of the cathode spot is a remarkable mixture of line and continuous spectra, apparently coming from the close proximity of regions with markedly different physical properties. The continuous spectrum in the radiation from the spot, discovered by Stark [6], is usually assumed to come from the cathode surface itself or from the part of the discharge adjacent to it. The fact that there is a continuous spectrum in the radiation from the spot was long interpreted as a proof that the cathode is at a high temperature, the more so since, as in temperature radiation of solids, the spectrum extends over and beyond the whole visible range. Smith found [4] that the region of continuous spectrum emission in a mercury cathode arc does not lie more than 10^{-3} cm from the cathode surface. Although the distribution of intensities in the continuous spectrum from the cathode spot is markedly different from the distribution characteristic of black or any other known bodies, a number of authors have attempted to determine the temperature of the spot by using optical brightness pyrometers. In this way temperatures were found for a mercury cathode lying between approximately 1000 and 2000°K [25, 8]. Similar measurements of the cathode-spot temperature in a copper arc [26] gave values of 2000-3400°K, depending on the current, which varied from 4 to 25 A. The fact that the spectral distribution is so different from temperature radiation, and the difficulties of explaining why the temperatures found for mercury were so high, finally caused the temperature version of the origin of the continuous spectrum from the cathode spot to collapse of itself.

A possible way of accounting for the continuous spectrum from the spot is provided by the peculiarities in the line spectrum. It has been shown in a number of observations that the line spectrum from the cathode region of the arc always contains bright lines from the cathode metal [4, 20]. This property of the arc spectrum is very significant, since it provides the most reliable indication that the vapor of the metal in the cathode plays an essential role in the arc-discharge mechanism. Comparing this fact with many other observations leads to the important conclusion that a necessary condition for the existence of a cold-cathode arc at moderate pressures in the surrounding medium is that the arc produce continuous and intense destruction of the cathode itself. As will be seen from what follows, this property of the arc in turn provides the explanation for a number of peculiarities in the behavior of the cathode spot on a cold cathode.

In contrast to the continuous spectrum from the spot, the line spectrum in all probability comes mainly from the negative luminescence region. Bright lines for the usual arc spectrum of the cathode metal predominate; spark spectrum lines are also observed [4, 20]. St. John and Winans found [93] that the intensity of the spark lines increases upon application of a magnetic field, which sets the spot into rapid directed motion. If the field intensity is increased, new spark lines appear, corresponding to higher and higher excitation energies, including lines from singly and doubly ionized mercury atoms as well as a number of forbidden lines. A peculiarity of the line spectrum from the cathode spot as compared with the spectrum from other parts of the arc is that the lines are considerably broadened and there are brighter regions in the continuous spectrum along side of them [93]. These bright regions were found to be symmetrically located with respect to some lines, but unsymmetrically located with respect to others. Starting with Ishida and Hiyama's [27] careful investigation of the Stark effect in the strong electric fields occurring in the cathode drop of a glow discharge, St. John and Winans came to the conclusion that the broadening of the lines and the adjacent parts of the intense continuous spectrum must be related to the effect of the electric field, which, in the region of the cathode spot, changes from small to enormous values. This applies above all to the asymmetric bands in the continuous spectrum. The symmetric bands may be the result of splitting of two different lines with a shift in opposite directions. Starting with the idea that the quadratic Stark effect is predominant, St. John and Winans calculated the possible values for the field gradient in the cathode-spot region from results of measuring the bandwidths. The most probable value for the maximum gradient coming from this calculation was $6 \cdot 10^6$ V/cm, which cannot be considered improbably large.

Thus the broadening of the spectrum lines in the radiation from the cathode spot and the bright parts of the continuous spectrum adjacent to them may actually be the result of the strong electric fields which exist beyond doubt at the cathode of the arc. Some authors [8, 93] consider the weak continuous background extending over the whole visible range to be the type of spectrum (pointed out by Finkelburg [28]) emitted by almost-free electrons when they move in ion fields. A spectrum of this sort always occurs in a highly ionized plasma having an exceedingly large particle concentration and, as a rule, it shows line broadening due to the Stark effect.

In trying to give an explanation for the cathode-spot spectrum we must keep in mind another possibility which was foreseen many years ago by Issendorf [29]. According to the existing concepts [30-33], the positive ions approaching the cathode are neutralized before the ions make thermal contact with the metal surface. Thus it is to be expected that recombination radiation will occur right at the surface of the cathode. A luminescence coming from neutralization of ions at the electrode surface was actually observed recently in some special experiments made on helium ions [34]. Spectral studies on the cathode spot undoubtedly have a great future. If this method is developed, it will very likely be possible to get data on the concentration and degree of ionization of mercury atoms in the different parts of the cathode spot, as well as to clear up the mechanism by which atoms are excited and ionized in the negative luminescence. An experiment of this sort was made for the arc column by Nottingham [20]; it was found that stepwise excitation processes play a considerable role in the column. It may be predicted that, in the cathode spot of the mercury arc with its enormous particle concentration, the ionization will be found to occur exclusively by processes of cumulative type.

6. Evaporation and Atomization of Mercury from the Cathode-Spot Region

The question of evaporation of mercury in the cathode spot first arose in connection with a study of the energy balance at the mercury arc cathode. In previous balance studies, evaporation of mercury was considered one of the principal ways of removing the energy given to the cathode by ion bombardment. This partly explains the attention that was given for a number of years to measuring the rate of evaporation of the mercury in the cathode spot.

Apparently, the first measurements of this sort were described in Schäfer's dissertation in 1910. Work published later gives the evaporation rate found by Schäfer as 36.7 mg/cal. Subsequently, similar measurements were made by Güntherschulze [7], who calculated the cathode evaporation rate from the mercury loss in the cathode reservoir of a mercury tube, the entire loss being ascribed to evaporation from the cathode spot. With currents of from 1.5 to 10 A, the mercury losses increased in proportion to the current. The mean rate of evaporation found from the measurements was 7.2 mg/cal. Compton and Voorhis [30] assumed that the greater part of these losses came from extensive atomization of the mercury by the cathode spot, which had not been considered by Schäfer and Güntherschulze. This conclusion was confirmed by Issendorf [29], who provided his apparatus with means for making a separate determination of the losses from evaporation and atomization of the mercury. An analysis of Issendorf's experiment shows that the separation was far from complete and that the technique used to separate the vapor from the drops inevitably led to errors in finding the evaporation rate. Further, the correction for mercury evaporation from the whole of the cathode surface and the walls of the vessel was many times greater than the amount of mercury evaporated from the cathode spot which he was trying to find. For these reasons, Issendorf had to restrict himself to finding the upper and lower limits of the evaporation rate in the spot; the limits he found were +1.3 and -0.3 mg/cal. The negative sign of the lower limit means that the evaporation from the spot may be so small that it is less than the condensation of mercury onto the spot. The amount of mercury atomized in the form of drops increased nonlinearly with increase in current: approximately from 9 to 15 mg/cal for currents from 2 to 10 A.

A different technique was subsequently used to measure the evaporation rate of the mercury. In order to eliminate random motion of the drop and losses from mercury atomization and evaporation from the part of the cathode surface not involved in the discharge, Kobel [35] made experiments with a spot confined to a small portion of the mercury surface bounded by a tungsten capillary. The tungsten and the mercury were generously cooled with water, and it was possible with this setup to find the mercury losses and the pressure produced by the arc on the cathode in the region of the spot simultaneously. For a current of about 30 A, the evaporation rate was found to be 0.017 mg/cal. Although this is almost exactly the same as that found by Tanberg [36] for the evaporation rate of copper in an arc with a copper cathode, Tonks [37] considered that it was much too low because of the inevitable condensation of vapor in the capillary in the conditions under which the experiments were made. By making similar experiments with a spot fixed at the wetting boundary between mercury and a platinum wire in a more open type of cathode, he found an evaporation rate of about 0.25 mg/cal.

The results obtained from these and a number of other experiments on the evaporation rate of different cathodes are given for ease of comparison in Table II with references to the papers they are taken from. It may be seen by comparison that the figures for solid cathodes are as a rule considerably less than the figures for mercury. This difference is not so much a matter of difference in atomic weight and other constants of the cathode materials as of the error coming from atomization of the mercury. It may be readily seen from the table that as the method of measurement was improved (principally by taking account of and eliminating atomization of the mercury), lower and lower values were, as a rule, obtained for the mercury evaporation rate. When the atomized droplets were eliminated to the greatest extent, as in Kobel's work, the

TABLE II. Results of Measuring the Amount of Metal Evaporated and Atomized by the Cathode Spot in Mercury and Other Metallic Arcs

Year of publication	Source	Metal	Experimental conditions	Weight of metal evaporated, mg per cal
1910	–	Hg	Evaporation and atomization	~37
1922	[7]	Hg	Same	7.2
1928	[29]	Hg	Evaporation	−0.3 to +1.3
1930	[35]	Hg	"	0.017
1938	[37]	Hg	"	0.25
1930	[36]	Cu	"	0.015
1934	[50]	Cu	"	0.007
1934	[50]	Zn	"	0.01
1938	[49]	Cu	"	~0.08

values were practically the same as those found in experiments on solid cathodes. This evolution of the subject leads us to a revision of our original ideas on the role played by evaporation of the cathode material in the energy balance at the cathode of the arc. If we make calculations using the data of Tanberg and Kobel, the energy expended in evaporating the cathode will not be more than 1% of the total energy liberated at the cathode surface. Accordingly, evaporation losses may be completely neglected in the cathode energy balance in an arc discharge.

Some other essential conclusions from these measurements belong to the next section. However, before we pass to these we need to give more detailed consideration to the atomization of mercury by the cathode spot.

As has already been pointed out, the question of the mercury drops arose originally in connection with the need to find the mercury losses from evaporation. Later, however, the question of the drops took on independent interest. The fact that drops are ejected by the cathode spot on mercury should have some intimate connection with the spot mechanism on liquid cathodes, to which, among others, Loeb [2] turned his attention. He assumed that the mercury droplets are formed as a result of local overheating of the metal at the cathode spot followed by explosive boiling of the metal. The drops ejected from the cathode spot can then in turn serve as sources for the formation of new cathode spots when they fall on the negative electrode [38]. Interest in the drops became especially great after it was found that they were able to cause backfires in mercury tubes when they fell on an anode heated by the discharge [39]. In finding the role played by the drops in the backfires and discovering ways of combatting them, it was not enough to have overall data on the mass of mercury atomized by the cathode spot. More detailed information was required on the dimensions, velocity and direction of flight of the drops, and their ability to be reflected from metallic and glass surfaces and to produce backfires. An answer was found to some of these questions in the work of Sena and co-workers [39, 40] and later in Udris' work [41]. It is sufficient here to point out simply the basic data on the process by which the mercury drops are generated in the cathode spot, as found by Udris. By studying the traces left by the drops on smoked glass held at different distances from the cathode, he found that the drops follow a statistical distribution in dimensions and velocities. Both of these distributions have the property that, as the diameter of the drop is reduced, the number and velocity increases monotonically. The most probable initial velocity for a large number of drops under observation is about 3 m/sec, and the maximum velocity reaches 8 m/sec, which corresponds to a distance of flight of up to 3 m above the level of the cathode. The greater part of the mercury is removed from the cathode in the form of drops with a diameter greater than 20 μ. Assuming all the mercury atomized to have what was found to be the mean diameter of 60 μ, and using the overall data of Issendorf, this author calculated that for a current of 6 A the spot ejects on the average about 10^5 drops every second. It is interesting that the data of Golubchin and Sena [40], supported by Udris' observations, show that the mercury drops are not only scattered by a freely moving but also by a fixed spot. If the spot is held fixed, there is a marked reduction in the dimensions of the drops but the number is still enormous. Hence it follows that the amount of mercury evaporating as measured by Tonks is obviously too high. This again confirms the idea that the value given by Kobel must be considered closer to the truth, since in his experiments the mercury atomization losses were held to a minimum.

7. Forces Acting in the Cathode Region of the Arc

The question of the rate of evaporation of the metal in the cathode spot is intimately related to the data on the large mechanical forces which the cathode experiences in a vacuum arc.

Until 1930 investigation of the forces was confined to an arc with a dense gaseous medium filling the space between the electrodes [42]. The result of the experiments was to show that repulsive forces exist tending to increase the length of the arc. Measuring the forces gave consistent values of the order of 1 dyne/A. Accordingly, it came as a surprise when Tanberg [36] showed in his experiments on a copper arc in vacuum that there was a pressure on the cathode of about 17 dyne/A. A force of this magnitude acted on a vane suspended in front of the cathode, tending to push it away. After the experiments, the vane was found to be covered with powdered copper. All this led Tanberg to the conclusion that the force that he had observed was due to the reaction of the stream of vapor emitted from the cathode at high velocity. Having found the loss of cathode material by weighing, he was able to calculate the velocity of the stream. It proved to be unexpectedly high, of the order of magnitude of 10^6 cm/sec. Assuming the stream to be of purely thermal origin, Tanberg calculated the temperature of the vapor in the cathode spot; it was about $5 \cdot 10^5$°K.

At approximately the same time, some similar measurements were made on the mercury arc by Kobel [35]. The discharge pressure that he found on the mercury cathode averaged 40 dyne/A, giving values for the velocity of the vapor stream between $1.6 \cdot 10^6$ and $4.3 \cdot 10^6$ cm/sec. The good agreement between the results of these two experiments produced a lively discussion. The high value found for the velocity and the temperature seemed absurd to the majority of physicists. In addition to other difficulties of explanation, Compton [22] called attention to the disparity between the temperature and the rate of evaporation of the metal. He made the assumption that the experimentally determined amount of evaporated metal M_0 is in essence the difference between the true amount evaporated, M, and the mass of metal ΔM returning to the cathode along with the ion current. If K is the electron current fraction of the total arc current I at the cathode and if h is the electrochemical equivalent of the cathode material, $\Delta M = h(1 - K)I$, and the total amount of evaporation from the cathode is

$$M = M_0 + h(1 - K)I. \qquad (1)$$

The difference between M and M_0 pointed out by Compton gave some reason to think that it is necessary when calculating the reaction force of the vapor to use the total mass M of evaporated metal instead of just the fraction M_0 found in the experiments. With this method of calculation it is no longer necessary to postulate either high velocities for the atoms or the exceedingly high temperatures corresponding to them. Compton himself gave another interpretation for the phenomena under discussion in which the discharge pressure on the cathode came from the emission of ions which were neutralized at the cathode surface. It may be shown from simple considerations [43] that ion bombardment of the cathode by itself cannot produce forces unless all the energy acquired by the ions in the accelerating field of the electrode is completely transferred to the cathode. On the basis of a series of data [30], Compton came to the conclusion that, when ions are neutralized at the cathode, the cathode receives only a part of the energy. Introducing into the discussion the accommodation coefficient g for the ions at the cathode, he found the fraction of the energy remaining in the neutralized ions,

$$10^7(1-K)(1-g)IU_c \text{ [ergs]},$$

and the corresponding total momentum, equal to the desired force,

$$F_p = [2\cdot 10^7 hU_c(1-g)]^{1/2}(1-K)I \text{ [dynes/A]}. \quad (2)$$

Substituting the value $U_C = 10$ V for the cathode drop in the mercury arc and h= 0.00209, using the reasonable values g = 0.9 and K= 0.8, we get $F_P = 41$ dyne/A, which agrees with the value given by Kobel. It may be seen from this calculation that the mechanism given by Compton for the origin of the force can explain the pressure exerted on the cathode by the discharge without considering the reaction force of the re-evaporated atoms. One of the objections raised against this interpretation of the phenomena was that of Slepian and Mason [44], who pointed out that it was impossible in this way to explain the high velocities of the stream of metal vapor found by Tanberg and Kobel, which they calculated to be equivalent to 70 V. Actually, for the emission mechanism given by Compton, the velocity of the reflected atoms would correspond to only an insignificant part of the cathode drop and in any case could not exceed a value equivalent to U_C. That Compton's interpretation of the forces and the nature of the fast metal-vapor streams is inadequate is also shown by the data on the emission forces experienced by the anode [50] and the data given by a number of authors on the existence of anode streams [45, 53].

In comparing the known facts dealing with the velocity of molecular beams in the arc, something that strikes the eye is the following curious state of affairs. Direct measurements of the velocity of the streams made by Haynes [45] and his predecessors [46-48] using optical methods gave values for the velocity of the order of 10^5 cm/sec, i.e., approximately an order of magnitude smaller than those found from data taken on the forces acting on the electrodes in a vacuum arc. They approached the velocities found from measuring the forces in high-pressure arcs, which may come from the fact that the optical measurements were as a rule also made with a dense gaseous medium present. However this may be, the difference causes one to suspect that some error is being made in the indirect method of finding the stream velocity. That such errors are possible is shown by the data obtained by Robertson [49] from an investigation of the forces in an arc with copper electrodes in an atmosphere of nitrogen and hydrogen over a wide range of intermediate pressures (approximately 0.3-500 mm Hg). Robertson tried to discover the reasons for the difference between the results obtained when measuring the forces in high- and low-pressure arcs. By using in his force measurements the sensitive method of pendulum deflection and reading the angle with a mirror, he found the discharge pressure force on the cathode as a function of the current and the gas pressure in the medium. At pressures above 5 mm Hg the forces that he observed did not differ materially from those which had been found in the high-pressure arc in previous investigations. As the pressure was reduced from 5 mm Hg to the minimum pressures used in the work, about 0.3 mm Hg, the force increased considerably but smoothly, approaching the values found by Tanberg and Kobel for the vacuum arc. This served to bridge the gap between the results of previous force measurements on the arc at atmospheric pressure and the new force measurements on the vacuum arc. It also put an end to the doubts as to the correctness of the quantitative results of the force measurements. There was something new in finding how the force varied with the current, the relation having turned out to be approximately quadratic. With a functional relationship of this type the ratio of the force to the current does not remain constant, as had previously been assumed. More than that, in the experiments with nitrogen at low currents (7 A) the force in general became negative, which corresponded to a tendency of the arc to contract. This result suggests that even though the quantitative determination of the force itself does not raise any particular doubts, it would be well to proceed with great caution in giving an interpretation of the forces.

Some possible sources of the forces were discussed by Robertson in the same paper. Even earlier, Tonks [43] had calculated that the partial pressure of electrons nkT_e at possible values of the electron density n and the temperature T_e existing in the negative column of the arc corresponds in order of magnitude to the forces observed experimentally. It is true that the pressure was calculated for 1 cm^2 of surface, while the forces developed in the cathode spot are applied to the extremely small part of the cathode surface occupied by the spot. In this case they would require electron concentrations a million times greater than the concentration used by Tonks in his calculations. If such concentrations are possible, then Tonk's conclusions as to the role played by electron pressure remain in force.

It thus follows that in trying to give an interpretation of where the forces in the arc come from the matter is not as simple as it seemed at the start. As for finding the second quantity needed to calculate the stream velocity, namely, the mass of cathode metal evaporated, it seems that a considerable error is always made in making the measurements. Thus, from Robertson's data the amoung of copper evaporated from the cathode was four times larger than the value found by Tanberg under similar conditions. In the indirect method of finding the velocity at which the vapor flows out under the emission forces, one has to operate with uncertain initial data. It is possible to see in this one of the reasons why such calculations differ from the results of direct measurements of the vapor stream velocity.

The interesting observations made by V. Hermoch [53] indicate another possibility of reconciling the contradictions. By using a mirror scanning method to photograph the luminous vapor streams in a pulse discharge, he observed on his pictures, in addition to the propagating luminescent front, a number of inhomogeneities inside the luminescence itself and propagating at a considerably higher velocity. These inhomogeneities correspond to oscillations in the density of the vapor in the stream, and thus show that the vapor flows out of the metal unevenly. As the author correctly assumed, the velocity of propagation of these inhomogeneities, which he called partial streams, would give a more accurate value for the rate of flow of vapor out of the metal in the steady state of the arc than the rate of motion of the front face of the luminescence as measured in previous papers. He measured the velocity of the partial streams for 17 metals having the most widely different physical constants. In absolute value, the velocity for anode streams was between $4 \cdot 10^5$ and $9 \cdot 10^5$ cm/sec, depending on the anode material. In general, the velocity increased with increase in boiling point of the metal. The velocity of the cathode streams was on the average 10% higher than the velocity of the anode streams. Thus, the measured velocities were of the same order of magnitude (10^6 cm/sec) as those calculated from the emissive forces, in view of which it may be assumed that an end has been put to the long discussed difference between the values for the stream velocity derived from different observations. As for the question of where the streams come from, it has apparently been definitely solved in favor of their being of thermal origin, as shown by the existence of anode streams, as well as by the sum total of experience from investigating electrical erosion of electrodes (see next section).

Summing up everything that has been said about spectra, forces, and evaporation in the cathode region of the arc, we can at the present time feel convinced that a region containing a large density of evaporated metal atoms will exist at the cathode in a normal cold-cathode arc. Apparently, evaporation of the metal plays an essential role in the mechanism of the arc, being a necessary condition for the formation of large space charges at the cathode. Hence it follows that the arc may be regarded as a form of discharge which is maintained by intensive destruction of the cathode itself. This conclusion is in complete accord with the discussion given below of the results of investigating electrical erosion of solid cathodes.

8. Electrode Erosion in Arc and Spark Discharges

The question of the electrical erosion of metals has several different aspects. By reducing the erosion of working contacts and electrodes, it is possible to increase the reliability and prolong the working life of various types of switching apparatus and rectifier installations [54-59].

Further, electrical erosion has acquired considerable importance as a good method of working metals. The question is discussed in detail from this point of view in [60, 61]. Here we shall be interested in the question of metal erosion solely from the point of view of the physical processes taking place at the electrode surfaces and in the region of the discharge adjoining it.

To systematize the results of the observations of different authors in the field of electrical erosion of metals, we must first draw a distinction between two sharply defined groups of experiments: experiments with the electrodes very close together (thousandths of a millimeter) and experiments where the discharge space is comparatively large. Both groups of observations deal principally with a pulse discharge at comparatively high gas pressures in the medium.

A typical case of erosion of electrodes close together is the erosion in the electrical working of materials as well as the destruction of relay contacts. It was found by Germer and co-workers [54-56] that the destruction of contacts observed in various types of relay installations results from the arc discharge formed when the contacts are brought together before they come into direct contact. Here it is almost exclusively the anode that is subject to destruction; each such discharge forms a small shallow depression with a wall around it in the metal. A small bump or projection is usually formed on the cathode at the same time. By combining electrodes made of gold, platinum, and silver, Germer showed from color microphotographs of the electrodes that the bump consists to a considerable extent of metal from the anode. The amount of metal carried over is proportional to the energy developed in the discharge and is about $4 \cdot 10^{-14}$ cm^3/erg. In these experiments the discharge through the test contacts came from a condenser or other capacity C, previously charged to some potential difference U_0. The energy required to transfer this amount of metal was only 2-3% of the total condenser energy $CU_0^2/2$. The form of the depression in the anode depended on the anode material. On a platinum anode, the depth of the pit was about a fourth of the diameter. The volume of the pit increased in proportion to the energy of the condenser discharge and was about $4 \cdot 10^{-12}$ cm^3/erg. Melting this amount of metal required an expenditure of $^1/_3$ of the total energy stored in the condenser. The conclusion was drawn from this that $^1/_3$ of the discharge energy is used in melting the anode metal, while the remaining $^2/_3$ is used in evaporation. The destruction of the electrodes is greatly reduced by putting a resistance or inductance in the circuit. In the experiments described, the distance between the contacts at which breakdown occurred was 10^{-3}-10^{-4}cm, depending on the applied voltage and the type of electrode. The extensive destruction of the anode observed at these distances was subsequently used for spark treatment of metals; particularly satisfactory results were obtained with discharges in a liquid medium.

As the distance between the electrodes is increased, the nature of the erosion changes sharply. This is shown first of all in the so-called erosion inversion phenomenon [62], which consists in a marked reduction in the degree of destruction of the anode and an increase as a result of this in the role played by cathode erosion. In the papers dealing with these experimental conditions [63-65], it was found that both electrodes undergo erosion, but the nature remains different in the two cases. The marks left by the arc on the anode are in the form of small craters of regular circular form, showing that there is a marked contraction of the discharge at the anode, giving current densities of the order of 10^4 A/cm^2 or even higher. The nature of these marks does not change if the anode is moved rapidly relative to the cathode, from which the conclusion was drawn that the arc column is, as it were, fastened to definite points on the anode and moves with it [65]. The anode craters on metals with melting points up to 600-700° C always show traces of melting, the size of the melted regions increasing with increase in current and duration of the arc. For not too short discharges the dimensions of the craters are appreciably greater than the dimensions of the luminous "anode" spot photographed during the discharge [63]. The depth of the craters is usually about $^1/_3$ of the radius. In contrast to this, the marks left on the cathode by the arc are of a more superficial nature and are irregular in form, showing a tendency to branching. They also show signs of melting when the metal in the cathode belongs to the low melting group. The superficial nature and the extended form of the marks found in experiments made with a moving cathode [65] show that the discharge does not stay in them but moves freely over the cathode and has a high degree of mobility. It follows from this that reliable information on the current density at the cathode of the arc cannot be obtained from the size of the eroded parts of the cathode. Both the anode and the cathode parts of the eroded surface become multiple at a current growth rate in the pulse greater than 10^7 A/cm^2, which, however, is not associated with branching of the actual discharge path in the gap.

The signs of metal melting in the eroded parts long ago suggested the idea that the erosion is of thermal origin, similar to the idea argued for by Zingerman [66]. By investigating the temperature distribution in the metal around the part of the surface to which some amount of energy is supplied from the path of the discharge, Zingerman concluded that the source of the erosion is the heat liberated at the electrode from the discharge. A quantitative verification of the role played by the thermal factor in cathode erosion and a refinement of the mechanism by which energy is transferred to the cathode from the discharge is given by Libin [67]; he discusses this problem for the conditions existing in large gas gaps. The experimental part of the paper gives an investigation, made by direct weighing of the electrodes, of the destruction rate of various metals in a pulse discharge in inert gases. As in Germer's experiments, the rate increases with increase in the amount of energy developed in the discharge, and decreases if resistance is put into the circuit. It turned out to be markedly different for some of the metals studied. As the author pointed out, the results of his measurements support the idea that cathode erosion is caused by melting as a result of positive ion bombardment.

Recently the opinion has been expressed that the special features in the erosion of electrodes at small spacings is due to the effect of the so-called torches, i.e., fast streams of metal vapor issuing normal to the metal surface from the points where the discharge path touches the electrodes. The destructive forces of the torches and the direct part that they play in the erosion of electrodes was demonstrated by Mandel'shtam and Raiskii [68]. These authors made a detailed investigation of the relation between the erosion rate and the size of the gap and the electrode geometry as well as the direction and intensity of the torches. A convincing argument is presented in this paper in favor of the thermal nature of the streams and the explosive nature of the metal evaporation, which serves to explain the high velocity and sharp directivity of the streams. In view of the results of this paper it seems probable that the extensive erosion of the anode at small electrode separations is due precisely to the destructive effect of a cathode torch, which is usually of very high intensity.

Thus, it may be seen that the cause of electrode erosion is the predominantly thermal and mechanical effect exerted on the electrodes by the charged and neutral particles moving at high velocity in the discharge gap.

9. Division of the Cathode Spot

As a result of observations made on the high-current arc in mercury rectifiers and other practical equipment, it has long been known that it is possible to have several cathode spots on the mercury simultaneously. Hence the idea has become fixed that at large currents the cathode spot can divide into a number of parts. Güntherschulze [11], after observing the spot with a rotating mirror, concluded that division of the spot begins at currents above 10 A, and that it is a purely random phenomenon. More detailed information on the division of the spot is to be found in Schmidt's dissertation [69], which gives a study of this phenomenon made by taking high-speed motion pictures of a mercury arc cathode. Counting the spots on a large number of frames showed that as the discharge current is increased from 16 to 75 A the number of spots on the cathode increases on the average in proportion to the current. The most probable current for one spot was about 6.7 A. This integrated survey of the data practically exhausts the existing quantitative information on division of the cathode spot.

The division process itself was very clearly demonstrated in the pictures given by Tonks [8] and Froome [12,13], who unfortunately did not give the matter any particular attention. The impression is created by these pictures that division of the spot plays a more important role in the arc processes than is usually assumed. However, the question of the governing laws, causes, and results of spot fission has now drifted completely out of the field of view of physicists.

10. Motion of Cathode Spot

a. Random Motion. One of the peculiarities in the behavior of the cathode spot on homogeneous metals with no external magnetic field is that it continuously executes a random motion as is well-known to anyone who has observed a mercury cathode arc. The random motion of the spot on mercury is assumed to come from local explosive boiling of the cathode produced by large amounts of heat liberated in small volumes of the metal [2,35,69]. This purely fanciful conclusion has never been verified experimentally, nor does it seem convincing in view of the observations which show that the same kind of motion occurs on solid cathodes, including the difficultly fusible ones. An investigation of the random motion of the spot has only begun in recent years. The paper by Schmidt already quoted is especially devoted to this problem. Here for the first time the question is asked of whether or not the motion of the spot is random in the strict sense of the word, and a quantitative study of the motion is made under various experimental conditions. One thing that can serve as an indication of random motion is that the mean square of the distance ρ of the spot from some initial point O on the cathode is a linear function of the time at which the observation is made:

$$\overline{\rho^2} = 4Dt.$$

By analogy with the problem in the kinetic theory of gases, D may be formally designated as the diffusion coefficient of the cathode spot. From a study of a large number of frames showing pictures of the spot moving over the cathode, Schmidt showed that a linear relation exists between $\overline{\rho^2}$ and t for the simplest experimental conditions using a symmetric arc. This seemed to show that the motion was completely random. This conclusion, however, was at variance with the observation that the probability of finding the spot in different parts of the cathode was not constant, especially if the discharge was asymmetric or the currents were large. It was found that the parts of the cathode where the discharge was most likely to be were those with the shortest path to the anode. Attempts to relate this fact to Steenbeck's minimum principle did not meet with success. The observed deviation from a purely random type of motion was attributed to the effect of the magnetic field of the positive column of the arc, which, apparently, is in accord with the data discussed below on the effect of a field on the motion of the spot.

As a direct continuation of this work, the motion of the spot at high currents, when there were several independent spots on the cathode, was investigated [70]. The results suggested that there was a force of the nature of a repulsion acting between them. Unfortunately, the fundamental question of the causes of the random motion of the spot was not even touched on in this work.

The tendency observed by Schmidt for the cathode spot to seek out the parts of the cathode lying closest to the anode has recently been investigated by Kloss [71], who established a number of interesting relations governing the behavior of the spot with direct and alternating current in an arc having either one or two anodes.

b. Directed Motion. Considerably more interest has been shown by physicists in the directed motion of the cathode spot observed when a magnetic field is applied tangential to the cathode. In all probability, the reason for this increased interest in the phenomenon has been the unusual direction, from the point of view of the laws of electrical engineering, of the deflection of the spot by a magnetic field at low pressures in the surrounding medium. Even Stark [72] studied an arc in a magnetic field and turned his attention to the intriguing fact that in a low-pressure arc the cathode spot moves in a magnetic field in a direction opposite to that predicted by Ampere's law. This type of motion subsequently received the name "retrograde motion." This discrepancy suggested the idea that the charges were moving in some unusual direction in the cathode region of the arc. On the strength of this, people became convinced that the retrograde motion was a key to an understanding of the mechanism of the cold arc. It is not surprising that investigating retrograde motion became one of the main directions of attack on the cold arc, as is quite eloquently attested to by the far from complete list of papers [73-98] that we give relating to the question. As we shall see below, the hopes that inspired these investigations have so far not been justified. More than that, even the question of the causes of the retrograde motion proved to be so complicated that recently it has been said [95] that there is no possibility of solving it until the mechanism of the cold arc is discovered.

The data in existence at the present time on the directed motion of the spot are as follows. First, it was found [84, 89, 93] that the rate of the retrograde motion increases nonlinearly up to approximately 100 m/sec with increase in current and field strength, after which saturation of the curve seems to set in. Measurements in fields as high as 20 kOe [88, 93] have shown, however, that for field intensities above 10 kOe the rate increases to approximately 200 m/sec. After an initial sharp rise it continues to increase more slowly with increase in field intensity, showing a tendency to saturate again. Further, a study was made of the effect of various gases introduced into the discharge in different amounts in addition to the mercury vapor [79, 84, 89]. It was found that increasing the mercury vapor pressure or introducing a gas tends to slow up the motion. If the amount of gas introduced is quite large, the motion stops completely, while if the gas pressure is increased further a change in the direction of motion is observed. Thus, at high enough pressures the direction of motion of the spot becomes the same as the direction of the ponderomotive forces acting on the charges. The critical gas pressure at which the sign of the motion changes is a function of the current, the field intensity, and the nature of the gas. As a rule, increasing the field intensity and reducing the current increases the critical pressure [84, 89]. There are indications that the extent to which any given gas affects the rate of motion is determined by the first ionization potential of the gas [84].

The retrograde type of motion of the cathode spot has been observed on all metals that have been tested [9, 79]. It is interesting to note that retrograde motion does not occur if the distance between the electrodes is very small [86, 91]. Also included among the essential facts is the observation [82] that, if tantalum and tungsten cathodes were heated to temperatures high enough for thermionic emission of electrons to occur, the spot could have only its normal motion in the direction prescribed by Ampere's law.

Passing now to a review of the points of view and the theories relating to the nature of the retrograde motion known at present, it is a good idea to divide them into two groups depending on the method of approach to the problem. To the more numerous group belong the theories in [75, 80, 81, 86, 89, 97], in which the motion of the spot is considered to be the direct result of primary deflection of some of the charged particles in the cathode region of the arc by a magnetic field applied externally. Since the deflection always occurs in the direction of the ponderomotive force, the authors of these theories are obliged to resort to more or less complicated, but always artificial, constructions imposed by the thankless job of getting from the normal Ampere deflection of the particles to the diametrically opposite motion of the cathode spot.

To the second group must be assigned the theories in [76, 92], which relate the retrograde motion of the cathode spot to phenomena taking place in the surface layer of the cathode itself, invoking familiar thermomagnetic and galvanomagnetic effects such as those of Righi-Leduc and Ettingshausen. According to these theories, in the metal in the spot, the effect of the tangential magnetic field on the heat fluxes and electric currents being propagated to the interior is to form a large temperature gradient in a direction perpendicular to the current and the magnetic vector. Accordingly, the region where the electron emission is a maximum moves across the lines of magnetic force. As is easily seen, by their very nature these ideas can only apply to a thermal electron type of emission, while the retrograde motion is observed nowhere except in a cold arc. Further, the effects that form the basis of these ideas are too small [77] and cannot explain why the retrograde motion is observed in all metals and moreover only at low pressures in the surrounding medium. In view of all this there is some point in stopping to give more detailed consideration to just the theories which give the reasons for the motion of the spot in terms of phenomena occurring in the discharge itself.

In trying to give an explanation for the retrograde type of motion of the cathode spot, Tanberg [75] suggested that the rapidly moving streams of metallic vapor torn loose from the cathode force the positive ions to move against the electric field, which would cause them to be deflected in the desired direction by the magnetic field. Similar to this is Rothstein's assumption [85] that the conductivity in the dense vapor region at the arc cathode is of a hole nature, which accounts for the current being deflected in the wrong direction. In both[1] cases, accounting for the retrograde motion of the spot required postulating motion of the charges in an electric field in something other than their proper direction, which in itself serves to show the futility of attempts of this sort. A more plausible theory is that of Longini [80], who related the retrograde motion to displacement of the positive space-charge region as a result of the positive ions and electrons being deflected by different amounts in the magnetic field. However, the change in direction of the motion of the spot at sufficiently strong fields, predicted by the author, was in fact not observed, which raises serious objections to his theory.

There is no point here in recounting all the ideas that have been advanced since they have been subjected to suitable criticism in their time [83, 89]. It will be sufficient to consider some of the new theories of the retrograde motion of the spot that have been advanced in recent years. St. John and Winans [89, 93] suggest considering the cathode spot as a peculiar mixture of the positive space charge at the cathode, of the negative surface charge which it induces in the metal, and of the electron cloud occurring above the cathode-drop region (Fig. 2). The electrons emitted normal to the cathode surface are deflected to the right by the magnetic field, i.e., in the direction predicted by Ampere's law. In the region Q they produce intense ionization of the metal vapor. The new positive ions formed here are accelerated by the electrical forces (F) resulting from interaction with the charges in the spot. The authors find that the resultant of these forces is in such a direction as to force the ions to move over the space charge from inertial forces and strike the cathode on the opposite side of the spot, in the region P, causing new emission centers to be formed. This sort of purely ballistic motion of the ions in a space containing a large density of neutral and charged particles does not seem very likely. However, there are more serious objections to the amount of liberty taken constructing the force diagram and the trajectories followed by the ions. Thus, for example, the question arises of why, in the force diagram, the ions are deflected to the cathode in the region P, instead of executing an oscillatory motion above the space charge. It is easily seen that making small changes in the magnitude of the individual forces in the diagram will give prac-

[1] The author errs in classing Rothstein's assumption with Tanberg's since the expected motion for holes is in the direction of the retrograde motion under proper circumstances — Ed.

tically any direction to the resultant, but here the diagrams should be to a first approximation symmetric to the right and left of the spot. A small amount of asymmetry can arise only as a result of the relatively small space charge of the primary electrons, but even this by its nature cannot make the ions fall into the region P. To all this must be added the fact that the ions in the plasma are in general not subject to any action from the electrode field or the space charge which balances it. The radial fields existing in the plasma should cause the ions to move in the direction of decreasing concentrations, i.e., from the center to the periphery of the cathode spot. We are thus led to the conviction that the scheme given by St. John and Winans for the motion of the ions is a last resort. Actually the ions are incident in the space-charge layer solely from the effects of thermal motion as a result of which a large number of ions will strike the cathode at the point P', corresponding to the maximum concentration, and this corresponds to a displacement of the spot in the normal direction given by Ampere's law.

An obviously direct contradiction to the above point of view is the theory published in 1958 by Ecker and Müller [97]. In this paper, which pretends to give a quantitative explanation of the laws governing the retrograde motion, a mechanism (see Fig. 3) is given for the motion of the spot. Here the primary reason for the motion of the cathode spot is taken to be the deflection by the magnetic field of positive ions moving from the ionized region Q

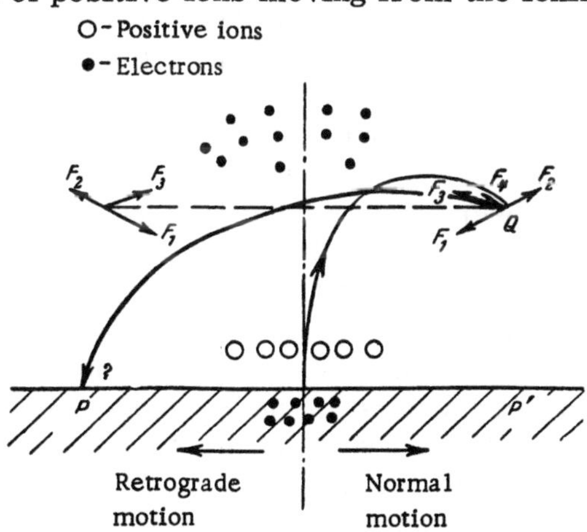

Fig. 2. The origin of the retrograde motion of the cathode spot as given by St. John and Winans.

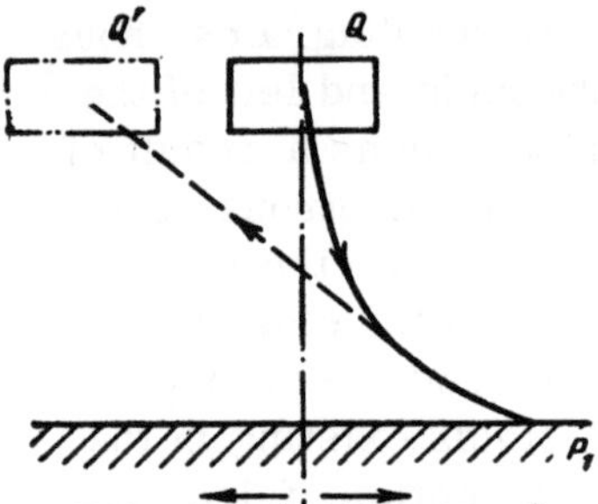

Fig. 3. Origin of the retrograde motion as given by Ecker and Müller.

to the cathode. As a result of the deflection, the ions strike the cathode at P_1, displaced with respect to Q in the direction normally given by the ponderomotive force. After making calculations to find some averaged path of the ions (solid curve, Fig. 3), the authors do not consider it necessary to make any special study of the electron trajectories. Instead it is assumed that the electrons emitted by the cathode at P_1 move along a tube of force produced by the ions, which in its lower half repeats the path followed by the ions going in the opposite direction. Here, in defiance of elementary facts, the authors neglect the effect of the magnetic field on the motion of the electrons, assuming the curvature of the electron trajectories produced by the field to be small in comparison to the curvature of the ion trajectories! It is further assumed that after going through half the cathode drop the electrons leave the tube of force and continue to move, as a result of their inertia, in the same direction, producing ionization of the vapor at the end of the mean free path λ_e in the region Q'. As a result of these unnatural constructions and simplifications, the authors get a displacement of the spot in the desired retrograde direction. By calculating the time required to form the space charge in the newly formed region of the cathode drop, they find the rate of the retrograde motion. The most surprising thing in this paper is that the relations found from these really quite fantastic ideas agree exactly with what is found experimentally. It seems completely absurd that calculations could give correct values for the rate of motion of the spot without taking into consideration the slowest process of all, namely heating the metal in the cathode to the boiling point, since a cathode spot cannot occur in a new location unless this occurs. It is even more difficult to admit that neglecting this slowest of all the processes, as well as some secondary phenomena, such as division of the cathode spot, would make it possible to reproduce theoretically in all their details functions of such a complicated nature as the way the rate of motion of the spot varies with current and magnetic field strength. For the reasons given, the results found in this paper do not inspire confidence, in spite of the surprising way in which they agree with experimental data.

In concluding this section, a short sketch should be given of the ingenious theory of the retrograde motion proposed by Robson and Engel [91, 94]. These authors relate the motion of the spot to the curvature of the path followed by the charge in an external magnetic field. In their opinion, the positive column of the arc, when deflected by the field in the normal direction given by Ampere's law, always makes a sharp bend at the cathode, as a result of which the strength of the inherent magnetic field of the arc is greatly increased in the direction of the bend. The authors assume that under certain conditions the inherent field in this direction can be greater than the external field. Since the direction of the inherent field is moreover opposite to the direction of the external field, the inherent field causes the cathode spot to move in the retrograde direction. The falsity of this idea was shown recently by Smith [95], by making an analysis of the resultant magnetic field distribution around the cathode spot. Smith pointed out further that the retrograde motion of the spot is observed even without any substantial curvature of the arc. While realizing the futility of other attempts made to explain the retrograde motion of the spot, he at the same time apparently gave the correct explanation for the normal motion observed at high pressures. According to the picture he sketched, the forward motion is a peculiar stepping process which is the result of bending of the positive column, after which it joins up with the cathode.

The above review shows that in spite of the large number of attempts that have been made to give an explanation for the ordered motion of the cathode spot in the magnetic field, there is so far no satisfactory theory to account for the retrograde motion of the spot. The experimental data in existence are almost exclusively limited to finding how the rate of motion varies with the different parameters of the discharge and the field strength.

11. Anchoring of the Cathode Spot

It was noted long ago that, if pieces of metal that are wetted by the mercury project out from the surface of the mercury cathode, the cathode spot sooner or later becomes fixed as a thin bright line running along the boundary where the metal is wetted by the mercury. This stops any visible motion of the spot, so that the process eventually came to be called anchoring of the spot. By photographing an enlarged image of the fixed spot on a moving film, Tonks [8] observed that the motion only seems to be stopped. In fact, the luminous line is a series of very small spots executing an odd sort of rapid motion along the boundary where the metal is wetted. From his measurements, each such spot takes a current of several tens of amperes, while the rate of motion reaches values of about 0.5 m/sec. Many metals have the ability to anchor the cathode spot, the metals themselves being to a greater or less degree atomized in the course of time to form metallic films on the walls of the vessel and on the surface of the mercury. Difficultly fusible metals such as Mo and W are relatively difficult to atomize. The ability of metals to anchor the spot depends on the degree of purity and the state of the surface. A special preliminary treatment of the surface is required for some metals. This introduces an element of unreliability into the anchoring action shown by metals. As the current is increased, the reliability of the anchoring action becomes smaller, unless definite conditions are maintained. According to Tonks' data, the spot cannot be reliably anchored unless the linear current density along the anchoring metal is less than 10 A/cm.

The rest of the papers in this field [99,100] are devoted principally to an investigation of possible engineering uses of cathode-spot anchoring and finding the most satisfactory ways of using it. It was found that if the heat liberated at the cathode can be removed satisfactorily, some forms of construction of the anchoring metal are capable of holding the spot up to currents of the order of 100 A. With all this, the anchoring is not distinguished for reliability, and the anchor device itself gradually fails during operation as a result of atomization of the anchoring metal which settles on the internal parts and walls of the apparatus. All this shows that the problem of anchoring the spot cannot be considered to be finally solved. The actual physical processes occurring in anchoring remain just as unclear at the present time as when the phenomenon was first discovered.

12. Conditions Under Which an Arc Occurs

From the point of view of making practical use of the arc discharge in various types of apparatus, as well as in the struggle with arc breakdowns, it is important to have some idea of what the conditions are under which a cathode spot occurs on metals and of what conditions are necessary to maintain it. This question may also be of fundamental importance in discovering the mechanism of the arc. The ways of producing a cold-arc discharge that are known at the present time fit into the following scheme:

a. exciting an arc by bringing contacts together;
b. exciting an arc by breaking the circuit;
c. breakdown of a gap in vacuum or in a gas with a fixed distance between the electrodes;
d. passing from a glow discharge to an arc or passing from a thermionic arc to a cold arc;
e. a cathode spot being formed as a result of mercury drops or dielectric particles falling on the electrode;
f. exciting an arc by passing current through a semiconductor dipped into mercury or some other fused or solid metal;
g. producing a cathode spot on metals brought into contact in a highly ionized gas.

It should be noted that on closer examination some of the above ways of producing an arc seem to be related or even practically identical in the sense of the physical principles that they are based on. This is true for example of the ways of producing an arc given in cases a and c. Nevertheless they are included as independent groups in the scheme in view of their great practical importance and because the fundamental similarity between them is not self-evident but only appears as a result of special investigations.

Our problem is just that of examining the existing data on the methods of producing an arc to find what processes they depend on, and thus get some additional indications as to the mechanism of the cold arc.

a. Exciting an Arc by Bringing Contacts Together. The phenomena that take place when metallic contacts connected to a voltage source are brought close together have been described in detail by Germer et al. [56-59]. By taking oscillograms of the voltage on the contacts as they were being brought together, it was found that, even before the contacts touch, an electrical discharge is formed between them with a striking voltage in the case of silver electrodes of about 15 V. It could be concluded from this that what occurred was an ordinary arc discharge. The voltage observed at the electrodes was not established instantaneously. The time required to set up the arc was about 10^{-8} sec regardless of the distance at which breakdown occurred. From a knowledge of the rate at which the contacts were being brought together and the time from the instant the arc started until the contacts came directly together, the distance between the electrodes at the instant the discharge appeared, and accordingly the field strength in the gap, could be calculated. The field strength calculated in this way for pure silver contacts at breakdown was equal to $2.3 \cdot 10^6$ V/cm. Considering the inevitable increase in field gradient at the irregularities in the metal surface, it is to be supposed that, when an arc is formed between contacts that are being brought together, cold electron emission plays an essential role. It should be emphasized that in these experiments the possibility of thermal emission of electrons taking any part in the breakdown was completely eliminated since the breakdown occurs before the contacts touch and takes place with obviously cold electrodes.

b. Exciting an Arc by Breaking the Circuit. In this sense, the case of exciting an arc by breaking the circuit is considerably less clear. It was long thought that a cathode spot is formed when the circuit is broken as a result of a sudden increase in temperature over small areas of the metal caused by the large transient contact resistance just before the electrodes separate. This idea was subjected to a review by L.A. Sena [101], who showed theoretically that, when a circuit is broken, field gradients can occur at the electrodes which are sufficient to produce cold-electron breakdown of the short gap formed. He made some simple calculations to show that as a result of charging the interelectrode capacity from the voltage source connected to the contacts, and of the continuous increase in distance between the contacts, the field strength in the gap reaches a maximum at some critical distance which, calculated under definite assumptions, is $3 \cdot 10^{-7}$ cm. Here the field strength may reach a maximum value of $3 \cdot 10^{-7}$ V/cm or more. Thus, actually, when the circuit is broken, an electric field will appear at the metal surface which is sufficient to break down the gap.

It thus follows for cases a and b that exciting an arc by bringing metallic contacts together or by breaking a circuit may in essence be considered the ordinary breakdown of a fixed gap in gas or in vacuum. Accordingly, the results given below of studies made on the breakdown of small fixed gaps will give a better understanding of the processes occurring in making and breaking circuits.

c. Breakdown of Gaps in Gas or Vacuum. Electrical breakdown of a small vacuum gap between tungsten electrodes has been investigated in detail in [102], which is marked by a high degree of clarity in the experimental conditions. The authors of this paper point out that before the experiments the electrodes were heated to a temperature of 1800°C, and that the working pressure was held to a level of 10^{-9} mm Hg during the measurements. The authors observed and measured cold-emission currents before breakdown. They established that there is a sudden change in the current as a function of voltage in accordance with the Fowler-Nordheim equation [103]. On this basis the authors came to the conclusion that the currents were due to cold emission. Comparing the slopes of the experimental and theoretical curves for tungsten showed that the field increases by a factor of 11 at the cathode as a result of inhomogeneities in the metal surface, which is about at the limits of probability. Breakdown occurred with a delay of the order of $0.3 \cdot 10^{-6}$ sec after the cold-emission current had reached some value depending on the applied voltage and the distance between the electrodes, which lay in the range of 1-15 mA for the conditions investigated. Using their data as a basis, the authors developed a theory of breakdown according to which arc discharge between electrodes close together in vacuum occurs as a result of appreciable evaporation of the anode produced by predischarge cold-emission currents followed by ionization of the metal vapor to form a space charge which in turn strengthens the cold-emission current. In a study of this paper, definite doubts as to whether the authors correctly interpreted the results of their experiments are inspired by the strange form of the oscillograms showing the predischarge currents and the breakdown itself. Further, a number of purely speculative conclusions are drawn in this paper regarding the evaporation of the anode and the role played by metal vapor in the breakdown mechanism; this causes one to treat the theory developed with circumspection.

Before passing to breakdown in gas gaps, it is well to point out, first of all, what the fundamental problems are that confront physicists in this field. Paschen's law [104] states that the breakdown voltage U_Z of a gas is a function of the product of the gas pressure p and the distance d_a between the electrodes. The basis for the law was the idea that the ionization processes in a gas depend solely on the mean particle energy, which in turn is determined by the ratio of the field gradient in the gas to the pressure.

Subsequent investigation on breakdown in gases showed departures from Paschen's law which consisted in a relation not envisaged in the law between the breakdown voltage and each of the variables p and d_a in the high-pressure range [105-107] and with small separations [55]. The breakdown voltage actually observed was considerably less than what was expected, especially at small separations, where this effect completely changes the nature of the relation between the voltage and the separation. Obviously, under these conditions some unexpected source of electrons comes into play. Germer and Haworth [54, 55] suggested that the reason for the sudden decrease in breakdown voltage at small electrode separations is cold emission of electrons from small irregularities in the metal of the cathode. They assumed that the electrons formed in this way can, by ionizing the gas in the gap, create a space charge which appreciably increases the electric field at the cathode. Since the voltage enters as an exponent in the Fowler-Nordheim equation, the increase in field will in its turn cause a considerable increase in the cold-emission current, causing the gap to break down at a comparatively low voltage. This theory was then developed quantitatively in [108] and an experimental verification for breakdown of a gap in air gave satisfactory results.

It is easily seen that the theories that have been given for arc breakdown of vacuum and gas gaps all in essence have as their basis the same idea that cold-electron emission plays a dominant role. Since there is no medium present to form a space charge in a vacuum gap, the authors of the theory conceived the idea of introducing one by assuming that the metal of the cathode evaporates from the effects of the predischarge cold-emission currents. This practically rubs out any dividing line between the theories for breakdown in vacuum and gas gaps. In the vacuum theories all that is needed to get breakdown is to create a medium ahead of time, which is supposed to correspond with the relatively long breakdown delay time.

d. Transition from a Glow Discharge to an Arc. Transition from a Thermionic Arc to a Cold Arc. The question of producing a cold arc by passing to this form from other forms of discharge can be of particularly great importance in solving the problem of the mechanism of cold cathode emission.

The transition from a glow discharge to a thermionic arc has been investigated [109-114] and treated theoretically by Engel and Staenbeck in [115]. The picture of the transition resulting from these studies is completely clear. As the current is increased, and the glow discharge changes to the anomalous form with a higher cathode drop, there is an increase in the energy liberated at the cathode. If, in the process, the cathode temperature reaches values at which thermal emission of electrons begins to play an appreciable role, the operating voltage of the discharge begins to drop. Further increase in current brings about an increase in cathode temperature and a sudden increase in the role played by thermal electron current. As a result of this process, the voltage drops to the values characteristic of an arc discharge. The whole gradual transition from a glow discharge to a thermionic arc has been followed in detail experimentally and given a quantitative theoretical treatment [113, 115], so that the question of the mechanism of the transition may be considered solved.

There is far from being this amount of clarity in the processes by which a glow discharge changes into a cold arc. A considerable role in forming present-day ideas on the mechanism of the transition has undoubtedly been played by the idea, advanced by Mackeown, that the anomalous glow discharge is unstable as a result of cold emission of electrons from the cathode [116]. The essence of the idea was that if the cathode drop is suddenly raised in an anomalous discharge, various portions of the cathode surface can, in a strong electron field, emit electrons in quantities sufficient to produce appreciable ionization of the gas in the cathode-drop region of the glow discharge. As a result of this, regions of increased charge concentration are formed opposite the active portions of the cathode. Since they have high mobility, the electrons leave these regions more quickly than the positive ions. The increased space charge thus formed at these points further increases the cold emission from the active parts of the cathode. This mutual strengthening process going on between the cold emission of the cathode and the space charge can, in Mackeown's opinion, end up by forming an arc discharge. This idea was later developed by Boyle and Haworth [117] into a semiquantitative theory based on the Fowler-Nordheim equation. The check which they made on the theory for the transition from a glow discharge to an arc in air at pressures of 50-1300 mm Hg gave positive results. In these experiments, the discharge was investigated between palladium electrodes. To make the measurements the current pulses were superimposed on the normal glow-discharge current to convert it first into the anomalous form and then into an arc discharge. By finding the field intensity at which the change occurred, the authors were able to get a comparison between experimental and calculated values. The most important evidence in favor of the theory is not so much the agreement between the measured and calculated values of the field strength as the solid basis given to the fact that, for all pressures, there is just one critical field strength of about 10^6 V/cm at which the glow discharge changes into an arc.

The transition from one form of discharge to the other has been treated from a somewhat different point of view by Bauer [118], who investigated the phenomena in the region of transition from a glow discharge to an arc, including the transition from the thermionic to the cold arc. This range of transition phenomena may be represented schematically as a triangle with the glow discharge at one corner, the thermionic arc at another corner and the cold arc at the third corner. Using the equation derived by Murphy and Good [119] for the electron emission of metals under the combined effect of an electric field and high temperature, Bauer was able to obtain volt-ampere characteristics of an arc with tungsten electrodes and compare them with the experimental curves. At low pressures in the medium (xenon), the phenomena observed were typical of the transition from a glow discharge to a thermionic arc, while at high pressures (20 atm and higher) they were typical of the transition to a cold arc. Bauer gave a qualitative description of the discharge contraction mechanism in the transition to an arc, in which the most important role is played by the strong dependence of the cold-emission current on the field strength. The results of Bauer's work give a good explanation of the earlier observations on arcs with and without cathode spot, and how they change from one to the other [120].

It may be seen from this short review of the results of studies made on the transition from a glow discharge to a cold arc that the transition may be satisfactorily accounted for by assuming that the arc is produced by a cold-emission mechanism. It must be kept in mind that as a matter of fact the transition can occur in such a way as to involve a number of factors that cannot be taken account of exactly, such as liberation of gas from the metal, as has been pointed out by Maxfield et al. [121]. However, beyond any doubt the secondary phenomena are by themselves in no position to provide an explanation for the transition to the arc, and accordingly they do not contradict the conclusion that the transition occurs by a cold-emission mechanism.

e. Other Methods of Exciting an Arc. Since there is no clear understanding of what happens when foreign objects fall onto the cathode, or when an arc is excited by passing current through a semiconductor, it is too early to give any detailed discussion of the processes involved. We shall limit ourselves to a short review of the observations that have been made in this field.

It was noticed long ago that mercury drops falling on the hot anode of a rectifier are a cause of backfires. This phenomenon has been investigated in detail [39-41] and the results indicate that the reason for the formation of a cathode spot on the anode is that there is a region of high vapor density at the metal surface. Apparently we are dealing here with the trivial case of transition of a glow discharge in mercury vapor to an arc. The formation of a cathode spot at the boundary between a dielectric and a metal, described in [122-124], belongs to a somewhat different class of phenomena. The experiments show that there are some conditions which make it much easier to produce the spot between mercury and the dielectric, but so far there is nothing but conflicting information as to what these conditions are, or what role is actually played by the dielectric, so it is just too early to discuss the question.

At the present time, great practical importance has been achieved by the method of starting the arc where current is passed through a semiconductor partially immersed in mercury. This method was patented as early as 1904 and was used later by Slepian and Ludwig [125], who concluded that the strong electric field produced at the mercury meniscus on the semiconductor plays an important role in the arc excitation mechanism. Ideas on the mechanism of cathode-spot formation at the semiconductor igniter have been advanced by other authors [126], but they are neither clear nor convincing. Experimental studies on this method of excitation have been made [127 and 128].

In conclusion, we should mention the possibility of forming

a cathode spot on metals in contact with plasma, an example of which is provided by the formation of cascade discharges, as well as the re-ignition of the working arc in rectifiers after it has been extinguished as a result of sudden changes in anode current [123]. It is to be expected that the processes taking place here are similar to the transition from a glow discharge to a cold arc.

In summary, we wish to emphasize that the majority of the known methods of exciting an arc may be satisfactorily explained qualitatively, and in a number of cases even quantitatively, on the assumption of cold emission of electrons by the cathode. This explanation provides a common feature in all the observed cases of arc excitations and enables them to be considered in essence as involving the same process by which the cathode spot is formed in a strong electric field which only develops under the various conditions existing at the start of the experiment.

13. Conditions for Arc Decay

As was done above for exciting an arc, we can now give the following classifications of the conditions under which the cathode spot decays and the arc is extinguished:

a. extinction of the arc from reduction or change in direction of current;
b. extinction resulting from shunting the arc with a plasma having high charge concentration;
c. decay of the arc resulting from destruction of the cathode;
d. spontaneous extinction of the arc.

a. Extinction of Arc from Reduction or Change in Polarity of Current. It was early noted that if an arc with a metallic cathode (of the mercury or copper type) is suddenly disconnected from the voltage source for even a fraction of a millisecond, the discharge stops instantly; this is not the case in an arc with carbon electrodes. It was concluded from this that the cathode emission in metallic arcs does not come from the cathode being at a high temperature. The conditions that tend to extinguish the discharge have been investigated in detail in a number of papers. Quite early studies of this sort showed [129] that the arc decays if the off-time is as short as about 10^{-7} sec. Later, the measurements were repeated by Mierdel [130], who found a considerably smaller value for the extinction time of the arc, of the order of 10^{-9} sec. This value has been used up to the present time in making various types of calculations and arriving at conclusions relating to the nature of the process occurring during decay of the cold arc [85, 123, etc.]. Meanwhile, an acquaintance with the technique of investigations used by Mierdel and the assumptions that he made in calculating the extinction time of the arc leads to the conclusion that his value in no way comes from the results of his actual measurements and is probably much too low. In his experiments, Mierdel extinguished the arc with a damped oscillatory discharge from a condenser which had previously been charged from a voltage source of the opposite polarity. By measuring the voltage and the capacity of the condenser he was able to determine the conditions under which decay of the arc occurred. A preliminary check showed that the extinctions only occurred for values of the discharge current close to zero. Assuming that the arc is always extinguished by the first negative half-wave of the oscillatory condenser discharge, and neglecting damping, the author wrote the condition for extinction of the arc at any discharge current I_a in the form $U_0 = R_V I_a$, where U_0 is the initial voltage on the condenser and R_V is the characteristic impedence of the discharge circuit used. This made it possible for him to calculate the unknown value of the inductance in the circuit, which was made up of the inherent inductance of the wires and the discharge. This gave the period of the oscillations T, which Mierdel found to be 10^{-7} sec for the conditions under which the arc was extinguished. It may be seen that these measurements gave values for the extinction time agreeing in order of magnitude with the values derived from earlier experiments. The figure 10^{-9} sec derived from them by the author was arrived at by making some rather doubtful corrections. In arriving at the final figure for the extinction time of the arc, he points out first that it is only the first negative half-wave of voltage that extinguishes the arc, i.e., the extinction time is not more than a quarter-period (since the current has to drop to values close to zero). After this he arbitrarily reduced it by another order of magnitude under the assumption that the discharge can only decay at a current near zero. Obviously, an approach of this sort to the calculation of the arc extinction time introduces an element of arbitrariness into the question. It is first of all unknown whether or not extinction depends on the subsequent current oscillations and the damping of the oscillatory discharge. This possibility is indicated by the very way in which the results of the experiment depend on the type of commutating device used. Further, the period of the oscillations is not known accurately, and neither is the value of the current at which extinction of the discharge occurs. All this leads to the conclusion that the arc extinction time of 10^{-9} sec given by Mierdel needs to be reconsidered.

Judging from the references in the literature, Mierdel's data giving the exceedingly small arc extinction time of the order of 10^{-9} sec were later substantiated [131]. Unfortunately, the author of the present review has been unable to study this work. However, the extinction time measurements made recently by Mantrov [132], using Mierdel's method, have again given values of the order of 10^{-7} sec as found in the first of the papers cited [129]. Some additional data on the extinction characteristics of the arc are contained in [133], where a study was made of the extinction current as a function of the duration of the extinguishing pulse, and dynamic volt-ampere characteristics were taken on the discharge. This investigation established that the current at which the arc decays is different from zero and depends on the duration of the extinguishing pulse, becoming less as the pulse becomes shorter. For short interruptions of current, the voltage across the discharge dropped to 5-6 V, from which the author concluded that the electrons drawn out of the cathode cannot overcome the potential barrier.

The study of arc extinction from sudden reduction or change in polarity of the current has taken on great practical importance at the present time in connection with the problem of eliminating extinction of the operating arc in high-power mercury tubes. Extinction under these conditions has been investigated in [123, 132], where it has been possible to observe re-ignitions of the arc, which is why decay of the cathode spot does not always end up by extinguishing the arc.

b. Other Types of Arc Extinction. Reduction or change in polarity of the discharge current is not the only reason arcs are extinguished. A sudden temporary increase in current followed by a rapid drop-off will do the same thing. Phenomena of this sort have been observed in investigating the extinction of the operating arc in high-power mercury tubes [123]. A detailed investigation [123,134,132] led to the conclusion that in this case the arc is extinguished because it is shunted by a highly conducting plasma. If the voltage on the arc electrodes is reduced to values that are too low to maintain the cathode spot, the shunting effect causes the spot to decay. However, as in extinctions of the first type, the spot may, under favorable conditions, form again, giving re-ignition of the arc. Thus, in extinctions of the first type the spot decays because the current is reduced, while in the second case it may be considered to decay because the voltage is reduced below the critical value.

A special category must be formed to include the cathode-spot decay phenomena resulting from the spot destroying the cathode itself, or from disturbing the conditions on the metal surface that are required to maintain the arc. This situation arises for example in arcs with oxidized copper or silver cathodes [135], in which the cathode spot can only exist at currents up to 10 A as long as there is an oxide film on the metal. As soon as the film is destroyed by the cathode spot, the arc goes out and cannot be started again. The same category of phenomena includes extinction of the cathode spot which is formed when mercury drops fall on a hot negative electrode. It was found [136] that if the current in the arc formed under these conditions is limited by the external circuit so that it does not exceed 2-3 A, the cathode spot decays as soon as the mercury in the drop is completely evaporated by the cathode or by the discharge.

The cases of arc decay considered above are alike in that the decay results from something external acting on the discharge space or on the cathode surface. It is well known, however, that arcs can also be extinguished without anything acting from outside and without any appreciable change in the conditions existing in the discharge itself. Spontaneous extinctions of this sort are usually observed with a mercury cathode discharge at currents below 7-6 A and become more frequent as the current is reduced. A study of the distribution of time intervals between arc extinctions has shown that the phenomenon is apparently of a random nature and that in the narrow range of currents investigated, from 3 to 6 A, the mean duration of the arc decreases approximately exponentially with decrease in current [137,138]. These experiments made use of the primitive method of reading the time interval between extinctions with an ordinary second indicator, which kept the authors from making observations at the lower currents and was bound to put a serious limit on the accuracy of measurements. Further, using a glass vessel for the discharge tube made any rigid control of the cathode temperature impossible, and this could lead to substantial errors including, in all probability, the conclusion of the authors that the duration of the arc is a function of the cathode surface area.

Spontaneous extinctions of the arc show up the inadequacy of the existing phenomenological theory of discharge stability [139] as applied to the mercury arc. What the extinction consists of remains unknown up to the present time, and this makes it problematical whether or not any correct understanding can be gained of the types of arc decay just discussed. Actually, for all these arc-decay phenomena to be intelligible under articial conditions in all their essential details, we must first find what extinguishes an arc that has been operating for some time and kept under constant operating conditions. But this can only be achieved by making a detailed study of the physical processes that accompany spontaneous arc extinction. This question is intimately related to the question of the stability of the cold arc and should occupy a considerable place in subsequent investigations.

B. DEVELOPMENT OF IDEAS ON THE ARC-DISCHARGE MECHANISM

The fundamental problem in the theory of the arc discharge is to find the mechanism which, operating in the cathode region of the arc, forms the electrons and ions required to maintain the discharge. The question of the mechanism of the arc already has a half-century of history. It may be said without exaggeration that, from the days when Mitkevich [140] and Stark [141] independently came to the conclusion that the arc is thermionic in nature, almost as many theoretical models have been advanced as there are ways known at the present time for liberating electrons from neutral bodies. Nevertheless, this wealth of ideas has still not led us to the desired result in the form of a finished theory verified by experiment. The critical review of the existing ideas given below will help to show the reasons for this, as well as what the present state of arc theory is.

14. The Thermionic Theory of the Arc and Its Associated Difficulties

The original theory of the arc related the passage of current in the discharge space with the ability of the cathode to emit electrons at a high temperature, which could be produced by heating the cathode artificially or by the bombardment of the cathode with the positive ions resulting from ionization of the gas. The thermionic theory could account for all the phenomena observed in the arc discharge as long as it was confined to an arc with carbon electrodes in the atmosphere; examples are given by Compton [142] as well as in previous reviews [143]. Moreover, the first time the theory was applied, it seemed that it could be extended even to metallic arcs in which the entire mass of metal in the cathode remains relatively cold. For this purpose it was sufficient to assume that high temperatures exist in microscopic volumes located at the surface of the metal in the region where the discharge takes place. Some observations, however, threw doubt on the possibility of applying the thermionic theory to metallic arcs. Stolt's experiments [144] played a particularly important role in this; Stolt showed for the first time that the cathode spot can move at high speed over a copper cathode, so that it was impossible to have any great heating of the copper even at the small parts of the surface occupied by the spot. At the present time, since the almost fantastic values of current density in the cathode spot have become known, arguments of this sort have lost their force. A much more serious objection to the universal applicability of the thermionic theory was advanced by Slepian [145], who pointed out that the majority of metals cannot be heated to high enough temperatures to produce appreciable emission. This is especially obvious with respect to such metals as mercury, copper, and silver. In searching for a way out of the difficulty Güntherschulze [7] suggested that the boiling point of the metal in the cathode spot is so greatly increased by the increase in local vapor pressure that the metal can be heated to high enough temperatures for electron emission to occur. Güntherschulze saw some support for this guess in the fact that the cathode spot in the mercury arc emits a continuous spectrum in addition to the line spectrum. On the basis of his calculations he allowed an increase in pressure at the cathode spot of up to 2-3 atm. The futility of this attempt to rehabilitate the thermionic theory was demonstrated by Ramberg [143].

15. Cold-Emission Theory of the Arc

Somewhat earlier, Langmuir [146] pointed out a new possibility of explaining the arc mechanism, which was the starting point for one of the most popular and productive theories of the cold-cathode arc. On the basis of the equation which he had derived for the space charge formed in the plasma at the negative electrode, Langmuir came to the conclusion that the space-charge field at the cathode of the arc may be large enough to pull electrons out of the metal by lowering the potential barrier. A detailed theoretical check of this conclusion as applied to the mercury arc was undertaken by Mackeown [147]. The fundamentally new aspect of this work was that it took into account the effect which the electrons emitted by the cathode have on the space charge at the cathode and the field resulting from it. It should be noted that since there are no trustworthy experimental data on the extent of the space-charge layer, one of the ways of checking the validity of the cold-emission theory of the arc is still to calculate the field at the cathode surface from the current density data by means of the space-charge equation. It is known from Langmuir's theory that the space charge at the negative electrode is composed of positive ions moving toward it from the plasma, the density of the ion current being determined by the ion concentration in the plasma. The electrons emitted from the cathode to a greater or less extent balance out the positive space charge, so that the results of calculating the field strength depend on how large a part of the current is being carried by ions. The neutralizing effect of electrons may be practically complete if the electron and ion current densities are related as the square roots of the masses of the ion and the electron. Hence it follows that in the mercury arc the ion current fraction $(1 - K)$ in the total current balance at the cathode will in any case be greater than 1/600. The maximum possible value of $(1-K)$ may be found from a consideration of the ionization efficiency in the negative luminescence region. The principal ionization process in the negative luminescence region was long thought to be ionization occurring from single collisions of mercury atoms with electrons accelerated in the cathode drop, the reason being the apparent agreement between the values of the cathode drop and the ionization potentials for some cathode materials. In view of the low efficiency of these ionization processes at voltages in the vicinity of the ionization potential there was little probability of values of $(1 - K)$ much in excess of 1/600. If the more efficient stepwise ionization processes are brought in, the maximum possible value of (1 - K) increases to approximately 1/4 [22]. The energy balance method of [22,153] leads to similar values.

In view of this discrepancy, the lack of data on the composition of the current in the cathode-drop region introduces a considerable uncertainty into the problem of calculating the field, so that such calculations have to be made in such a way as to include all the possibilities that may occur in the actual composition of the current at the cathode. This fact was taken into consideration by Mackeown, who calculated that the space-charge field at the cathode surface can reach values of $5 \cdot 10^5$-$1.3 \cdot 10^6$ V/cm, if the ion current fraction in the total current balance at the cathode is 0.05-0.3. The starting point for these calculations was the value 4000 A/cm^2 for the current density found in 1922 by Güntherschulze in his mercury arc experiments and used in all papers up until 1948. The author of these calculations himself interpreted the results optimistically, counting on the possibility of getting appreciable emission currents in practice from the contaminated mercury surface at the field gradients found. However, from the point of view of Schottky's theory as well as from the quantum mechanical equation derived by Fowler and Nordheim [103], the gradients found by Mackeown were obviously not large enough to produce appreciable electron emission from the cathode. At the same time, the initial data which physicists then had at their disposal for calculating the field in the cathode-drop region were not such as to allow anybody to entertain the possibility of any much stronger field existing at a liquid[1] mercury cathode. On the other hand it did not seem highly probable that mercury that had been repeatedly distilled in vacuum would have surface films on it that could appreciably reduce the electron work function. All this was accepted by many workers as evidence for the inadequacy of the cold-emission theory of the arc.

16. Slepian's Thermal Theory

The chronological sequence of the development of arc theory must be interrupted to emphasize that the disillusionment with the cold-emission theory came too soon. The disillusionment was a direct result of the exceedingly low value for the current density

[1]With solid cathodes it is usual to assume that a considerable increase in field gradient is possible at the cathode because of the surface roughness.

in the cathode spot of the mercury arc which had been taken from Güntherschulze's experiments. In the 1920's, the figure of 4000 A/cm^2 published by Güntherschulze still looked quite impressive. It is easy to imagine that at that time the only way of getting out of the difficulty may have seemed to be the one pointed out by Slepian in 1926 [145,148]. According to the new theory, the role of the principal agent furnishing electrons to the cathode region of the arc was again assigned to the high temperatures, but now the scene of action was transferred from the cathode surface to the volume of the gas, to the so-called "ionization space," removed from the cathode surface by 1-10 mean free path lengths of the atoms in the gaseous medium. As one of the arguments in favor of his theory, Slepian advanced the important idea that since the gas has relatively small heat capacity and thermal conductivity, it should be more strongly heated at the cathode surface from the passage of current than any of the metal in the cathode. Hence it followed that if electrons are liberated anywhere as a result of high temperatures, it ought to be primarily in the volume of the gas located at the proper distance from the cathode for the temperature to reach its maximum value. Not long before this Compton [142] had made use of the theory of thermal ionization of a gas to give a description of the positive column in the arc. Based on this work, Slepian originally used the same ionization mechanism to explain the transition from a glow discharge to an arc; he assumed that a condition for the transition with increase in current density is to have sufficient thermal ionization in the dark space to produce an appreciable increase in current density. The critical ionization values are reached first in the hottest central region of the discharge, which disturbs the uniform current distribution. This increase in current density causes a further increase in temperature of the gas in the center and leads to a steadily increasing concentration of the current at a small portion of the cathode surface. Using the familiar Saha equation to calculate the degree of ionization of the gas, Slepian showed that instability of this sort will be given by temperatures of 3000-5000°K which may be assumed to exist in the discharge. Along with this, a satisfactory explanation was found for the process by which the current is kept within the limits of the cathode spot of the arc. A logical climax of this idea was to attempt an explanation of the mechanism of the arc in the steady state using the same picture of thermal ionization of the gas [145]. The main difficulty with extending the theory to the steady-state arc was that the large current density in the cathode spot requires enormous ion concentrations in the ionization space. There are only two ways of explaining this latter condition in terms of thermal ionization of the gas. One is to bring in exceedingly high temperatures, at which the degree of ionization of the gas approaches 1. In view of how close the ionization space is to the relatively cold cathode, this idea did not seem realistic to Slepian and his adherents [149]. As a result of this they chose the other way of solving the problem, which consists in postulating a very high concentration of neutral particles in the discharge space. It is therefore not simply a matter of chance that the thermionic theory has since been considered applicable only to high-pressure arc conditions. As will be seen from what follows, the new data on the current density in the cathode spot make this last idea even less realistic, since the ion concentration increases too slowly with increase in pressure due to the reduction in degree of thermal ionization of the gas at high pressures. The futility of this way of going about the problem is clearly visible from the simple fact that even at atmospheric pressure Slepian, when obviously forcing the matter, could get a calculated current density only of the order of 100 A/cm^2. Nevertheless, in 1940, Weisel, Rompe, and Schön [149] again turned to the theory of thermal ionization at the cathode of the arc. As a result of formal use of the mathematical apparatus, they came to the conclusion that electron emission from the cathode does not play an essential role in accounting for the processes taking place in the arc, and accordingly it does not need to be present at all when the arc is in the fully developed stage.

Maintaining that the current in the cathode-drop region is purely ionic is quite characteristic of Slepian's theory and is a logical consequence of the idea that the charges are regenerated in the ionization space. In this respect Slepian's theory may be regarded as the antithesis of the cold-emission theory of the arc, which regards electron emission at the cathode as being the principal agent active in carrying the current to the cathode-drop region. This difference can be of great importance in solving the problem of the most probable mechanism for the cold-cathode arc. Leaving the discussion of this question to a subsequent section, it is sufficient to point out here that in view of the rejection of cathode emission by Slepian's theory, it is difficult to understand what role is played by the cathode in the arc discharge, the more so since the theory regards the cathode as a cooling surface which limits the heating of the gas.

17. Further Development of Ideas on the Mechanism of the Cold-Cathode Arc

These and the more fundamental difficulties that have arisen in Slepian's theory in recent years have left the question of the arc mechanism open for a long time. During the last three decades, various authors have advanced a number of new ideas in this field, but not one of them has yet received the necessary development or experimental verification. Thus, in 1930, Druywesteyn [150] made the suggestion that insulating and semiconducting particles and films on the cathode surface may play a role in the cold-emission mechanism; this involves the charging of the particles by positive ions, thus forming strong electric fields. This theory, according to which one would expect the phenomena to be strongly dependent on the degree of contamination of the surface, was not confirmed by observations. A particularly convincing objection to it is that there is no difficulty in maintaining an arc on the extremely clean surface of mercury that has been repeatedly distilled in vacuum.

In contrast to this, Smith [76] associated the electron emission from the cathode with microscopic volumes existing in the metal in which the conduction electrons were maintained in a "hot" state in spite of the fact that the metal itself remained cold. In trying to substantiate his theory, he was forced to resort to highly improbable assumptions concerning the mechanism by which energy is transferred to the electrons in the metal by electrons returning from the gas and also concerning the energy exchange processes between the electrons and the metal.

Considerably more attention is deserved by Rothstein's idea [85], according to which the surface layer of the cathode in the spot is in a special vapor-like state with an exceedingly high particle density. Referring to the data from studies made on the conductivity of mercury in the supercritical region [151], the author assumes that it is possible to have conduction bands for the dense vapor layer. Then, between the liquid metal and the vapor there will be a region with ordinary metallic conductivity. It may be assumed that for a high enough temperature in the vapor-like part of the cathode, due to intense influx of energy along with the ions bombarding the cathode the region will be able to emit thermal electrons.

Ideas similar to this were advanced previously by Güntherschulze [7]. In some degree or other they all represent a return to the earlier thermionic theory of the arc although in a new sense. A tendency has been observed recently to combine the thermionic with the cold-emission theory of the arc (theory of the T-F arc). One of these attempts was undertaken by Bauer [118], who considered the primary emission from the cathode of the arc to be the result of simply combining the thermionic and cold-emission currents calculated separately from familiar equations. Lee [154] called attention to the unsatisfactory character of this primitive treatment of the question. The basis for the criticism was a more logical solution of the problem of electron emission by a metal under the combined action of an electric field and temperature [119]. An important result of this work is the idea that there is a continuous transition between thermionic and cold-emission arcs, depending on the relation between the cathode temperature, the field strength at the cathode, and the work function. It follows from the calculations that have been made that at a field strength of about $5 \cdot 10^7$ V/cm even heating the cathode to a temperature of the order of 3000°K does not appreciably affect the amount of emission, and in this case we are dealing with a purely cold-emission arc. On the other hand, reducing the field intensity to $1 \cdot 10^7$ V/cm is enough to change the state of affairs considerably, and heating the cathode to 2000°K will have a decisive effect on the emission. At lower field strengths, the range of the purely thermionic arc becomes broader. It is self-evident that cathode heating cannot be of any importance to the emission of a mercury cathode, so that in this case the T-F arc theory does not provide any way out of the difficulty.

A new and interesting possibility of explaining the arc mechanism was recently pointed out by Robson and Engel [152]. These authors assume that the electrons are pulled out of the cathode by means of excited atoms with a potential energy greater than the work function of the electrons at the cathode. There are several considerations that support this possibility. The intense luminescence from the cathode spot shows that there is a region with an exceedingly high concentration of excited atoms very close to the metal. The flux of vapor leaving this region is considered by the authors to be the difference between the large numbers of atoms evaporated from and returning to the cathode. Because of the small distances between the negative luminescence and the cathode, the majority of the returning atoms behave as metastable atoms, which accounts for the large emission output of the cathode. As for the electrons emitted, after they pass through the cathode drop, their energy is only enough to excite the mercury vapor. The necessary ionization is achieved by collisions with the excited atoms, the ion current only being necessary to produce the space charge and the cathode evaporation. Orientational calculations show that the theory can account for the high current densities of the order of 10^6 A/cm^2 observed in the cathode spot if the greater part of the returning atoms pull electrons out of the cathode. The requirement of an improbably high yield of the γ-process at the cathode, approximating 1, is a weak point of the the subject of discussion [156,157].

Quite recently a similar point of view of the role played by excited atoms in the arc mechanism was advanced by Hernqvist [18], who came to the conclusion that the cold-emission theory of the arc was inadequate. The reason for this was that he got a high value for the current density at the cathode, reaching the order of 10^9 A/cm^2, when he made a combined solution of Mackeown's and the Fowler-Nordheim equations. The analysis of this solution of the problem given in the next section shows that the objection raised to the cold-emission theory of the arc need not be taken seriously in view of the arbitrary value for the ratios of ion to electron current used in the calculation. The weakness of the derivation is also obvious from the fact that somewhat earlier, as a result of making a graphic solution of the same equations, Wasserrab [153] came to the conclusion that the theory is compatible with the new data on the current density. A more serious reason for doubting that cold-emission theory applies to the cold-cathode arc is Hernqvist's reference to the so-called D-type arc with a current density of the order of 10 A/cm^2 or less, previously described by Smith [17]. These unusually low values for the current density, as well as the fact that the arc occurs on oxidized molybdenum, which has a high work function, exclude any possibility here of cold-emission aided by the space charge and oblige us to look for some other mechanism by which the charges are regenerated at the cathode. Hernqvist is inclined to associate the mechanism with excited atoms coming from the negative luminescence onto the cathode surface. The paper cited discusses two ways in which excited atoms can liberate electrons. In those cases where the cathode work function is less than the difference between the ionization and excitation energies, electrons may be pulled out of the cathode by transferring the energy from the excited state. In the opposite case, an electron may be taken out of the atom. The atoms thus formed are capable of producing electron emission from the cathode as a result of the tunnel effect.

It may be seen from the above review of the development of ideas on the mechanism of the arc that at present there are a number of theories based on various processes by which electrons may be liberated from solids and gases, from thermionic electron emission to complicated stepwise processes involving excited atoms. Some of them have been widely known for a long time and have at their disposal quantitative data and a well-developed mathematical apparatus, while others have only been published in the last few years and are of a purely speculative nature. It is highly probable that a number of new ideas on the nature of the cold-cathode arc will be advanced in the near future. Indications of some new possibilities in this direction are given in [158]. In this connection the question of what criteria we have at our disposal at present to select the right ideas is of considerable importance. Is it possible that the ideas generally used for this purpose really provide a reliable basis for choosing our theories? These questions, as well as possible paths to be followed in subsequent investigations of the arc, are discussed in the next section.

C. THE MOST PROBABLE MECHANISM FOR THE COLD-CATHODE ARC. ADDITIONAL METHODS OF INVESTIGATION

18. The Energy Balance at the Cathode as a Criterion for the Correctness of Theory

Until recently one of the basic criteria of the correctness of a given arc theory was held to be whether or not the energy balance was maintained at the cathode. This view found expression in the fact that almost all authors, in attempting to substantiate their ideas, turned first of all to the energy balance. One is struck by the remarkable fact that the energy balance is almost always satisfactory, in spite of all the differences in the basic ideas underlying the arc mechanism and in spite of all the simplifications and errors made in setting up the balance. A striking example of this is provided by the next two papers. One of the first to turn to the evidence provided by the energy balance at the mercury arc cathode was Güntherschulze [7], who found that it supported his theory in spite of the fact that he was using Stark's data on the cathode drop, which were too low by almost a factor of two. Later, in attempting to substantiate his absurd arc theory, Smith called for help from the energy balance equation in his paper [76], although in his balance he omitted the extremely important source of energy brought to the cathode in the form of positive ions.

These absurdities suggest that in setting up the balance it is possible to be biased in deciding where the energy comes from and where it goes, so that this method of evaluating the theory does not deserve any particular confidence. This idea is supported by the extensive investigation of the energy balance at the mercury cathode undertaken by Compton [22] on the basis of Langmuir's theory, which is presented below with a certain amount of digression and in more general form. In any realistic theory of the arc, one of the most important sources of energy given to the cathode must be the positive ions which bombard the cathode at the velocities acquired in the cathode-drop region. If K is the electron fraction in the total flux of charge incident on the cathode surface, and g is the ion accommodation coefficient for the cathode in question, the energy given to the cathode per unit time will be

$$(1-K)(gU_c+U_i-\varphi_0)+X,$$

where U_c, U_i and φ_0 are, respectively, the cathode drop, the ionization potential of isolated atoms from the cathode (which make up the gaseous medium), and the cathode work function. By adding the term X, we retain the possibility of including other energy sources such as Joule heat. On the other hand, the electrons emitted by the cathode always carry off a part of the energy, which is given by the latent heat of evaporation of the electrons,

$$-K\varphi_1,$$

where φ_1 is the effective electron work function in the presence of the accelerating voltage; the minus sign is used since the energy is dissipated.

In setting up the balance, we must also include the total energy losses from evaporation of the cathode material as well as from heat conduction, convection, and radiation; we designate these by the general symbol σ. Setting the sum of all the terms equal to zero gives the cathode energy balance equation, which may be solved for the quantity giving the electron current fraction:

$$K=\frac{U'+X-\sigma}{U'+\varphi_1}, \tag{3}$$

where $U' = gU_c + U_i - \varphi_0$. If all the quantities on the right-hand side of Eq. (3) were known, the energy balance could be used to find the value of K characteristic of any theory, and this could serve as a basis for deciding which theory is the correct one. Actually, at the present level of knowledge, the majority of the quantities in the equation cannot be found with any satisfactory degree of accuracy. Matters are in relatively good shape for the quantity U', where the only uncertainty depends on the accommodation coefficent g. In Compton's opinion the value of g for mercury lies between 0.9 and 1, with the uncertainty in U' being only a few percent. There is a great deal more uncertainty in the values of φ_1, X, and σ. Thus, φ_1 can take on values from 0 to φ_0 depending on the way in which the space-charge field acts in pulling electrons out of the cathode. The quantities X and σ are even more important, since, depending on the ratio between them, K can take on values from 0 to 1. That this does not exaggerate the degree of uncertainty is shown by the following considerations. At the present time, in view of the low values of the evaporation rate of mercury found by Kobel [35], there is a tendency to neglect the value of σ in comparision with U', which makes K take on a value close to 1. The state of affairs changes considerably if we consider the possibility that the cathode throws off a large part of the energy that is received along with the greatly overheated mercury resulting from explosive boiling. These energy losses may be quite large, if the mercury in the spot can be instantaneously heated to high temperatures of the order of 1000° or more, the possibility of which is admitted by some authors. If this source of energy dissipation is included in σ, values of K can be found which are close to zero. This same result is found if we include the continuous rapid motion of the cathode spot and the active emission regions on the cathode surface that make it up. As a result of this motion, energy is uselessly dissipated on the cathode surface much faster than could be assumed by simply considering the thermal conductivity of the metal, which is what is usually included in the balance equation. However strange it may seem, no one has yet considered these trivial sources of energy dissipation in setting up the cathode energy balance.

At the same time, it is well known that, for any conditions under which the cold-cathode arc operates, there is continuous rapid motion of the cathode spot as a whole [69], or of individual parts of it [16], and ejection of material from the cathode in the form of a stream of more or less large particles [29]. It is clear from the analysis given that setting up the energy balance for the arc cathode at the present state of our knowledge involves a considerable amount of arbitrary choice of the quantities used, so that this method of solving the problem is obviously defective.

19. The Current Density in the Cathode Spot and Its Importance in the Cold-Arc Problem

There is another, considerably more promising way of solving the problem of the arc mechanism that does not require a knowledge of a large number of difficultly accessible quantities. In solving the problem in this way, greatest importance is given to the data on the extent of the cathode-drop region d and the area of the emission surface of the arc S_j, which determines the calculated current density j in the cathode spot. Knowing the value of d would permit a direct determination of the field strength E_c at the cathode surface, since approximate values for the cathode drop are known for many cathodes. However, as has already been pointed out in Part A, the values of d measured experimentally do not deserve much confidence. A less direct method for finding the value of the field strength E_c consists in using the space-charge relations derived by Langmuir and Mackeown and calculating the field from the data on the current density at the cathode of the arc. This method has already been used several times, but in the past it led only to false conclusions because exceedingly low values of j were used in the calculations. The progress that has been made recently in measuring the size of the emission surface permits a more reliable calculation of the current density at the mercury arc cathode. From Froome's data, values for j may be taken up to 10^7 A/cm^2, which is tens of thousands of times greater than the values originally used. The changes that have occurred in the situation require a serious re-evaluation of the conclusions drawn previously and of the theories now existing, as was correctly pointed out by Wasserrab [153]. The importance of the new data on the current density is clearly illustrated by the following simple considerations. If, along with Slepian and his successors, we assume that the current in the cathode-drop region of the arc is carried exclusively by positive ions, information on the thickness of the space-charge layer is given by Langmuir's relation

$$j_p = \frac{\sqrt{2}}{9\pi}\sqrt{\frac{q}{m_p}}\cdot\frac{U_c^{3/2}}{d^2}, \tag{4}$$

which holds for $d < \lambda_p$, where j_p is the ion current density, q and m_p are the charge and mass of the ions, and λ_p is their mean free path. For reasons discussed by Compton [22], the value of d cannot be much in excess of the upper values given by the inequality.[1]

Thus the value of d may be found from the data on the current density in the spot, after which the maximum field strength at the cathode surface may easily be found as $E_c = (4/3)\cdot U_c/d$. Substituting the numerical values of m_p and U_c for the mercury arc gives:

$$d = 3.44\cdot 10^{-4}(j_p)^{-\frac{1}{2}}; \tag{5}$$

$$E_c = 3.81\cdot 10^4 (j_p)^{\frac{1}{2}}, \tag{6}$$

where j_p is expressed in A/cm^2 and E_c in V/cm.

In Eq. (4), the current density j_p is essentially an independent variable, the value of which is given by the positive ion concentration, n_p, in the plasma and the rate of random motion, w, of the ions according to the kinetic theory equation

$$j_p = \frac{1}{4} n_p q w. \tag{7}$$

After finding w, assuming the improbably high temperature T = 6300°K for the mercury vapor, it is a simple matter to find the minimum positive ion concentration required to maintain the ion current and the equivalent mercury vapor pressure reduced to zero:

$$p = \frac{n_p}{L}, \tag{8}$$

where L is the Loschmidt number. The results of calculating d, E_c, n_p, and p for different values of j are given in Table III. It is not difficult to see from a comparison of all these values that the new data on the current density in the arc (10^6–10^7 A/cm^2) produce a substantial change in the relation between the facts in favor of

[1]This statement is based on the fact that the region in which the gas is strongly ionized by electrons begins at a distance no further than one mean free electron path from the cathode surface.

TABLE 1II

j, A/cm²	d, cm	E_c, V/cm	n, cm⁻³	P, atm
10^4	$3.44 \cdot 10^{-6}$	$3.81 \cdot 10^6$	$3.1 \cdot 10^{18}$	>0.115
10^5	$1.09 \cdot 10^{-6}$	$1.2 \cdot 10^7$	$3.1 \cdot 10^{19}$	>1.15
10^6	$3.44 \cdot 10^{-7}$	$3.81 \cdot 10^7$	$3.1 \cdot 10^{20}$	>11.5
10^7	$1.09 \cdot 10^{-7}$	$1.2 \cdot 10^8$	$3.1 \cdot 10^{21}$	>115

the cold-emission theory of the arc. Actually, increase in j_p requires a proportionate increase in ion concentration in the ionization space, which increases the difficulties facing the thermionic theory. A current density of the order of 10^6 A/cm² requires a concentration of ions so large that for the ordinarily allowable plasma temperatures there is no possibility of maintaining them by thermal ionization of the gas (or vapor). Actually, the degree of ionization of mercury vapor at T = 6300°K is not greater than 0.01, which would require pressures hundreds of times greater than those shown in the table. For a current density of 10^6 A/cm² this would amount to thousands of atmospheres.

In contrast with this, the chances for the cold-emission theory only improve with increase in current density, as a result of the reduction in extent of the space-charge region and the increase in field strength at the cathode. Even for $j_p = 10^6$ A/cm², E_c reaches values of the order of $4 \cdot 10^7$ V/cm². Calculations show (see Table IV) that at this field strength cold emission of electrons will play a substantial role in the current balance at the mercury arc cathode. Thus, Slepian's assumption that the current is purely ionic in the cathode-drop region is incompatible with the new data showing the high values of current density existing at the mercury cathode surface. On the other hand, it may be shown that the cold-emission theory of the arc is compatible with these data. The field strength E_c at the cathode due to the space charge may be expressed in terms of the ion and electron current densities j_p and j_e by means of Mackeown's equation [147], which does not raise any doubts. Expressing the electrical quantities in this equation in practical units, it may be written in the form

$$E_c^2 = 7.57 \cdot 10^5 U_c^{1/2} j_e \left[\left(\frac{m_p}{m_e} \right)^{1/2} \frac{j_p}{j_e} - 1 \right], \tag{9}$$

where U_c is the cathode drop, and m_p/m_e is the ratio of the ion and electron masses.

On the other hand, the relation between the field strength E_c and the cold-emission current density j_e is given by the Fowler-

Nordheim equation [103]:

$$j_e = 6.2\cdot 10^{-6}\,(E_c^2/\varphi_0)\,e^{-6.8\cdot 10^7\,\varphi_0^{3/2}/E_c}\,, \tag{10}$$

where φ_0 is the electron work function in volts. Solving these equations simultaneously gives

$$j_e = 6.11\cdot 10^9\,\varphi_0^3\left\{U_c^{1/2}\left[\left(\frac{m_p}{m_e}\right)^{1/2}\frac{j_p}{j_e}-1\right]\ln^2\left\{4.69\frac{U_c^{1/2}}{\varphi_0}\left[\left(\frac{m_p}{m_e}\right)^{1/2}\frac{j_p}{j_e}-1\right]\right\}\right\}^{-1}, \tag{11}$$

which makes it possible to calculate the value of j_e if U_c, φ_0, and j_p/j_e are known for any given cathode. As may be seen from (11), the electron emission current density will increase sharply with increase in work function of the metal in the cathode. This is due to the fact that, as φ_0 increases, a higher field strength is required to maintain the current. Since the field is related to the space charge, this is equivalent to requiring higher values of current density. In all probability, the magnitude of the cathode drop U_c is determined by the excitation potentials of the atoms in the gaseous medium (or vapor) and the cathode work function, while the ration j_p/j_e also depends on the boiling point of the metal. This ratio has a considerable effect on the electron current density, which must be taken into consideration when it is being calculated from Eq. (11). It was already pointed out in Sec. 15 that having no reliable methods of determining the ratio j_p/j_e makes it necessary to calculate j_e using all the possible values of j_p/j_e, running approximately from 0.01 to 0.25. The results of calculations made from Eq. (11) as a function of j_p/j_e for the mercury arc are given in Table IV, using the values $U_c = 10$ V, $m_p/m_e = 600$, and $\varphi_0 = 4.52$ V.

The same table gives values of the field strength at the cathode surface calculated from Eq. (9). These calculations show that as the ratio of the currents varies from 0.01 to 0.3 the field strength varies by only a factor of two, remaining equal in order of magnitude to 10^8 V/cm, while j_e drops from $4.57\cdot 10^9$ to $2.45\cdot 10^7$ A/cm^2. The reason for the difference between the results

TABLE IV

j_p/j_e	i_e, A/cm^2	E_c, V/cm
1	$5.22\cdot 10^6$	$8.64\cdot 10^7$
0.3	$2.45\cdot 10^7$	$1.02\cdot 10^8$
0.1	$1.09\cdot 10^8$	$1.24\cdot 10^8$
0.03	$6.5\cdot 10^8$	$1.63\cdot 10^8$
0.01	$4.57\cdot 10^9$	$2.34\cdot 10^8$

of Wasserrab and Hernqvist is to be looked for in this abrupt change in j_e as a function of j_p/j_e. Hernqvist in his calculations used the value $j_p/j_e = 0.02$, derived from formal considerations. The result of this was an exceedingly high value for the current density of the order of 10^9 A/cm^2, which led the author to the unjustified conclustion that the cold-emission theory is no good. As a matter of fact, there are a number of data which make it more plausible to assume high values for the ratio of the currents. The majority of scientists have recently become convinced that the metal vapor is practically 100% ionized in the negative luminescence region of the arc. This fact, as well as the low values for the cathode drop, indicates that an important role is played by stepwise ionization processes which have high yields. Under these conditions, the ratio j_p/j_e can take on values close to 0.25. Compton [22] and Wasserrab [153] used precisely this order of magnitude for the ratio in their calculations. In Table IV, this value corresponds to current densities of the order of $2 \cdot 10^7$ A/cm^2, which is close to those found experimentally.

In evaluating the results of the analysis just given, there are, in addition, two important facts to consider which speak in favor of the cold-emission theory of the arc. First, the field gradients for cold emission required by the theory are for some reason higher than those found experimentally. Thus, from Beams' data [24] cold emission from a clean mercury surface occurs at a field strength of about $1.8 \cdot 10^6$ V/cm. If the surface is dirty, this figure drops to $3.5 \cdot 10^5$ V/cm. Dyke and Trolan [155], working with extremely clean metal surfaces, observed cold emission at a field strength of about $2 \cdot 10^7$ V/cm. Second, it must not be forgotten that, as a result of imperfections in the technique of measuring the current density at the cathode spot, we are always dealing with values that are obviously low, as is clearly illustrated by the whole history of measurements of this sort. As was pointed out in Sec. 3, they are not based on a direct measurement of the cross section of the actual current channel at the cathode surface, but rather of the area of intense luminescence in the cathode region of the arc of the eroded surface of the metal, none of which gives an exact idea of the current density. Further, even if we take these areas to be equivalent to the cross section of the current channel, by their very nature these measurements can only give averaged values for the current density. The current may be so unevenly distributed that the instantaneous maximum values of the current density are an order of magnitude greater than the mean values found. The general conclusion must be drawn that the values of current density in the cathode spot which are not high enough from the point of view of theory cannot be used as a serious argument against the cold-emission theory of the arc, especially if we are talking about differences between theoretical and experimental values of only 1-2 orders of magnitude. However strange this may be, this obvious source of disagreement between theory and experiment is usually ignored by the opponents of the cold-emission theory, and even a difference of as little as one order of magnitude between the theoretical and experimental values for the current density in the spot is considered to be sufficient reason for rejecting the theory [152].

A no less important source of conclusions as to the nature of the cold arc is provided by the data on the conditions under which the cathode spot is formed for different methods of exciting the arc, including the processes discussed in Sec. 12 on breakdown of vacuum and gas gaps, transition from a glow discharge to an arc, etc. As has already been pointed out, all observations testify to the dominant role of cold emission in the formation of the cold arc. It must be concluded from this that the same emission mechanism is also dominant in the steady state of the arc, where the cathode current density, and accordingly the electric field strength, will be many times greater.

We can conclude from the arguments given above that there is less reason than ever to say that the cold-emission theory of the cold arc is inadequate. It is true, of course, that other processes can take part in liberating electrons from the cathode. In view of the high concentration of excited atoms in the negative luminescence region it is natural to assume that they too make some contribution. They may actually take on a decisive role in the recently observed arc in which the current density is unusually low. However, even if subsequent observations show that this form of discharge actually exists, it would scarcely be sensible on this basis to declare the cold-emission theory to be inadequate in the range of normal metallic arcs with high current density for which it was originally proposed. Since there are at present no data which embarrass the theory, it seems natural to let it take over the above range of phenomena, just as the thermionic theory has taken over the carbon and other hot arcs.

20. Conclusion

A review has been given above of several decades of experimental work on the cold arc, as well as a discussion of the present state of the mechanism of the arc. This leaves us with only the problem of summing up.

Even a cursory glance at the list of papers cited is sufficient to give some idea of the great attention physicists have from the beginning devoted to the problem of the arc. The great contribution they have made to this field of knowledge is also evident. Let us recall that the investigations of the arc made in the past have made possible the present wide use of the arc in rectifier engineering and other branches of the national economy. This makes all the more paradoxical the conclusion to which one inevitably comes in trying to answer a natural question: namely, what do we positively know about the processes in the cold arc and the mechanisms involved? No matter how the answer to this question is given, it will have to admit that we still know basically very little about this type of discharge. Actually, if we consider the basic question of arc theory, the mechanism by which the charges are regenerated at the cathode, we can only say that many ideas have been advanced at different times but not one of them has yet been finally established. Here, the attention of physicists has by choice been concentrated on the question of the mechanism by which electrons are emitted from the cathode, while the other aspects of the arc cycle have been for the most part neglected. In analyzing the existing theories of emission from the arc cathode it is easy to see that in looking for a solution to the problem, physicists have in fact considered almost all the known ways in which electrons can be drawn out of neutral matter. Nevertheless, the correct one has still not been found, although this is precisely the essence of the problem.

Apparently, the second part of the problem of investigating the arc, namely accounting for the remarkable properties and the peculiarities of behavior of the cathode spot, is in approximately the same state. In spite of the number of valuable observations made by various authors, we still do not understand such important properties of the arc as why the current is drawn together into the small region of the cathode spot, why the spot executes continuous random motion over the metal, and why the arc is extinguished spontaneously at small currents. A comparatively detailed study has been made of the directed motion of the cathode spot in a magnetic field, but even for this phenomenon there is no reliable explanation supported by special experiments. Practically nothing is known of the reasons for, or the laws governing, division of the spot, while the problem of the stability of the spot has not yet been even raised. It is in particular characteristic of the present state of information on the arc that the relation between the above recounted phenomena has so far completely eluded physicists.

Thus, the past several decades of investigation on the cold arc, in spite of a number of successes, have not produced the final solution to any one of the problems which this form of discharge presents. The resulting state of affairs causes one to think seriously about the sources of the difficulties and methods of overcoming them. If careful attention is given to the state of the investigations in the arc-discharge field, one is struck by the absence of any clear plan or even any system. Actually, this whole branch of physics is a collection of a large number of isolated pieces of work, conducted under conditions difficult to compare and, as a rule, giving equivocal results. As a result of this lack of system and the isolated nature of the various works, investigation in the various branches of the problem has in several cases been broken off at the point at which in reality a determined attack on the arc should just be starting. The lack of system has shown up in particular in the fact that no attention has yet been given to a question of first-order importance, namely, that of the stability of the cold-cathode arc as a special form of discharge. The present state of the question of the cathode-emission mechanism in the cold arc reminds one of the problem of solving an equation with many unknowns. In order for the problem to be understood we must first of all eliminate the uncertainties by finding the missing equations. The main source of uncertainty in the problem in this case is not lack of ideas to start with on the possible ways by which electrons are liberated from the cathode, but rather the absence of any criterion for telling which of them is correct. Under these conditions, new ideas that turn up only increase the number of unknowns. The touchstone of theory had usually been considered to be setting up the energy balance for the cathode region of the arc. However, as was shown above, this method of evaluating the theory must be regarded as inadequate at the present time. Since there is no accurate information on the effective sources of energy or the energy losses in the cathode region of the arc when the balance equation is set up, a door always remains open to biased conclusions. Accordingly, this method, for practical purposes, only creates the appearance of a quantitative check of theory. Further, the final criterion for the validity of a theory must be that its conclusion should agree with the results of experiment. Although the sum total of present data on metallic arcs supports the cold-emission theory, these data are still far from sufficient to consider the problem of the cold-arc mechanism as finally exhausted.

Thus, in further investigation of the arc the principal efforts should not be concentrated in looking for some new ideas for the cathode-emission mechanism, but rather on a planned accumulation of accurate data on the arc, and on using this as a basis for developing reliable quantitative criteria for checking and choosing among the existing ideas. In such a research program, a prominent position must be occupied by those problems in the physics of the arc which have so far not received sufficient attention. One of the most important of these is the problem of the stability of the arc.

Chapter 2

THE STABILITY OF THE MERCURY CATHODE ARC[1]

21. Statement of the Problem

Any self-sustaining or independent electric discharge is a complicated chain of interrelated processes forming some sort of closed cycle. The typical features of the cycle depend on the form of the discharge. The most general feature of the arc cycle is its ability to maintain itself, i.e., the cycle remains closed for relatively low input voltages. The processes occurring in the cycle are of a violent sort and are concentrated in the small portion of the cathode occupied by the cathode spot. One portion of the cycle of the arc, producing the metallic vapor from the cathode, must necessarily be evaporation and withdrawal of electrons from the cathode. The processes that close the cycle are ionization of the metallic vapor from the cathode and formation of the cathode portions of the discharge; these in turn serve to regenerate the charged particles in the neutral medium.

Identifying the individual processes that take place is not sufficient to give a complete description of any form of discharge, however important their role in the discharge mechanism. To be complete, a description must be given of the whole cycle. Before beginning an investigation of a particular aspect of the arc cycle, it is natural to try to get some idea of its properties as a whole, primarily of its stability. It must be emphasized that the question raised here goes far beyond the framework of the familiar phenomenological theory of discharge stability developed by Kaufman [139] and since applied to the arc discharge by a number of authors [2, 115, 161, 162]. According to this theory, the individual properties of the discharge reduce to a falling volt-ampere characteristic, so that any set of operating conditions is considered stable if for

[1]Some of the results of the investigations described here have been published as seperate papers [163-167].

a given set of parameters in the external circuit there are no auto-oscillations, which usually occur in a circuit containing a gas-discharge apparatus with negative resistance. It follows from the original assumptions of the theory that it only holds for the determination of the current behavior in an apparatus with a known falling characteristic. But the falling characteristic does not give the cold-arc model any of the properties which this form of discharge actually has. The phenomenological theory does not touch on any of the questions of an internal sort which form the core of the arc-stability problem in the true sense of the word and which may be listed as follows:

1. Is equilibrium maintained between the individual processes in the arc cycle at every instant of time when the arc is in the steady-state current condition?
2. What coordinates these processes so that a closed arc cycle can be maintained?
3. Under what conditions does the arc cycle remain closed, and how does the cycle change with variations in the experimental conditions?
4. Can the cold-cathode arc be considered in general a stable form of discharge?

It may be seen from the review just given of the data on the arc discharge that there are no answers to any of these questions at present. Nevertheless, these questions on stability touch on the fundamental properties of the arc and thus can serve as a general basis for beginning a systematic study of the arc. These questions do not only list the characteristics of the arc cycle as a whole; they are a potential source of information on the individual processes taking place in the cycle. Actually, by having data on how the stability of the arc changes with variations in the physical properties of the cathode and the gaseous medium, we can get a rather complete idea of the processes occurring in the discharge.

Apparently, the stability of the cold arc has never as a matter of principle been subjected to any serious doubt, since the arc can maintain itself under ordinary conditions for a practically unlimited time. It must, however, be kept in mind that this property of the arc is still no criterion for its stability and is of little importance in this respect. Actually it does not exclude the possibility that the arc is maintained by processes that are incompatible with the criterion for discharge stability, e.g., voltage oscillations or even periodic momentary extinction of the discharge. Accordingly, for greater clarity in formulating the question of arc stability, we must state exactly what we mean by the stability of a form of discharge. A natural criterion for the stability of any discharge fed from a constant voltage source might be whether or not it can maintain itself for an unlimited time with constant current and voltage on the electrodes and constant geometry of its fundamental parts, i.e., their fixed location with respect to the electrodes. It is obvious that this criterion is intended only for conditions in which there are no changes in the amplitude and frequency characteristics exceeding the limits of the usual statistical fluctuations inherent in any discharge.

We will next consider a number of departures from the discharge stability criterion in the conditions under which the mercury cathode arc operates, as well as relate them to their proper causes. The grossest departures of this sort occur in spontaneous extinction of the arc and we will therefore consider them first.

A. STATISTICAL STUDY OF SPONTANEOUS ARC EXTINCTION

22. Method of Observation and Apparatus

The tendency of the mercury arc toward spontaneous extinction at currents below 7-8 A has long been known both from observations [137,138] and from years of experience in working with ionic current rectifiers with mercury cathodes. This has led to the idea that, under normal operating conditions for this type of discharge with a spot executing random motions over the cathode, 3-7 A is a transitional or threshold current range: at lower currents the discharge cannot be maintained, while at higher currents it is completely stable. Along with this, the random nature of the extinctions, as shown by the lack of any appreciable regularity in the distribution of time intervals between extinctions, has from the very beginning attracted attention. However, the laws governing phenomena of this sort are best discovered by applying statistical methods of investigation, based on systematizing and drawing general conclusions from a large number of observations. It is therefore natural to begin the presentation of the investigations with a description of a statistical study of the distribution of time intervals between extinctions of the discharge. Here the principal property of the arc will be the mean length of time ϑ that the arc continues to exist under various experimental conditions. By changing the experimental conditions according to a definite plan, a logical study may be made of the effect of various factors on the stability of the arc and their role in the arc cycle.

Carrying out such a research program requires an enormous number, on the order of 10^6 or more, of separate observations or readings of the length of time the arc runs. Thus, in carrying out this program such observations were made with small discharge currents for which the arc lasts for only a small fraction of a second. This range of currents is also the most interesting from the point of view of the actual physical processes that extinguish the discharge. The extreme smallness of the intervals measured and the need to take a very large number of readings required the development of special measuring apparatus sufficiently accurate and convenient to operate. A short description of the apparatus is given below.

A schematic diagram of the discharge circuit showing the necessary measuring equipment is given in Fig.4. The arc discharge under test was excited in the tube T, which has a mercury cathode, either by means of a semiconducting igniter or by breaking down the gap between the cathode and an auxiliary needle-shaped electrode. The discharge was excited by short voltage pulses separated by definite, regulated time intervals and produced by the special thyratron generator P. The pulses were formed by discharging a condenser through the thyratron and the load RL, which was resistive or reactive depending on the required pulse shape. The pulse lengths could be varied from 10^{-6} to 10^{-3} sec, and in the majority of cases the interval between them was large enough for spontaneous extinction of the arc to occur during the interval at the desired discharge current, allowing for the statistical distribution in the arc duration times ϑ_i. The power supply for the arc discharge under test was a large-capacity storage battery, and the current was limited by the bifilar rheostat R_1 with an improved screw pressure contact and by the system of noninductive resistances R_2 and R_k. The total inductance in the circuit was not more than 50 μH. To investigate the effect of inductance on the mean duration ϑ of the arc, additional coils with known inductance

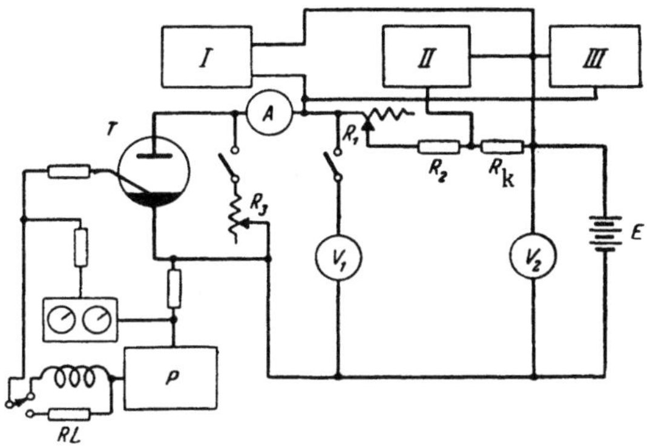

Fig. 4. Schematic diagram of connections to the tube T for the measurement of the mean duration of the arc.

were connected into the circuit. The voltage from the arc power supply was changed by disconnecting some of the cells from the storage battery or by means of a noninductive potentiometer connected into the circuit.

With the small currents used, starting at approximately 0.05 A, the duration of the arc was so short that a special method had to be developed to produce the desired currents and measure them accurately. To make a calibration current setting, the discharge tube T was shunted by the rheostat R_3 before making any measurements, and the values of R_1 and R_3 were adjusted so that the instruments connected in the circuit showed the required value of current for a voltage drop at R_3 equal to the operating voltage U_g of the arc under test. The operating voltage was found with an ordinary voltmeter, with the arc burning continuously at a current of 5-6 A. Because of a possible dependence of U_g on the current, and because of the oscillations and transient processes observed during discharge, the current values found in this way could be appreciably different from the true ones. With the strong dependence that exists between ϑ and the current, a substantial error may be introduced into the measurements by this difference. Therefore the calibration setting of the discharge current must be supplemented by measuring the true values; this has to be done with reference to the results of measuring ϑ. Since, as we noted above, oscillations are present, there is no point in talking about anything but the mean values of the true current found for each individual life cycle of the arc and for each series of discharges for which a value of ϑ is obtained.

The mean lifetime of the arc, ϑ, and the mean discharge current I were measured by means of an electronic "memory" circuit operating on the principle of charge buildup in a condenser. The circuit consisted of the three separate units I, II, III in Fig. 4. One of them (I) served to measure the time intervals ϑ_i from the moment the arc was excited until it was spontaneously extinguished, and it made an automatic summation $\sum_{i=1}^{N} \vartheta_i$ of the required number of readings. The total number of readings in each series of measurements was recorded by an electric pulse counter actuated by the pulse generator P. In view of the statistical distribution of the values of ϑ_1, the mean value of ϑ was found from at least 500 readings as $\frac{1}{N}\sum_{i=1}^{N} \vartheta_i$. At the same time, the unit II measured the total charge Q that had passed through the tube during the N readings. The mean value of the current was found by taking the ratio of the charge to the total time the current flowed through the tube, $\sum_{i=1}^{N} \vartheta'_i$. This last quantity was found in unit III by summing the individual conducting cycles in the tube ϑ'_i. In addition to the duration of the arc discharge it includes the duration of the various transient processes and discharges of other types. In the majority of cases, the two sums do not, for all practical purposes, differ from one another; in these cases unit III functioned as a monitoring apparatus. However, for a number of conditions, particularly in the low-current range, the duration of the arc discharge was only a fraction of the total time that the current had been flowing through the tube. In this case it was necessary to know both the quantities determined separately by units I and III.

A schematic diagram of these last two units is shown in Fig. 5. The voltage divider made up of resistances 7 and 9 is the dividing line between the two units and is connected to a constant 430-V source. The unit I is to the right of this line and consists of two vacuum tubes. The grid circuit of the input tube of each of the units contains the system of ballast resistances for the discharge tube, shown as R in Fig. 5. The cathode and grid of the in-

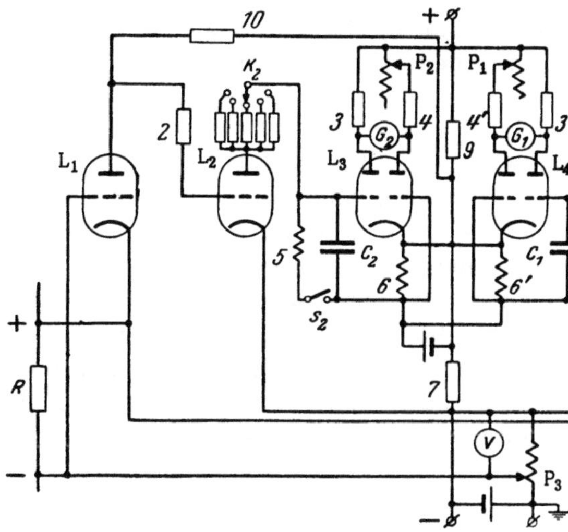

Fig. 5. Schematic diagram of the units I and III used to measure the mean duration of the arc and the length of time current has passed through the discharge gap under test.

put tube of unit I are connected to R in such a way that when the discharge current passes through the tube a positive potential appears on the grid. Further, a negative bias measured by the voltmeter V may be applied to the grid of this tube through the potentiometer P_3. During the measurements, this bias was maintained at a level U_1 such that the tube was conducting only when an arc form of discharge occurred in the tube T at a sufficiently low operating voltage U_g. For any given value of $\mathcal{E}$ the emf of the arc power supply, the value of U_1 will be somewhat less than the difference $\mathcal{E} - U_g$. When the tube is conducting, there is gradual charging of the condenser C_1, which is connected between the grid and the cathode of the double diode L_4. The potential difference U_c of the plates of the condenser is read by means of the galvanometer G_1, which is connected in a balanced bridge circuit between the plates of the double triode. The galvanometer is set to zero at zero potential difference by means of the variable resistance P_1. With the bias U_1 fixed in this way, and with a quite small time constant in the grid circuit of the input tube, the time required to charge the condenser C_1 will, to a sufficient degree of accuracy, be equal to the lifetime of the arc during each excitation cycle. Since the various types of leakage have been eliminated, the circuit can "remember" separate readings and add a large number of readings by accumulating charge. After the desired series of charges has been recorded, the indication on the instrument can be reduced to zero by simply short-circuiting the condenser plates with the switch s_1. Since the potential difference on the condenser plates was never more than a very small part of the power supply voltage (of the order of 1 V), the value of U_C increased linearly with the length of time the arc had been operating. At the low charging voltages used, and since the condensers were of a special oil type, there was practically no dielectric absorption in the condensers. Thus it was possible to calibrate the galvanometer readings directly in fractions of a second. In order to get a sufficiently wide time scale for the circuit, provision was made for changing the charging time constant of the condenser over wide limits by means of a set of resistances selected by the switch K_1.

Much of what has been said about unit I may also be said about unit III, which occupies the left-hand side of the diagram in Fig. 5. The principal difference between them is that the ballast resistance is connected directly between the cathode and the grid of the input tube of unit III without any additional negative bias. Further, the cathode and grid are connected to R in such a way that if discharge current flows in R the input tube is cut off. As long as the discharge tube is in the nonconducting state, the input tube is conducting and the grid of the intermediate stage L_2 receives a negative potential with respect to the cathode. In this case the condenser C_2 cannot be charged. As soon as the discharge tube begins to conduct current, the tube L_1 cuts off and a positive potential appears on the grid of L_2. Accordingly, the condenser C_2 starts to charge as soon as the discharge gap becomes conducting, and, to a sufficient degree of accuracy, the duration of the charging is equal to the length of time current flows in the tube, regardless of the type of discharge occurring there. It may be seen from Fig. 5 that the output or measuring stage of unit III is an exact copy of the output stage of unit I.

As a basic feature of operation of these circuits, the indications of the instruments connected to the output were completely independent of the magnitude of the discharge current, so that it was possible to make an accurate determination of the duration of the discharge for any values of current. The instrument readings are current-independent since the input tubes of the circuit were connected to the ends of the ballast resistance of the arc and were completely conducting or else completely cut off as the current varied over extremely wide limits. Changing the current beyond these limits could have no additional effect on the operation of the circuit.

In this respect, exactly the opposit requirements were placed on unit II, which played essentially the role of a coulometer. The indications of the instrument connected to the output had to be proportional to the product of the discharge current and its duration. By maintaining proper conditions, this requirement was satisfactorily met by the simple circuit shown in Fig. 6. The output stage is no different from the stages already described in units I and III. However, in unit II the condenser C is charged through a diode, the source of emf being the potential difference IR_k appearing across the calibrated constant resistance R_k when the discharge current I being measured flows through it. In the circuit shown, the charging current i_c of the condenser through the diode and the limiting resistance R' may be found from the equation

$$i_c = \frac{IR_k - U_0 - U_c(t) - U_a(i)}{R'}.$$

Here U_c is the potential difference on the condenser plates, U_a is the potential difference on the diode, and U_0 includes the contact potential difference of the diode as well as the constant negative bias used to eliminate electron currents which result in harmful leakage in the circuit. The value of $U_c(t)$ is variable, increasing with time as the condenser is charged, while U_a depends on the charging current i_c. These terms introduce an ambiguity into the measurements of the charge that has passed through the tube. In order that the charge on the condenser

$$q_c = \frac{1}{R'} \int_0^t [IR_k - U_0 - U_c(t) - U_a(i)]\, dt$$

be a unique linear function of the product of the current I by the time t that the current has been flowing, conditions of measurement must be maintained so that the terms $U_c(t)$ and $U_a(i)$ may be neglected in comparison to IR_k. This requirement could easily be met. The use of an extremely sensitive galvanometer G limited the maximum possible value of the potential difference U_c to tenths of a volt. On the other hand, the current-limiting resistors R' were large enough so that the working range of diode currents was completely limited to the low-current exponential part of the diode characteristic, corresponding to a negative potential on the anode with respect to the cathode. The part of the volt-ampere charac-

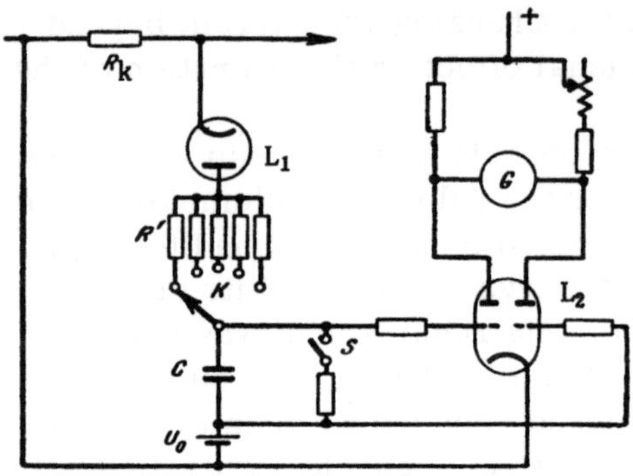

Fig. 6. Schematic diagram of the coulometer.

teristic of the 6Kh6 diode used is shown in the semilog plot of Fig. 7. It may be seen from the figure that the potential difference U_a on the diode changes by only 0.6 V when the current changes over the whole range of operating currents, i.e., from $2.4 \cdot 10^{-7}$ to 10^{-4} A. In order to reduce the effect of the nonlinear terms even further, this current range was divided into three parts, within each of which U_a could change by only 0.2 V. This was done by using the set of limiting resistances R' selected by the switch K. By connecting the resistances in various ways, the charging time constant of the condenser, and with it the sensitivity of the circuit, could be varied over wide limits. Since the minimum value of IR_k during the measurements was 18 V and the sum of the terms U_c and U_a was not more than 0.6 V, the maximum error in the current measurements was about 3%. The error was even smaller in measurements of the duration of the arc and the conducting period of the discharge tube under test.

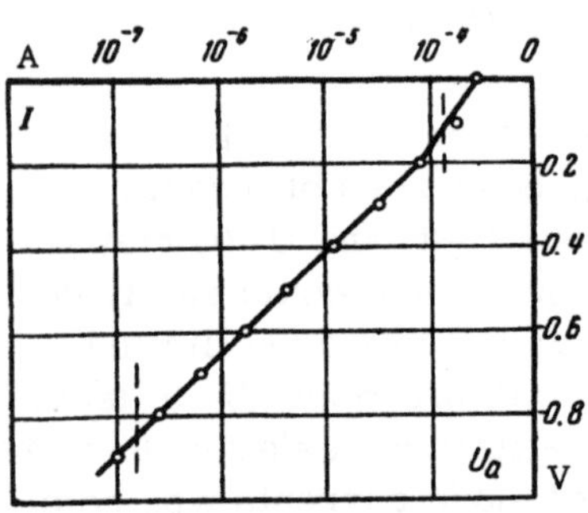

Fig. 7. Volt-ampere characteristic of 6Kh6C diode in the low-current range.

Before measurements were made, the circuits just described were calibrated for each value of the resistances used to measure the charging time constant of the condenser. For this purpose, the readings from all three units were compared with the readings of an electric second meter for different values and durations of the current. The second meter made it possible to read intervals of time with an accuracy of 0.01 sec. It was started automatically at the same time as the units in the circuit by closing the circuit (Fig. 4) at the resistance R_3 equivalent to the discharge, with a previously set value of current. Along with calibration, a check was made of the linearity of the readings of the circuit and whether or not the calibration curves were independent of current. Typical calibration curves of units I and II are given in Fig. 8; the different kinds of points are for readings made at different current values. Ideally they all fall on a straight line passing through the origin. In addition to the calibration, a check was made on the overall operation of the circuit. For this purpose, rectangular pulses of known frequency, duration, and current were fed to the input of the units. The only failure of the circuit to work properly was observed at very short pulse durations less than 10^{-5} sec.

The time spent in adjusting and calibrating the circuit was compensated by the convenience which the circuit provided. It is sufficient to note that obtaining the desired mean currents and arc lifetimes for 1000 discharges usually required no more than five minutes, with the possibility of error reduced to a minimum. Of course, a well-stabilized voltage source was used to feed the circuit, and the calibration was checked from time to time.

The measurements were made with three different types of discharge tubes. Type 1 was a cylindrical glass vessel with a mercury cathode and a plain molybdenum anode suspended on a flexible wire (Fig. 9). The distance between them could be varied from 1 to 40 cm by winding the wire up on the shaft of the ground

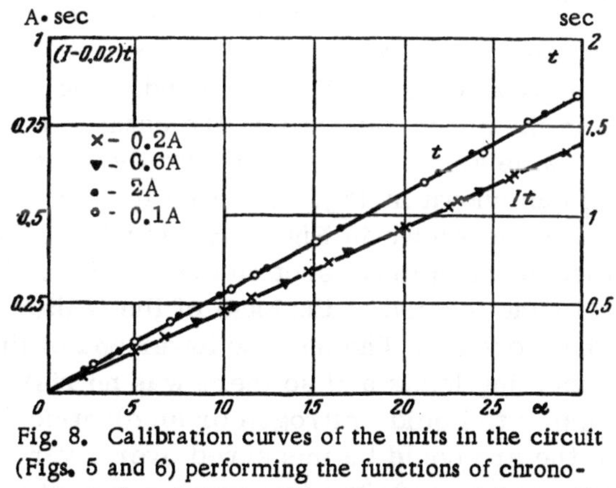

Fig. 8. Calibration curves of the units in the circuit (Figs. 5 and 6) performing the functions of chronometer and coulometer. The abscissas are the readings of the output instruments G, and the ordinates are time (right) and measured values of the charge (left).

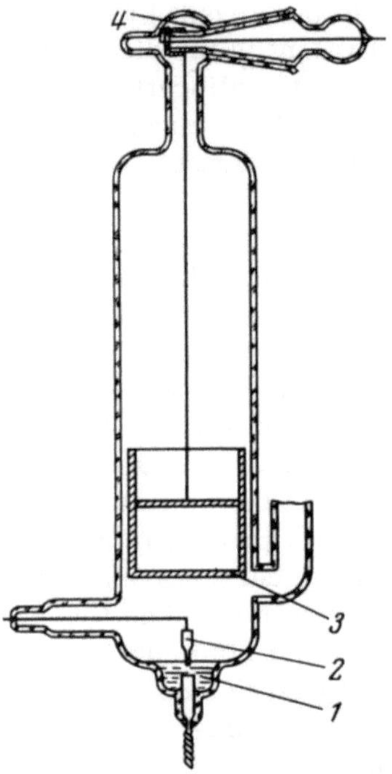

Fig. 9. First type of experimental discharge tube. 1) Mercury cathode; 2) igniter; 3) movable anode; 4) shaft of stopper used to wind up flexible wire holding the anode.

glass stopper in the upper part of the tube. The diameter of the tube was about 10 cm, and the area of the cathode was somewhat less than 10 cm^2. The pressure of mercury vapor in the tube was kept constant by putting the tube in a tank with flowing water at a constant temperature. In spite of the low thermal conductivity of the glass, the heat exchange between the cathode and the water was entirely satisfactory at currents up to 3-4 A, since the arc lasted for only a short time in this range of discharge currents. The first type of tube was used to determine the dependence of the stability of the arc on length, as well as for experiments on an arc of normal length.

In contrast to this, tubes of the second type (Fig. 10) were intended for studies on the stability of a short arc. Here the distance between the mercury cathode and the anode was only a few millimeters and could be varied by vacuum distillation of larger or smaller amounts of mercury from an auxiliary reservoir connected to the exhaust system above the tube. The anode of sheet molybdenum and the semiconducting igniter passing through a hole in the anode, as well as the cathode lead, were mounted on the same glass stem sealed into the upper wide portion of the glass bulb as shown in Fig. 10a. One form of this tube, the lower part of which is shown in Fig. 10b, made it possible to investigate the stability of the cathode spot on the thin film of mercury coating a nickel cone located at the base of the igniter and partially immersed in the mercury. The narrow lower end of these tubes was free of connecting leads and so there was no risk in putting the discharge gap into liquid nitrogen or an electric furnace, and the stability of the arc could be measured over a very wide range of temperatures, from -180°C to the temperature at which the mercury boils violently. In order to get any desired intermediate

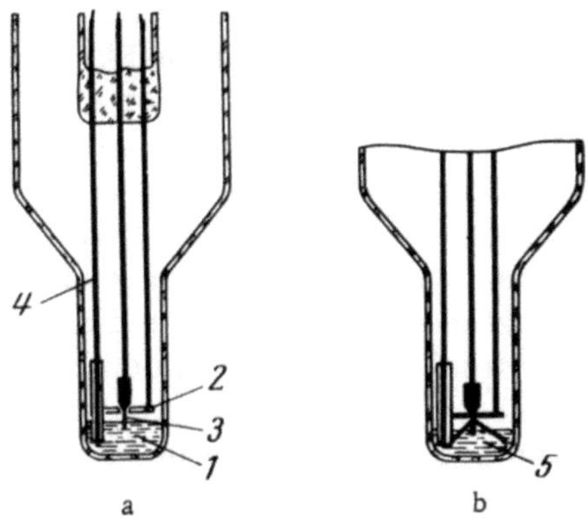

Fig. 10. Second type of experimental discharge tube. a) Vertical tube for measurements with a homogeneous mercury cathode; b) tube with the cathode in the form of a thin mercury film on a nickel cone; c) inclined tube for measurements in a magnetic field. 1) Mercury cathode; 2) anode; 3) igniter; 4) cathode lead; 5) nickel cone wetted by mercury; 6) aluminum cooler; 7) alcohol to improve thermal contact between tube and cooler; 8) thermal insulation; 9) copper strip; 10) dewar flask containing liquid nitrogen.

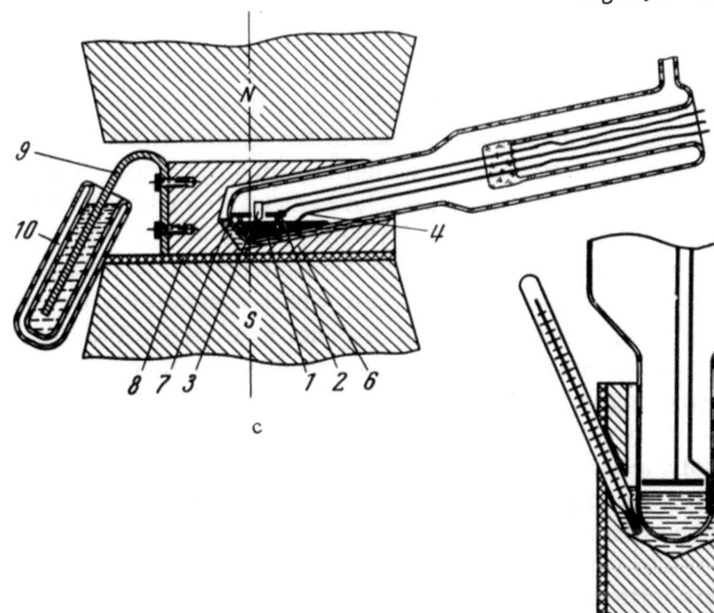

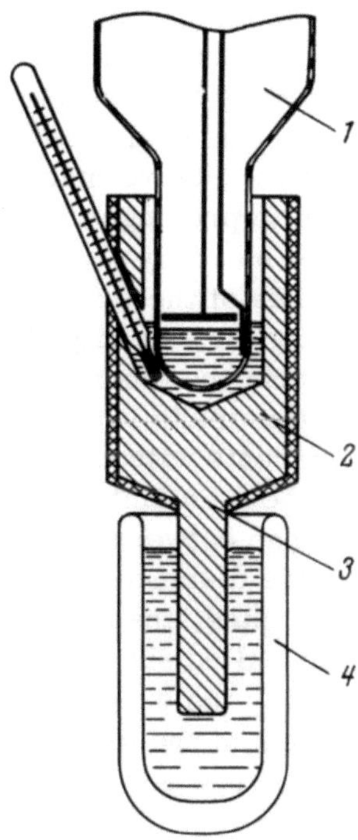

Fig. 11. Arrangement of cooled mercury cathode in a vertical tube. 1) Lower part of tube; 2) aluminum cooler; 3) aluminum heat conductor; 4) dewar flask containing liquid nitrogen.

temperatures in the negative range, use was made of a cooling device; the principle of operation of this device is shown in Fig. 11. A solid aluminum cylinder with a drilled hole holds the lower end of the tube containing the discharge space. The hole was filled with alcohol to improve the thermal contact between the tube and the cylinder. The aluminum cylinder was covered on the outside with a thick layer of asbestos as thermal insulation from the surrounding air. An aluminum or copper rod fastened to the lower end of the cylinder was immersed in a dewar flask containing liquid nitrogen. By varying the length and diameter of the rod, or the nitrogen level in the dewar, any temperature could be produced from 0 to -180°C.

For experiments of the same sort with the discharge in a longitudinal magnetic field, an inclined version of the tube was used (Fig. 10c). This tube, together with the cooler, was placed between the poles of a large magnet. In this version of the experiment, the hole for the tube was drilled in the cooler at an angle equal to the angle of inclination at which the tube operated. The cooler was in the form of a bar, and the heat was conducted from it to the liquid nitrogen by means of a bent copper strip. The steady-state temperature of the cooler was measured by means of an ordinary alcohol thermometer, for which a special hole was provided in the cooler and filled with alcohol to improve the thermal contact. Before making any measurements, the tubes were subjected to prolonged heating at 450°C, after which the required amount of mercury was distilled over in vacuum.

Tubes of the third type (Fig. 12) were also intended for investigating the stability of a short arc. They had a larger cathode surface than the other tubes, and the heat control was greatly improved; this made it possible to take measurements at relatively large currents without departing appreciably from the temperature that was being maintained at the cathode. In one tube of this type, the distance between the electrodes was about 2 cm, while in another the distance was reduced to approximately 0.3 cm. In the latter case, the experimental conditions approximated those in the so-called short arc, in which the behavior is completely determined by the properties of the cathode spot. The maximum axial length of the tube was not more than 4 cm. Accordingly, it could be placed between the poles of a large electromagnet, as

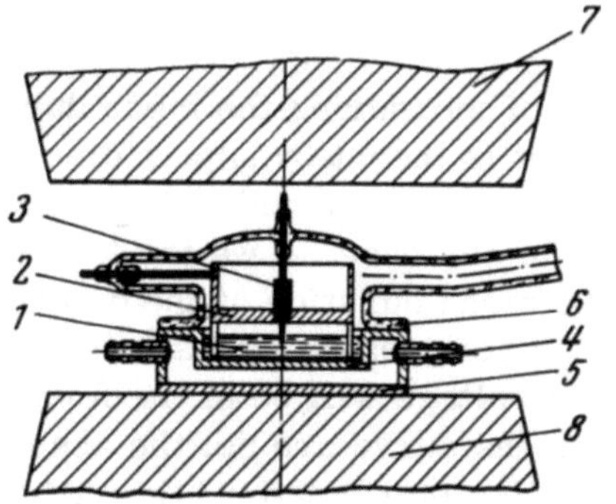

Fig. 12. Third type of experimental discharge tube between the poles of an electromagnet. 1) Mercury cathode; 2) anode; 3) igniter; 4) steel cathode reservoir; 5) bottom of water jacket; 6) ground joint; 7 and 8) pole tips of magnet.

shown schematically in Fig. 12. The anode and the bottom of the cathode were made of nonmagnetic chrome nickel steel, which eliminated field distortion and ensured good heat exchange between the mercury and the water circulating in the cathode cavity. If the temperature of the water was held at some fixed level, the mercury vapor pressure in the tube could be rigidly controlled. The glass anode portion of the tube was connected to the cathode reservoir with a flat ground joint, which was also cooled by the water flow. The discharge was excited in the center of the cathode by a semiconducting igniter passing through a small hole in the anode, or else by breaking down the vacuum gap with a high-voltage pulse. To prevent the cathode spot from going over to the walls of the cathode reservoir, the walls were protected by a glass ring mounted in the tube.

The mean duration of the arc in mercury vapor was measured while the tubes were being continuously evacuated by diffusion pumps. It was also possible to study the behavior of the discharge in an atmosphere made up of mercury vapor and an inert gas, for which purpose a suitable amount of the gas was admitted to the tube. The internal conditions in the discharge could also be changed by applying a longitudinal magnetic field. A quite uniform field of this type with intensities up to 10^4 Oe could be produced in the discharge space by means of an electromagnet. The field strength was determined by measuring the emf in a miniature coil placed in the uniform field at the level of the cathode and rapidly rotated by a synchronous motor.

23. Methodological Notes

Before passing to the results of measuring the mean duration of the arc, we must first decide what is meant by an arc discharge in measurements of this sort. The following considerations show that this question has to be settled in our case. If we try to use the above-described method to find the mean duration of the discharge, ϑ_u, at an operating voltage which is not greater than some value U, the curve giving ϑ_u as a function of voltage will be of the form shown in Fig. 13. The ordinates on the curve are not the absolute values of ϑ_u, but the ratio of ϑ_u to the total length of time ϑ' that the current passes through the tube, regardless of the form or properties of the discharge. This curve is characterized by some threshold value of the potential U_m which separates the low values of ϑ_u/ϑ' from the region where the ratio rises rapidly. To an accuracy of 2-3 V, the value of U_m is the same as the operating voltage of the arc as found with ordinary apparatus at higher currents. Within the limits of this threshold value of the potential, the ratio ϑ_u/ϑ' increases continuously from low values to 1. This condition proves to be due to the voltage oscillations inherent in the arc, and introduces an element of uncertainty into the problem of measuring the lifetime of the arc, especially since the form of the curves in question and the threshold values of the potential U_m change with the conditions of the experiment. In order to reduce to a minimum the effect which this uncertainty has on the results of measuring the mean lifetime of the arc, ϑ, there was some point in assigning to the arc region all those forms of discharge for which the potential difference between the electrodes in the tube is less than 40 V. In Fig. 13, this corresponds to the region of the curves to the left of the vertical dotted line. This region contains all the portions of the curve where the value of the ratio ϑ_u/ϑ' changes rapidly. With the arc defined in this way, the inevitable voltage oscillations on the arc-discharge electrodes are not reflected in the results of the measurements. and the transitional form of the discharge is included as described below.

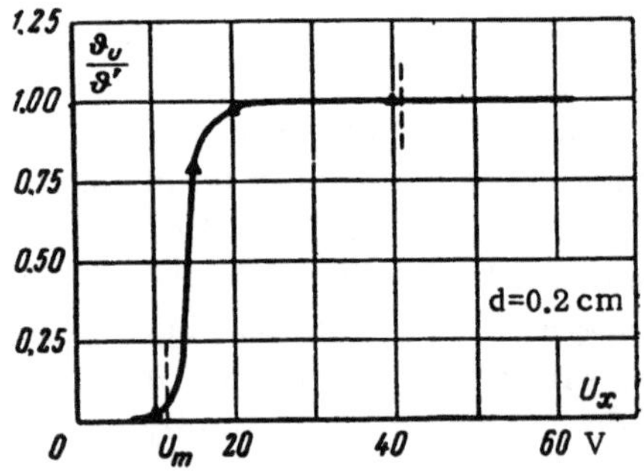

Fig. 13. Ratio of chronometer circuit readings, ϑ_u/ϑ', as a function of the threshold tripping voltage of the circuit.

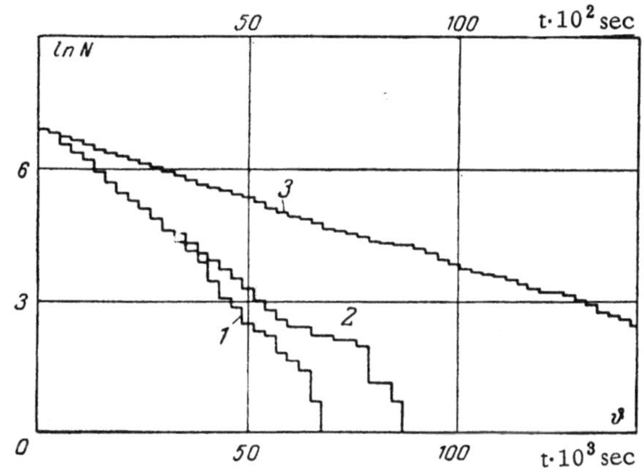

Fig. 14. Statistical distribution of arc-duration readings for three series of measurements. 1) I = 1.5 A, H = 0; 2) I = 0.3 A, H = 7 kOe; 3) I = 0.55 A, H = 7 kOe (abscissa scale at top).

To find the true nature of the way in which the arc extinguishes itself and to refine the technique for investigating the phenomenon, it was also necessary to find what sort of distribution holds for the time intervals ϑ_i from the instant the arc is excited until it is spontaneously extinguished; this required taking a quite large number of readings of ϑ_i and subjecting them to the usual statistical treatment. The results of a statistical treatment of three series of observations with 1000 readings each are shown in Fig. 14. The abscissas in the graph are the time intervals ϑ_i and the ordinates are the natural logarithms of the number of readings, N_t, with a lifetime greater than ϑ_i. The observations were made on a short arc in the third type of tube, both under ordinary conditions and in a magnetic field with a strength of about 5500 Oe. The histogram 1 shows the distribution of the time intervals at a current of 1.5 A under normal conditions for an arc unperturbed by a magnetic field. In spite of a certain lack of large intervals, the distribution is quite well described by the function

$$N_t = N_0 e^{-\frac{t}{\vartheta}}, \qquad (12)$$

i.e., it approaches a purely random distribution. There is even more reason to consider that there is a random distribution of the intervals shown by histograms 1 and 2. They refer to the series of measurements made at currents of 0.3 and 0.55 A in a longitudinal magnetic field. Without disturbing the random nature of the distribution of the intervals, the magnetic field produces a marked increase in the duration of the arc. Each of the random distributions found has a uniquely characteristic mean duration of the discharge, ϑ, that may be defined as the tangent of the angle which the straight line makes with the ordinary axis. It must be kept in mind that in order to present the results of treating all three series of readings on one graph, the scale of abscissas for series 3 had to be reduced by a factor of 10. The values of ϑ found from the data of Fig. 14 are given in Table V.

The method of finding the mean lifetime of the arc from the distribution of the intervals ϑ_i is too laborious and is only to be used as a check, The data given below were found by simply adding the separate ϑ_i readings, done automatically by the electronic circuit already described, and then dividing the resulting sum by the total number of readings N in each series. The resulting arithmetic mean value for the arc lifetime ϑ is the most complete characteristic of the distribution in question. The standard fractional deviation of the quantity measured will not be more than $1/N^{\frac{1}{2}}$. In order for this quantity not to be much in excess of the maximum error of 0.03 introduced by the apparatus itself, each value of ϑ was based on at least 500 readings.

In order to systematize and generalize the observations better, separate discussions are given below of the data on the effect which the parameters of the external circuit and the internal conditions of the discharge have on the quantity under investigation.

In measuring ϑ, the methodological question inevitably arises as to whether or not the form of the igniting pulse has any effect on the results of the measurements. In order to get an answer to this question and then set up the most favorable experimental conditions, it was necessary to make several series of preliminary measurements with different pulse forms, other conditions being the same. For the sake of clarity, the results of

TABLE V

I, A	1.5	0.3	0.55
H, Oe	0	5 500	5 500
ϑ, sec	$12 \cdot 10^{-3}$	$13 \cdot 10^{-3}$	$317 \cdot 10^{-3}$

these auxiliary observations must be given briefly before presenting the basic data of this section.

One of the conditions for exciting the arc in the discharge tube is simply that the discharge current in the igniting pulse must exceed some threshold value of about 0.4 A. So that the ignition of the arc would be of a completely regular nature, the parameters of the circuit used during the measurement were such as to produce pulses in which the maximum value of discharge current in the igniting pulse was 5-10 A. Providing these ignition conditions does not, however, ensure that the arc will be regularly picked up by the main anode in the tube under some experimental conditions. Thus, for example, in a magnetic field the discharge was not picked up in a considerable percentage of the cases when the pulse was a half sine wave, as is characteristic of a forming circuit consisting of a capacitor, a thyratron, and an inductive load. These failures to pick up the arc came, in all probability, from reverse current in the thyratron, and they markedly changed the statistical distribution of the arc life cycles. They distorted the results of measuring the mean lifetime of the arc. No failures to pick up the arc or associated distortions of the data were observed when using a purely aperiodic pulse, produced by discharging a capacitor through a thyratron and a quite large ohmic resistance. On the basis of these considerations a purely aperiodic igniting pulse was used in making most of the measurements. The duration of the pulse was unimportant, provided, obviously, that it was less than the natural duration of the arc under study. The discharge time constant of the capacitor through the load resistance in the pulse circuit was about 10^{-3} sec in the first experiments, while later it was reduced to $5 \cdot 10^{-6}$ sec, with pulse lengths of the order of 10^{-5} sec.

After these preliminary remarks, which deal principally with the choice of experimental conditions, we can pass to a presentation of the fundamental material contained in this section.

24. The Mean Arc Duration ϑ as Effected by External Circuit Parameters

a. Variation of ϑ with Current and Voltage. The results of measuring the mean lifetime of the arc as a function of the discharge current I for different values of emf of the source $\mathcal{E}$ are shown by the family of solid curves in the semilog plot of Fig. 15. The curves are plotted from data taken in the first type of glass

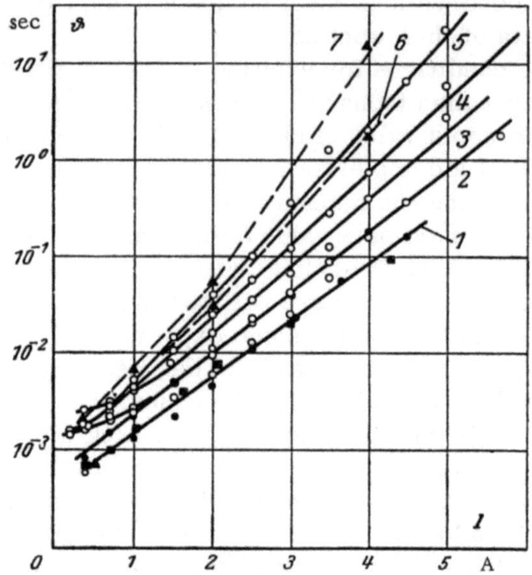

Fig. 15. Mean arc duration ϑ as a function of current for different emf's of the source feeding the arc. 1) 11.5 V; 2) 30 V; 3) 60 V; 4) 90 V; 5) 120 V; 6) 60 V, $L = 1 \cdot 10^{-3}$ H; 7) 60 V, $L = 5 \cdot 10^{-3}$ H.

tube with an arc length of about 5 cm and at a bath temperature of 18-20°C. The individual curves are for different emf's between 11.5 and 120 V. The circles give the experimental values of ϑ found for an igniting pulse length of the order of 10^{-3} sec. As was to be expected, both sets of values are everywhere the same, other conditions being equal, except in the low-current range, in which ϑ becomes comparable to the duration of the igniting pulse for the first group of curves. The departures from linearity shown by the curves at low currents result from the fact that the igniting pulse is too long in this case. It should be noted that the spread in the points is much greater than would be expected from the large number of points used in finding the mean values. This is apparently due to inaccuracy in reading the current and to the oscillations in the current. The data of Fig. 15 were taken before unit III was put into the circuit of Fig. 4, and after this addition accurate mean values were obtained for the discharge current.

It is obvious, in spite of the spread, that the points are definitely grouped around inclined straight lines. The slope of the lines increases monotonically with increase in $\mathcal{E}$ and in the majority of cases the ordinate intercept is only slightly displaced. The irregularity of the displacements suggests that the displacement is attributable principally to errors. The displacement is significant only in the lower curve, but this is the case for which there is most reason to think that there was a large error in measuring the current.

In order to check these ideas, some measurements of ϑ were later made at two greatly different values of $\mathcal{E}$, 90 and 225 V, using the method already described for measuring the mean discharge currents. Here the observations were extended to currents as low as about 0.05 A, where some peculiarities were to be expected in the behavior of the arc. The results of these measurements are shown in Fig. 16. In contrast to Fig. 15, the abscissas are accurate mean values of the discharge current, measured with the electronic circuit, and the scale is expanded to show the low-current range. Our attention is attracted first to the fact that each of the curves consists of two linear segments with a definite break at about 0.5 A. A simple exponential relation between ϑ and I is maintained with great accuracy in each of the segments. In view of the two-exponential form of the function, it is convenient to write it in the form

$$\vartheta = \vartheta_k e^{\psi_k (I - I_k)}, \tag{13}$$

where ψ_k is the slope of the segment, and I_k and ϑ_k are the coordinates of the beginning of the segment or the point where the break in segments occurs. Each segment is assigned a subscript k; in what follows the first low-current segment will always be given the subscript $k = 0$. Then the parameter I_0 is the minimum or threshold value of the arc current, and ϑ_0 is the duration of the arc at this current. For the conditions under which the mercury arc normally operates, an independent discharge with a low voltage on the electrodes is recorded with our method of measurement only at currents greater than 0.07 A, which is thus obviously the threshold arc current I_0. The condition for fitting the ends of the exponential segments gives an equation relating the various parameters of the relation

$$\vartheta_{k+1} = \vartheta_k e^{\psi_k (I_{k+1} - I_k)}. \tag{14}$$

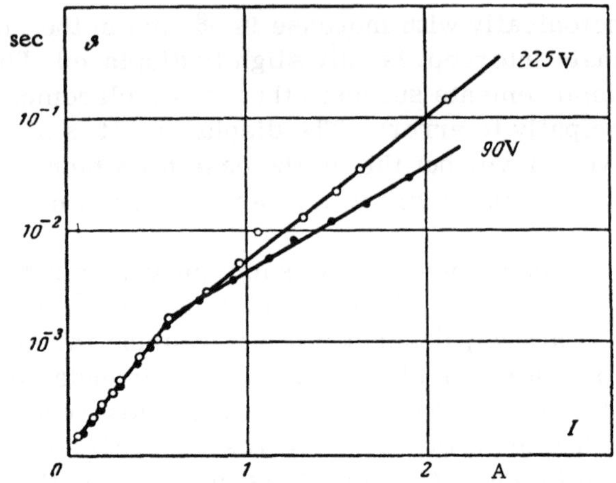

Fig. 16. Mean arc duration ϑ as a function of the accurate mean value of the current I for $\mathcal{E}$ equal to 90 and 225 V.

Writing the function in this form has the advantage that it includes the possibility of having more than two exponential segments. Further, it presents a number of convenient features for generalizing and systematizing the observations. Here the problem is to trace changes in the parameters in (13) when changes are made in some of the external circuit constants or in the internal conditions of the discharge. It is obvious from Figs. 15 and 16 that increasing the emf of the discharge power supply $\mathcal{E}$ only increases the slope of the high-current part of the curve. This corresponds to increasing ψ_1 in an equation such as (13) without changing the other parameters. For the conditions in this particular experiment, ψ_1 as a function of $\mathcal{E}$ may be represented by the approximate formula

$$\psi_1 = 1 + 8.6 \cdot 10^{-3} \mathcal{E}, \tag{15}$$

from which it may be seen that, for values of $\mathcal{E}$ near the operating voltage of the arc, the duration of the arc is practically independent of the emf.

b. Effect of Circuit Inductance. The reasons the effect of source emf on ϑ is less at low values of $\mathcal{E}$ are quite obvious from the graph (Fig. 15), where the dashed lines give the results of measuring ϑ when additional inductances L, equal to $1 \cdot 10^{-3}$ and

$5 \cdot 10^{-3}$ H, are connected into a discharge circuit with $\mathscr{E} = 60$ V. It may be seen from comparing the dashed and the solid curves that putting inductance into the discharge circuit is equivalent to increasing the source emf. The relation between the equivalent values of L and $\mathscr{E}$ depends in a complicated way on the current and the value of L. In particular, at a current of about 1 A, an inductance of 10^{-3} H produces an effect equivalent to increasing the emf by 50-60 V. This may only mean that the inductive emf plays some part, along with the power supply, in maintaining the arc. But an inductive emf can only occur when the discharge current changes. Thus, the relation between ϑ and L shows that there are oscillations in the discharge current and that they are related to spontaneous extinction. The arc seems to be trying to go out, but is kept from doing so by the additional emf produced by the inductance that is always present in the circuit. The effect of inductance is greatly magnified when the source emf is low, which accounts for the reduction in the effect of emf on ϑ observed under these conditions.

An opportunity to make a detailed study of the effect of circuit inductance on arc duration is provided by the data of Fig. 17,

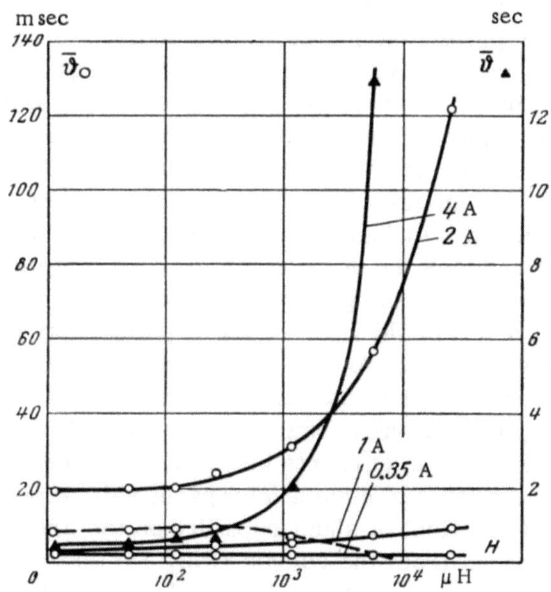

Fig. 17. Variation of ϑ with circuit inductance for a number of currents.

where $\log_{10}L$ is plotted as abscissa and ϑ plotted on a linear scale is the ordinate. The solid curves show the variation of ϑ with L for $\mathcal{E}$ = 60 V and for four different values of discharge current; for the current of 4 A, the ordinate scale was reduced by a factor of 100. For currents up to 0.5 A, the inductance has no appreciable effect on the duration of the discharge, and at a current of 1 A the effect is still very small. With further increase in current, the effect of inductance increases rapidly, but even in this case an inductance of less than 100 μH has no effect. This may be due to the fact that the discharge circuit itself already had an inductance of this order. It is not impossible, moreover, that with an inductance less than 100 μH the effect of the inductance is insignificant in comparison with the power-supply emf. It is somewhat unexpected to find the arc duration decreasing with increase in the inductance at low source emf's as shown by the dotted curve in Fig. 17 for $\mathcal{E}$ = 11.5 V and 2 A.

c. Effect of Capacity. In order to complete the presentation of the effect of external circuit constants on the arc, we must consider how ϑ varies with the capacity in parallel with the electrodes. The results of measurements of this kind for the second type of tube with d_a = 0.3 cm are given in Fig. 18, where the abscissa is the shunt capacity plotted on a logarithmic scale. For convenience in comparing the results found at different discharge currents, the ordinates in the figure are the ratio ϑ_c/ϑ, where ϑ_c is the mean arc life for some value of shunt capacity, and ϑ is the arc life under the same conditions but without the capacity. For currents less than 0.5 A, the capacity has no appreciable effect on the stability of the discharge. At larger currents, shunting the tube with a large capacity reduces the duration of the arc; the higher the capacity and the current, the greater the reduction. Apparently, for each current a threshold value of the capacity may be found at which the curve starts to drop off. As shown by the lowest curve, which was taken in a longitudinal magnetic field with a strength of about 5500 Oe, if a strong enough field is applied, ϑ is also reduced by the capacity at currents below 0.5 A. Under these conditions, the curve begins to drop at comparatively small values of shunt capacity. The effect of shunt capacity on the arc at different currents may be seen from the curves (Fig. 19) showing ϑ as a function of current for capacities of the order of 10^{-1} and 10^{-4} μF.[1]

[1]The interelectrode capacity including the distributed capacity of the wires was not more than 20 cm.

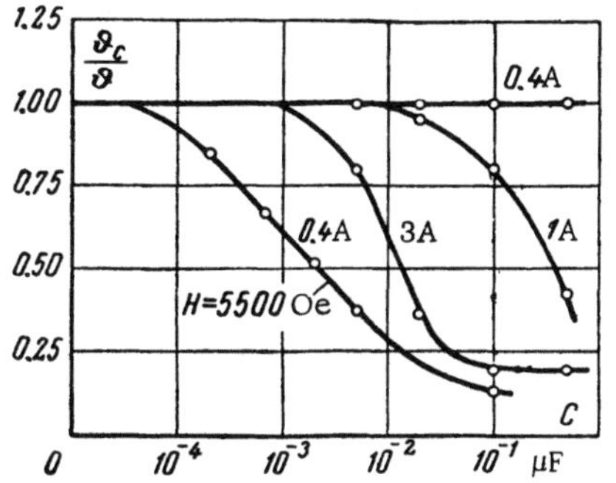

Fig. 18. Relative values of ϑ as a function of the capacity in parallel with the discharge tube.

It follows from this part of the work that, for small values of inductance in the circuit, the mean arc life ϑ as a function of current may be represented by a curve consisting of two exponential segments. The boundary between them occurs at currents of about 0.5 A. In the low-current segment and in the transition region the external circuit constants have no appreciable effect on ϑ. It is characteristic of the effect of the external circuit constants that changing them only changes the slope of the curves in the upper, high-current segment. Thus, for example, the slope of the upper segment increases if the source emf and the inductance in the circuit are increased. In this sense, emf and inductance have an equivalent effect. On the other hand, capacity can only reduce the slope, and in this sense it exerts an effect opposite to that of the emf.

25. Mean Arc Duration ϑ as Affected by Conditions Inside the Discharge

a. Effect of Arc Length and Cleanness of Mercury Surface on the Arc Duration ϑ. Before we consider the effect which conditions within the discharge have on the stability, there is some

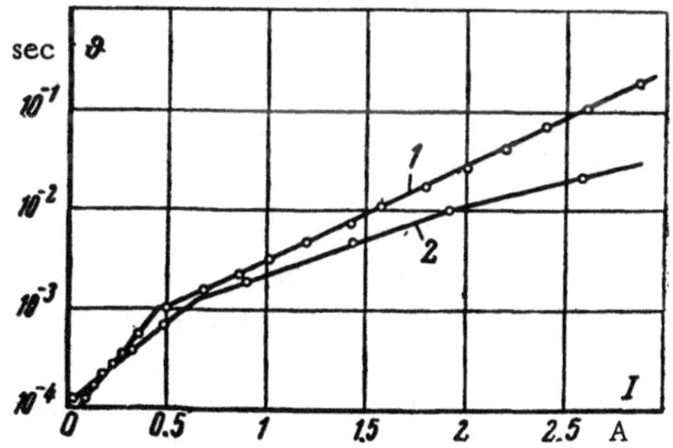

Fig. 19. Effect of capacity on the ϑ vs. current curves for d_a = 2 cm. 1) C = 10^{-4} μF; 2) C = 10^{-1} μF.

point in first determining whether or not the arc duration changes as the distance between the electrodes is varied. A partial answer to this question may be obtained by a study of Fig. 20, which brings together the results of measuring ϑ for interelectrode distances d_a = 1, 12, and 27 cm. With the usual scattering, they all fit on the same inclined straight line, from which it may be concluded that changing the interelectrode distance from 1 to 27 cm does not have any appreciable effect on the duration of the arc. It should be noted that in this series of measurements, made during the initial period of the work, the mean discharge currents were not accurately determined. For this reason, the data of Fig. 20 cannot be considered sufficiently reliable. With later, more accurate measurements, it was possible to find some change in the form of the ϑ vs. I curve when the distance between the electrodes was reduced. The changes are only appreciable at small distances, when passing over to short arc conditions; as d was reduced, the slope of the initial low-current part of the curve decreased and the slope of the high-current part increased. As a result, the break in the curve becomes less sharp. A change of this sort is clearly visible in Fig. 21, which shows ϑ vs. I curves for the third type of tube with distances between the electrodes of 2 and 0.3 cm. It may be assumed that for larger distances the duration of the arc will be practically independent of its length. This must be considered as evidence showing that the arc is extinguished as a result of processes taking place in the cathode spot and its immediate surroundings. This is also indicated by the fact that the duration is strongly dependent on the cathode material, but the study of arc stability on other metals does not fall within the scope of this work. It need only be noted here that the arc lasts appreciably longer when the mercury surface is dirty, but not enough to justify special measures for keeping it clean. Thus, for example, repeatedly distilling the mercury produced no appreciable change in the results of

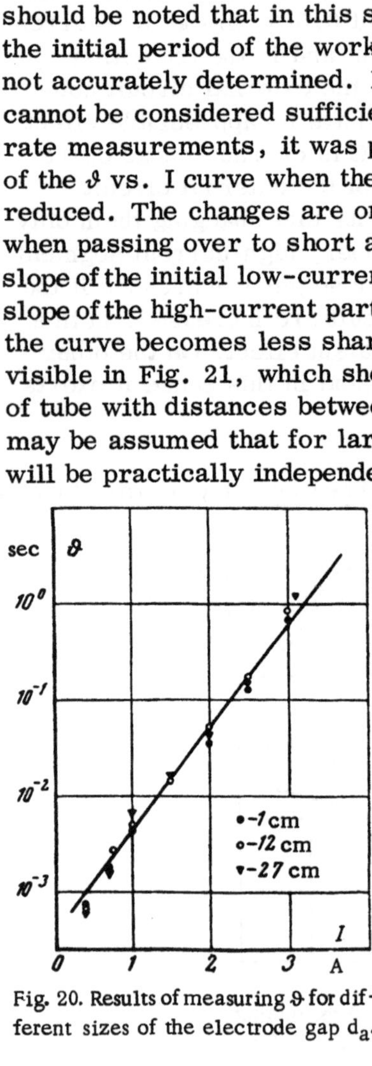

Fig. 20. Results of measuring ϑ for different sizes of the electrode gap d_a.

the measurements. Other conditions being equal, a noticeable increase in arc duration was produced only when, during prolonged measurements, the mercury surface became covered with films of dirt visible to the eye.

b. Effect of Temperature and State of Aggregation of the Cathode on ϑ. Results of Measurements on an Arc with an Anchored Spot. All the above data were taken with the cathode and tube walls maintained at the temperature level 18-20°C. It was found even during these measurements that if the temperature departed from this level there was a change in the arc duration ϑ; this showed that ϑ was a direct function of the temperature. Special measurements were later made over a wide range of temperatures; they showed that, in an arc with a liquid-mercury cathode, the duration of the arc actually does always increase with increase in temperature, in spite of the assertions that there is a maximum value of ϑ in the vicinity of 20°C [137, 138]. A number of interesting effects were observed during these measurements, so that there is some point in stopping to consider them in more detail.

The results of measuring ϑ from -180 to +130°C are shown in Fig. 22 as a series of curves[1] plotted in the usual (I, ϑ) coordinates. Each of the curves is for a definite temperature. The solid curves give the results of measurements with a liquid-mercury cathode, while the dotted curves are for a solid cathode. Throughout the whole range of temperatures studied, the curves have the characteristic break dividing them into two linear seg-

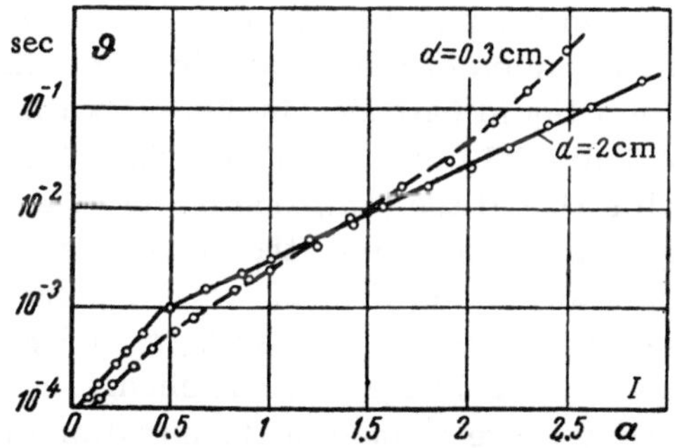

Fig. 21. The function ϑ(I) for small interelectrode distances.

[1]Here and in the rest of the book data are given for measurements made with a source voltage of $\mathscr{E}$ = 120 V and a circuit inductance or L < 50 μH.

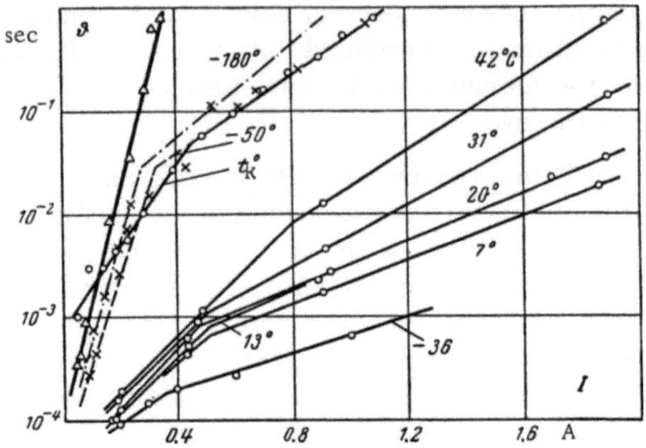

Fig. 22. Results of measuring the mean arc duration ϑ for different temperatures and different cathode states. The curve t_k is for an arc with a boiling cathode. The curves for -50 and -180°C are for a solid cathode. The heavy line at the top shows the results of making measurements with an anchored spot.

ments with different slopes; thus the break must be related to some general law. The slope of the low-current segment is always greater than the slope of the high-current segment. Under all conditions, ϑ for an arc with a liquid cathode increases monotonically with increase in temperature, reaching a limit when the cathode starts to boil. The increase occurs both by the motion of the whole curve upward and by an increase in the slope of both segments, which means a change in the parameters ϑ_0, ϑ_1, ψ_0, and ψ_1 in Eq. (13). The approximate values of the parameters for different temperature conditions in a liquid cathode, including boiling, are given in the upper part of Table VI for a short arc ($d_a = 0.3$ cm). It may be seen from the table that if the temperature is changed from -36°C to the boiling point of mercury, the parameters ψ_0 and ψ_1 increase by approximately a factor of 2, while ϑ_0 and ϑ_1 increase by two orders of magnitude. Since, however, ψ_0 and ψ_1 enter into the exponent of a power, a relatively small change in these quantities can produce a very large change in the duration of the arc at high currents. It is easily seen from the table that the various parameters do not all change in the same way. While the changes in ψ_0 and ψ_1 occur predominantly in the intermediate temperature range, the increase in ϑ_0 and ϑ_1 occurs almost exclusively at

TABLE VI

T_k, °C	ψ_0, A^{-1}	ψ_1, A^{-1}	$\vartheta_0 \cdot 10^4$, sec	$\vartheta_1 \cdot 10^3$, sec
Boiling	9.2	4.45	15	50
42	6.5	4.1	0.8	—
31	6.5	3.5	0.72	1.0
20	4.5	2.8	0.55	0.6
—36	4.1	2.06	0.55	0.25
—50	20	4.45	—	30
—180	21	5.2	3	30
Anchored spot	25.6	—	7	—

high temperatures and probably comes principally from boiling of the mercury.

A central question in the effect of temperature on the arc is what happens to the stability when the cathode passes from one state of aggregation to another. As shown by the dotted curves (Fig. 22), plotted from the data on a solid mercury cathode at -50 and -180°C, freezing the mercury produces an unexpected discontinuous increase of approximately 3 orders of magnitude in the duration of the arc. Perhaps the most interesting thing about this effect is that, within the accuracy of measurement, the increase in stability is such that the arc has the same duration along the high-current part of the curve as when the cathode is boiling. This can scarcely be the result of chance coincidence. Thus, as the temperature is reduced, ϑ decreases monotonically over a very wide range of temperatures corresponding to a liquid cathode; but when the cathode solidifies, ϑ suddenly increases to the values shown at the start by an arc with a boiling cathode. Further reduction in temperature to -180°C produces only a relatively insignificant increase in ϑ. This remarkable behavior is shown graphically in Fig. 23, which gives a curve of ϑ as a function of temperature, plotted for a current of 1 A from the data of the preceding figure. The values of the various parameters in Eq. (13) for a solid cathode

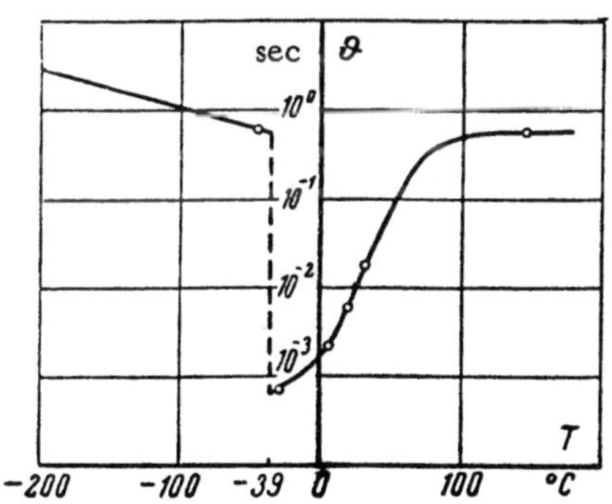

Fig. 23. Change in ϑ with a cathode temperature for a current of 1 A.

are given in the lower part of Table VI. A careful comparison of the data in this table shows that high values of the parameter ϑ_0 are especially characteristic of an arc with a boiling cathode, while if the arc has a solid cathode our attention is attracted by the exceedingly high values of the parameter ψ_0, which gives the slope of the low-current part of the curve. This difference shows that the boiling cathode arc has relatively high duration even at the smallest currents, while the high duration of the solid cathode arc is achieved as a result of the large slope of the initial part of the curve.

In view of the sudden change in the nature of the curves observed with change in state of aggregation of the cathode, the data on the lifetime of an arc with a cathode spot anchored in place on a thin film of mercury wetting a nickel plate are of some interest. The results of these measurements are shown in Fig. 22 by the heavy straight line. In contrast to the curves obtained for other conditions, we have in this case an exact exponential relation between ϑ and I over the whole range of ϑ investigated, i.e., at least over a range of 4 orders of magnitude. The values of the parameters in the equation for this case are given in the bottom line of Table VI. The arc with the anchored spot has the highest values of ψ_0 and ϑ_0, i.e., the highest stability.

In interpreting the relation found between the arc duration ϑ and the temperature we must consider the possibility that ϑ is affected by two or more factors at the same time. The suddenness of the discontinuous increase in arc stability when the temperature is reduced below -40°C, and the fact that it coincides with the freezing point of mercury, gives us confidence in ascribing the whole change to the change in state of aggregation of the cathode. This conclusion is supported by the fact that the increase in stability always lags somewhat behind the decrease in temperature and coincides with the instant at which the mercury crystallizes. A similar discontinuous increase in arc duration when the cathode solidifies has been found for a number of other metals with low melting points, such as bismuth, lead, indium, etc. Thus, this effect may be considered one of the general laws governing the behavior of the cold arc.

Somewhat less well defined is the reason the duration of the arc gradually increases when the temperature is increased in a liquid cathode. This could be the result of both the increase in the

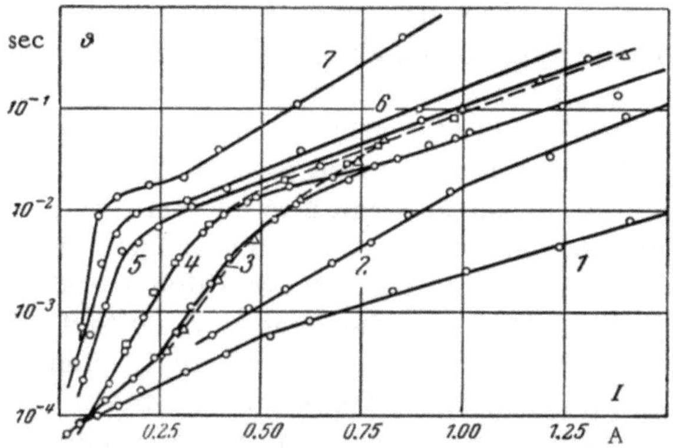

Fig. 24. Results of measuring ϑ with helium (solid curves) and neon (dotted curves) present. 1) p_{He} = 0; 2) p_{He} = 1 mm Hg; 3) p_{He} = 6 mm Hg; 4) p_{He} = 20 mm Hg; 5) p_{He} = 63 mm Hg; 6) p_{He} = 100 mm Hg; 7) p_{He} = 200 mm Hg ;---Neon: p_{Ne} = 1 and 6 mm Hg.

equilibrium mercury vapor pressure in the discharge space and the simple fact that the cathode temperature is increased. It can only be assumed that in the present case the first factor predominates over the second. In order to substantiate this conclusion and give it definite form, a detailed study must be made of the effect of the surrounding medium on the stability of the arc, with the temperature and state of aggregation of the cathode kept constant.

c. Effect of a Gaseous Medium on the Stability of the Arc. There is particular interest in the effect of inert gases on the arc. The results of measuring the duration of a short arc in an atmosphere of mercury vapor and helium are shown by the solid curves in Fig. 24. Each of the curves corresponds to a definite pressure. The lowest curve is for a pure mercury discharge. When helium is introduced, the curves undergo changes showing an increase in arc duration over the whole range of currents studied. At the same time, however, the nature of the change in the curves is not the same in the low- and high-current part of the curves. While the change in the low-current part of the curve consists principally in an increase in slope, the high-current part is displaced upward without appreciable change in slope, as long as the helium pressure stays below 200 mm Hg. Since the slope of the low-current part of the curve increases continuously with increase in gas pressure, the break in the curve becomes sharper and sharper. At the same time, the break point moves uniformly toward large times and low currents. Our attention is also attracted by the fact that the magnitude of the shift in the high-current part of the curve is practically constant for helium pressures from 5 to 50 mm Hg.

Similar changes in form and location of the curves are noted in the case of neon, the results for which are shown by the dotted curves in Fig. 24. The difference in the effect of helium and neon on the stability of the arc is of a purely quantitative character. The two gases cause the same changes in slope of the first part of the curve at pressures that are inversely proportional to the atomic weights. The effect of the gases may be included in Eq. (13) by increasing the parameters ψ_0 and ϑ_1 and reducing I_1. When using a short arc, as in the present case, the value of the first of these parameters may be written as a function of the gas pressure p using the approximate empirical formula

$$\psi_0 = 4.5 + 1.5(Ap)^{1/2}, \tag{16}$$

where A is the atomic weight of the gas and p is expressed in millimeters of mercury. As for the other two parameters, ϑ_1 increases by a factor of 20 or 30 when the gas is present, while I_1 decreases from 0.5 to 0.1 A as the pressure of the medium is increased.

The heavy inert gases exert an effect on the arc stability similar to that described, as may be gathered from the data of Fig. 25, which shows the results of measurements made with argon and xenon. The quantitative difference between the effects which various inert gases have on the arc is simply that, as the atomic weight of the gas is increased, the stabilizing effect becomes less and less sharp. As a matter of fact, for equal gas pressures the curves for the heavier gases are on the whole always located above those for the light gases. In all probability this effect comes not so much from the actual atomic weights of the gases as from their critical potentials.

If the changes in the $\vartheta = f(I)$ curves produced by inert gases are compared with the temperature changes discussed previously (Fig. 22), it may be seen that the latter are actually to a considerable extent due to the increase in mercury vapor pressure in the discharge space. An indication of this is the characteristic

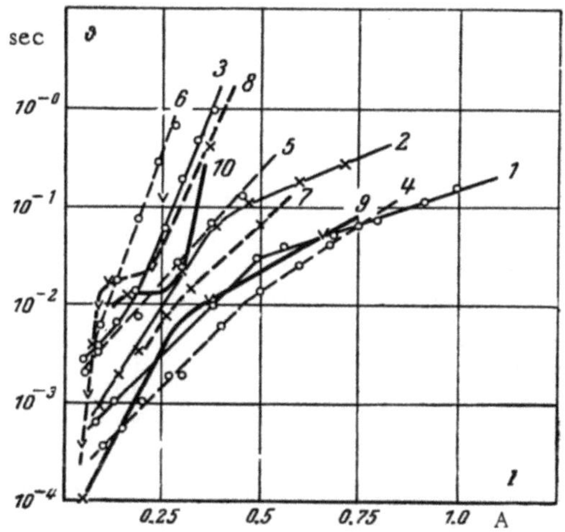

Fig. 25. Results of measuring ϑ with argon, xenon, hydrogen, and air present. Xenon: 1) p_{Xe} = 1 mm Hg; 2) p_{Xe} = 20 mm Hg; 3) p_{Xe} = 100 mm Hg. Argon: 4) p_{Ar} = 1 mm Hg; 5) p_{Ar} = 50 mm Hg; 6) p_{Ar} = 200 mm Hg. Hydrogen: 7) p_H = 1 mm Hg; 8) p_H = 4 mm Hg. Air: 9) p = 11 mm Hg; 10) p = 4 mm Hg.

upward displacement of the high-current part of the curves. However, the simultaneous displacement of the whole low-current part of the curve may be explained in this case as a direct result of the effect exerted by the cathode temperature on the stability of the arc, which can in turn be related to the reduction in thermal losses at the cathode spot.

Air as a medium exerts a considerably stronger effect on arc stability than inert gases, as is shown by the heavy curves (see Fig. 25), which are for air pressures of about 1 and 4 mm Hg. The same marked increase in arc stability is produced by hydrogen, although here the operating voltage of the discharge is greatly increased. The curves for hydrogen are shown by heavy dashed lines in the same figure. A comparison of the curves suggests that the sudden reaction of the arc to introducing air into the discharge space may be due precisely to the hydrogen contained in the air. Among the peculiarities of the reaction of the arc to introducing hydrogen and air must also be included the gradual increase in stability with time which is one of the reasons for poor reproducibility of the measurements.

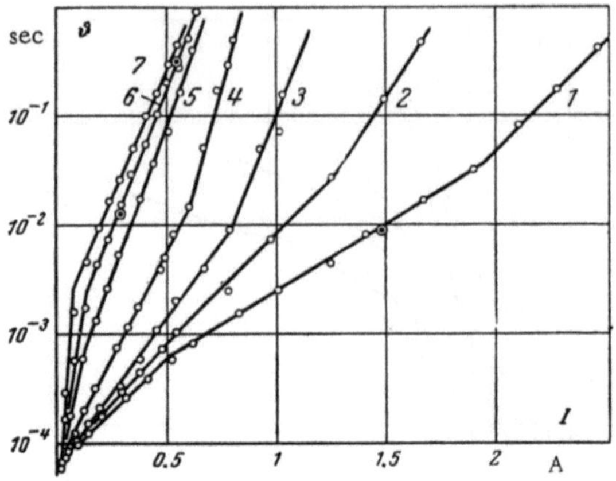

Fig. 26. Curves of $\vartheta(I)$ for various longitudinal magnetic field strengths H in a short arc (d_a = 0.3 cm). 1) H = 0; 2) H = 450 Oe; 3) H = 1000 Oe; 4) H = 2070 Oe; 5) H = 3270 Oe; 6) H = 5500 Oe; 7) H = 7250 Oe.

d. Stabilizing Effect of a Magnetic Field. One of the most satisfactory ways of varying the internal conditions of the arc is to apply a longitudinal magnetic field. This type of field has the minimum effect on the motion of the cathode spot and the voltage drops in the discharge, so that the changes in conditions produced by the field will be of a comparatively simple sort. The changes consist principally in distorting the trajectory of the electrons, the net effect of which is to reduce their transverse diffusion coefficient. The results of measuring the mean arc duration as a function of current and magnetic field strength are given in Figs. 26 and 27 as a family of $\vartheta = f(I)$ curves plotted to our usual semilog scale. The lowest curve in both figures is the ordinary curve of ϑ vs. current taken for the normal arc unperturbed by an external magnetic field. In making this series of measurements, the residual field in the electromagnet was completely eliminated by means of a small demagnetizing current. The other curves illustrate the same relationship, but with a longitudinal magnetic field. The data in the first figure are for a short arc with an interelectrode distance of about 0.3 cm, while in the second figure the distance is about 2 cm. The curves are plotted from the readings given by the integrating circuit. In contrast to this, the circles with the dot inside (Fig. 26) give values of ϑ derived from the results of finding

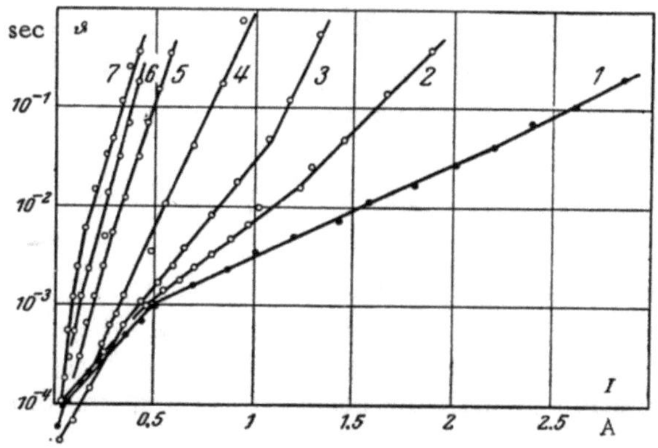

Fig. 27 Curves of $\vartheta(I)$ for various values of H with d_a = 2 cm.
1) H = 0; 2) H = 450 Oe; 3) H = 1030 Oe; 4) H = 2090 Oe;
5) H = 4030 Oe; 6) H = 5640 Oe; 7) H = 7680 Oe;

the statistical distribution of the arc lifetimes and borrowed from Table V. These points are located very close to the curves for the conditions to which they correspond. This is probably the best justification for the method of measurement used and illustrates the excellent reproducibility of the results. In its basic features, the actual effect of the field on the arc is as follows:

1. Application of the field, just as the introduction of an inert gas, increases the duration of the arc over the whole range of currents studied.
2. As in the inert gas experiments, the increase is achieved in the low-current part of the curves by an increase in slope.
3. In contrast to the effect of inert gases, the field causes a sharp increase in the slope in the high-current part of the curve; the slope reaches a maximum value at H = 4 kOe.
4. Under the action of a strong field, the break point in the curve is displaced toward low currents.
5. The stabilizing effect of the field is more marked at large interelectrode distances than when the distances are small.

It follows from the changes in form of the ϑ vs. current curve in the presence of a field that the effect of a magnetic field

TABLE VII

	d = 0.3 cm		d = 2 cm	
	H = 0	H = 7250 Oe	H = 0	H = 7680 Oe
I_1, A	0.5	0.08	0.5	0.08
ϑ_1, sec	$6 \cdot 10^{-4}$	$2.6 \cdot 10^{-3}$	10^{-3}	$2.4 \cdot 10^{-3}$
ψ_1, A^{-1}	2.8	10.8	2.2	16.4

is to change almost all the parameters in Eq. (13). The effect of the field in the low-current part of the curve is felt principally by the parameter ψ_0, which for a short arc (d_a = 0.3 cm) may be written as a function of the field strength H using the empirical equation

$$\psi_0 = 4.5 + 1.17H^2, \tag{17}$$

where the field strength is expressed in kilooersteds. It is still impossible to say with confidence whether or not the field has any effect on ϑ_0. If any such effect occurs, it is in any case small, so that the value of this parameter may be taken from Table VI even for an arc under the influence of a magnetic field.

As a result of the complicated behavior of the curves in the high-current region an analytic expression for ψ_1 as a function of H cannot be obtained as in the case of ψ_0. Table VII gives for comparison only the values of the various parameters for the two limiting cases: no field at all and the highest field. If we consider the quantitative side of the question of the effect of the magnetic field on the arc, our attention is drawn to the almost astronomical figures giving the increase in arc duration when a strong field is present. Thus, from the data of Table VII, at a field strength of the order of 7000 Oe and at an interelectrode distance of 2 cm the parameter ψ_1 increases by a factor of 8 as a result of the presence of the field. At a current of 2 A, this corresponds to an increase in arc duration by a factor of approximately 10^{12}.

In view of the large increase in arc stability observed when the cathode solidifies or a magnetic field is applied, the question arises as to whether these effects will be superimposed when a field acts on an arc with a solid cathode or whether they are mutually exclusive. The answer to this question is given by the curves of Fig. 28, which gives a comparison of the effects of a field with a strength of about 7 kOe on the stability of an arc with a liquid and with a solid cathode. The solid curves on the graph are for an arc

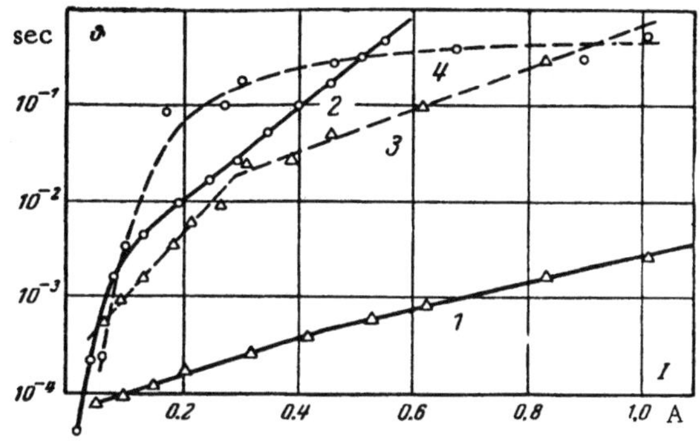

Fig. 28. Effect of a magnetic field on the duration of an arc with a liquid and with a solid mercury cathode.
Liquid Cathode (T = 20°C): 1) H = 0; 2) H = 7 kOe.
Solid Cathode (T = −50°C): 3) H = 0; 4) H = 7 kOe.

with a liquid cathode; the lower curve is for a normal arc while the upper is for an arc under the action of a magnetic field. Similarly, the dotted curves illustrate the effect of a field on an arc with a solid cathode. In the liquid cathode arc, application of the field produces a marked increase in arc duration, increasing with increase in current, while in the solid cathode arc the effect of the field is less marked and decreases practically to zero when the current is raised to 1 A. This latter effect could be due to the movement of the spot out of the region of the cathode lying directly under the anode.

The reaction of an arc to a field applied when the discharge space is filled with a quite dense gas is of considerable importance to an understanding of how a magnetic field stabilizes the arc. The results of a study of the effect of the field on the arc in helium and xenon atmospheres are shown in Figs. 29 and 30. The abscissas are the magnetic field strength H, and the ordinates are the relative values of arc duration, ϑ_H/ϑ, for the field strength used. The individual curves give ϑ as a function of H at different gas pressures. The value of ϑ is not increased by the field unless the field strength exceeds some threshold value H_p, which depends on the kind of gas and on the pressure. This value increases with increase in gas pressure and for the same pressures is higher for xenon than for helium. Because of the differences in form of the curves, and the fact that there is no clearly defined break in the

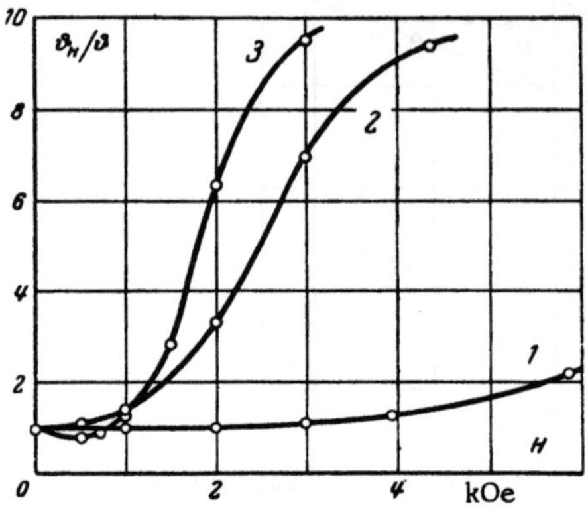

Fig. 29. ϑ as a function of longitudinal magnetic field strength H in helium.
1) p_{He}=300 mm Hg; I = 0.3 A
2) p_{He}=100 mm Hg; I = 0.3 A
3) p_{He}=100 mm Hg; I = 0.6 A

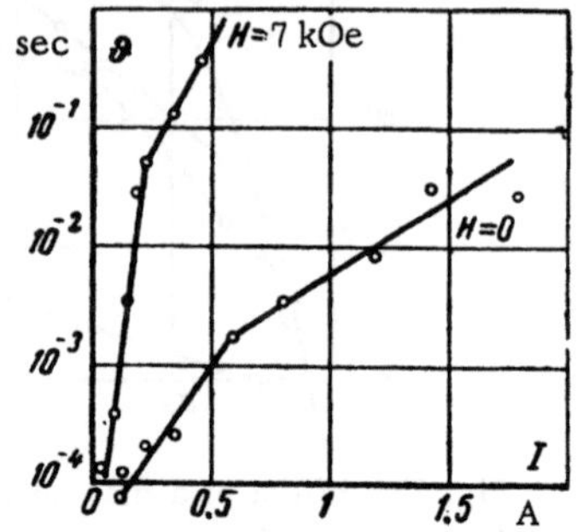

Fig. 31. Results of measurements of ϑ with an arc excited by means of a pulse breaking the vacuum gap, with H equal to 0 and 7 kOe.

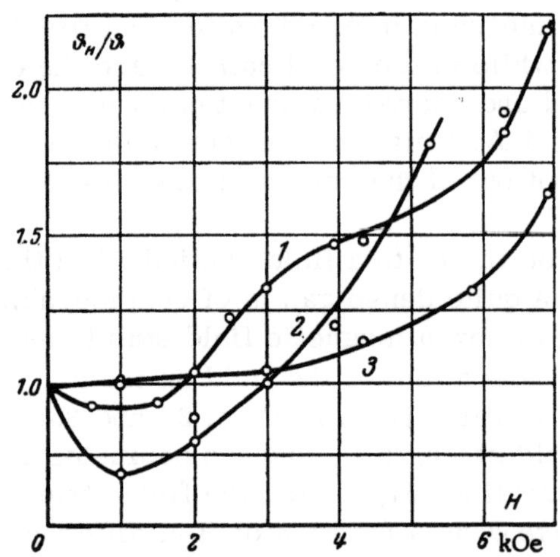

Fig. 30. ϑ as a function of H in xenon.
1) p_{Xe} = 25 mm Hg; 2) p_{Xe} = 50 mm Hg;
3) p_{Xe} = 100 mm Hg.

TABLE VIII

Gas	Helium	Xenon
Range of pressures studied, mm Hg	100-300	25-100
Range of variation of H_p, kOe	1.0-4.0	1.5-5.0
Calculated values of H_p, kOe	1.0-3.0	0.6-3.0

increase in ϑ, it is impossible to get accurate values of H_p from the data of the experiments. It is only possible to give approximate limits for the variation of the threshold field strength in the range of pressures of the two gases studied, as is done in Table VIII.

e. Effect of the Method of Exciting the Arc. Before we try to evaluate the results of the measurements just described, it should be emphasized that the special features observed in the $\vartheta = f$ (I) curves and, in particular, the deviation from a purely exponential form of the function do not come from errors in method of measurement, but give expression to inherent properties of the arc. In order to eliminate methodological errors it was particularly important to be sure that the break in the curves usually observed at currents of about 0.5 A does not come from using a semiconducting igniter in the experiments. For this purpose, a series of control measurements was made in which the discharge was excited by breaking down the vacuum gap between the cathode and an auxiliary sharp-pointed electrode above it, using high-voltage pulses of short duration for the purpose. Here, the sharp point was the negative electrode in the auxiliary discharge, so that the auxiliary voltage played no role whatever in forming the cathode spot on the mercury. In addition, the current at the point from the main power supply was limited by a resistance of such a size that a cathode spot could not be maintained on this electrode by the power supply. In spite of the greatly different conditions under which the arc was excited, and in spite of the fact that the contact between the cathode spot and the semiconducting igniter was eliminated in the control experiments, the curves obtained from these measurements showed the same features as before. Figure 31 shows two curves taken during the control measurements made with the arc excited in the way described above. The lower curve is for a normal mercury arc with no external field; the upper curve is for an arc with a longitudinal magnetic field of about 7000 Oe. In spite of the fact that there is considerably more spread in the points, partly due to failures of the arc to ignite, the curves clearly show the familiar break, the sharp increase in slope produced by the field, and the shift in the break point toward low currents and large times.

26. Discussion of the Results of Measurements

In summarizing, we must bring the fundamental results together for comparison and try to extract from them information as complete as possible on the arc cycle that we are interested in. The most general result of the observations is that there is a wide range of discharge currents over which spontaneous arc extinction occurs. For the normal low-pressure arc with a liquid mercury cathode, this range extends approximately from 0.07 to 7 A. Although the desired form of discharge can be produced even by a current corresponding to the lower limit of the above range, the arc cycle cannot be maintained for an unlimited time at any value of current in this range, and when it is extinguished remains extinguished. This tends to refute the idea that spontaneous extinction of the arc only occurs in a narrow range of currents, considered to be a transition range for the arc discharge. The measurements show further that, as the current is increased from 0.07 to 7 A, the mean duration of the arc increases continuously, approximately from 10^{-4} to 10^{3} sec and higher. In the various segments of this region an exponential relation exists between ϑ and the mean discharge current I. In the earlier papers there were some indications of the exponential character of the function ϑ (I) which were based on measurements of ϑ in the narrow current range of approximately 3-6 A [137,138]. Using a more highly developed method of measurement in the present work has enabled us to make observations over a considerably wider range of currents and to establish the fact that there are regular departures from the simple exponential law, consisting in a clearly defined break between separate exponential segments. With deviations of this sort, the function ϑ (I) may be represented by a curve consisting of two or more exponential segments intersecting at break points occurring at definite values of current. The break point that occurs most regularly is at a current of about 0.5 A. It is unaffected by changes in the external circuit constants and changes its position only when changes are made in the temperature or state of aggregation of the cathode, in the dimensions of the gap, or in the gas filling the gap. It is important that changes in the conditions under which the discharge takes place are reflected differently in the behavior of the two exponential segments of the curve intersecting at the break point. It may be concluded from this that the break in the curve at a current of about 0.5 A comes from inherent changes in the properties of the cathode region of the arc. In all probability, a change in structure of the cathode spot itself occurs in this range of currents. That changes of this sort necessarily occur follows from the observations of Froome [12,16], who came to the conclusion that the cathode spot on mercury consists of separate emission foci, each of which has a current of the order of 1 A. The focal structure of the cathode spot on mercury is supported by the observations given in the next chapter. According to the data given there, the low-current part of the curve may be considered to give the range of currents in which formation of a single spot is brought to completion. New spots are formed in the high-current part of the curve, with two or more spots existing simultaneously on the cathode for a certain part of the time.

Attention should be given not only to the two-segment nature of the function ϑ (I), but also to the observed differences between the effects produced on the function by the external circuit constants and the conditions existing within the discharge. As has already been pointed out, the change in form of the ϑ (I) curve produced by altering the external circuit constants is of an easily defined sort, appearing only as a change in slope of the high-current part of the curve that may be taken care of completely by changing the parameter ψ_1 in Eq. (13). In contrast to this, altering the conditions existing inside the discharge produces more profound changes in the curve, usually of a complex sort. Thus, when the cathode passes from the liquid to the solid state, a discontinuous increase is observed in the slope of the first part of the curve, and at the same time, the second part of the curve moved sharply upward. A similar transformation occurs in the curve if a gas is introduced or if the mercury vapor pressure in the discharge gap is increased. Heating the cathode to the boiling point raises the entire first part of the curve, while applying a magnetic field greatly increases the slope of both parts of the curve and the break point shifts toward low currents. If these internal changes are made in the discharge, all the parameters in Eq. (13) have to be changed.

These differences in the effect produced on the arc by external and internal factors can serve as the key to an understanding of the physical meaning of the individual parameters in (13) and to an interpretation of the exponential coefficient itself. Along with this, we are in a position to draw a number of preliminary conclusions as to the properties and stability of the arc cycle. In solving this problem it is convenient to imagine anything that affects the discharge as a superposition of two simple effects: a change in stability of the arc in the narrow sense of the word, and a change in the effectiveness of the particular restoring mechanism which operates every time the arc reaches the critical state or starts to decompose and whose action restores the discharge to its original state. That such a restoring mechanism exists is quite convincingly shown by the very relation between the arc lifetime and such parameters as the emf of the source and the inductance in the circuit. When interpreted in this way, the stability of the arc can depend only on the conditions existing inside the discharge. The effect of the external circuit can only extend to the restoring mechanism. We shall simply note for the present that the cause of the effect resides in the short-time voltage increases at the arc electrodes needed to activate the cathode processes in the arc cycle. Obviously, the effect of the external circuit constants will, in the final count, be that when they are changed, they either increase or decrease the rate at which the voltage rises at the arc electrodes, which will determine whether the probability of restoring the discharge increases or decreases. From this point of view it is possible to understand the difference in the effect of emf and inductance on the arc and of the effect of capacity on the arc. Since the effect of the external circuit is completely taken care of by the parameter ψ_1, which determines the exponential increase in arc duration with increase in current in the high-current part of the curve, ψ_1 is an index of the effectiveness of the restoring mechanism in this part of the curve. It is natural to conclude from this that the exponential coefficient gives the increase in duration of the discharge due to increasing the stability of the discharge in the proper sense of the word. In all probability, this applies to the same extent to the low-current part of the curve, which is controlled by the exponent ψ_0. The fact that ψ_0 is independent of the external circuit constants can only mean that in the low-current part of the curve the effectiveness of the restoring mechanism is not determined by the rate of voltage rise, but by the state of the cathode and the gas surrounding it. This last assumption is in complete agreement with the effect that the state of aggregation of the cathode and the surrounding medium are found to have on the duration of the discharge. The confirmation of this assumption, as well as the explanation of the actual mechanism that restores the arc in this portion of the curve, will be found later, when we investigate the oscillatory processes taking place in the cathode region of the discharge.

A fundamentally different role must be played by the parameters I_0, I_1, and ϑ_0. The coordinates of the special points on the $\vartheta(I)$ curve represent definite stages in the formation of the arc and different stages of stability under the conditions in the experiment. As was already pointed out in Sec. 24a, it is natural to consider the quantities I_0 and ϑ_0 as the threshold value of the current in an arc with a mercury cathode and as the normal duration of the arc at this current respectively. In contrast to this, I_1 and ϑ_1 can be considered as the current and duration of a completely formed arc cell. A reduction in the parameters I_0 and I_1 or an increase in ϑ_0 would be an indication of a true increase in stability of the discharge. We have observed an appreciable increase in ϑ_0 only when the cathode is boiling and when the spot is fixed on a thin mercury film. In all other cases, this parameter remains constant within the limits of accuracy of observation, from which it may be concluded that it is a quantity characteristic of the state of the cathode itself. A considerable reduction in I_0 and I_1 was observed when a gas was present and when a magnetic field was applied. This condition may be interpreted as a sign of a true increase in stability of the arc, brought about by the factors in question. The increase occurring simultaneously in the parameters ψ_0 and ψ_1 shows, however, that the effect of the field is complex. It may be thought of as the superposition of a stabilizing effect in the proper sense of the word and an increase in effectiveness of the restoring mechanism in the discharge when the field is present. This explains the exceedingly large increase in arc duration when the field is applied.

The data given in this section show that the mercury cathode arc is of limited duration over the quite wide range of currents of approximately 0.07-7-10 A, and that the mean duration gradually increases according to a definite law as the current is increased. From this we can easily calculate that even at a current of 10 A the duration of the arc will be at least 4 days. It is thus obvious that there is no point in investigating arc extinction at high currents. This is not simply because such observations would require an unjustifiably large expenditure of time, but because at currents of the order of 10 A and above, arc extinctions lose their value as an indication of instability in the discharge. Here, the first order of importance is yielded. Other manifestations of instability in the arc, considered below, are most important here.

B. OSCILLATIONAL PROCESSES IN THE CATHODE REGION AND THE RESTORING MECHANISM OF THE ARC

27. Voltage Oscillations in the Short Arc

In Part A, a discussion was given of one of the grossest manifestations of departure from the conditions required to maintain a stable discharge in a mercury cathode arc, namely spontaneous extinction. A statistical approach was used to investigate the extinction of the discharge. A quantitative description of the tendency of the arc to go out was provided by the data on the mean duration of the arc under varying experimental conditions.

Despite all the well-known advantages of statistical methods of observation they are still of a somewhat abstract nature. To gain a complete understanding of a phenomenon requires further investigation of its detailed structure and of the more important phenomena accompanying it; in this case, these include, first, the voltage oscillations in the arc as observed under all conditions. The oscillations are also of independent interest since they are one of the most important signs of internal instability in the arc. Thus, having started with an investigation of spontaneous extinction in the arc, we shall next investigate the oscillation processes occurring there, these being an externally less obvious, but nevertheless more consistent form of departure from the conditions required to maintain the stability of the discharge.

a. Method of Observation. If oscillograms are taken of the voltage at the electrodes of an operating arc, a number of oscillation processes are usually observed, differing from one another in amplitude, frequency range, and in the causative factor. For the questions under discussion here, there is special interest in the voltage oscillations located at the cathode. The frequencies lie in the range 10^4-10^7 cps, and are oscillations of the relaxational type. These oscillations may be observed in purer form in a short arc which has no positive column. In order to discover the nature and the basic laws governing the oscillations, and also to establish the reason for spontaneous extinction of the arc, we must turn to an investigation of the behavior of a short arc in the low-current range, where the discharge is of short duration.

Oscillograms of the voltage oscillations in a rapidly decaying arc were taken by two different methods. In one case, the oscillograph scanning circuit was triggered with a suitable delay after the arc had been excited without changing the low current under study, while in the second case the scanning was triggered at the instant the current was suddenly reduced to a fixed value below some initial value sufficient for prolonged operation of the arc. By changing the delay time, it was possible to follow the change in voltage on the electrodes of the arc from the instant it was excited, or from the instant the discharge current was reduced up to the point where it was extinguished. A method was also provided for simultaneously oscillographing the various changes in light flux emitted by the cathode spot. The simplest scheme for investigating the oscillations by the second method is given in Fig. 32. The tube T_1 with a mercury cathode and an anode located close to it was connected in a circuit consisting of a storage battery and a system of noninductive resistances R_1 and R_2 connected in series. The cathode-anode distance was 0.3 to 0.5 cm. A device in the center of the cathode for anchoring the discharge consisted of a small triangular piece of sheet molybdenum with one of the corners protruding slightly above the surface of the mercury. The cathode spot, excited by means of a contact needle, was very quickly anchored at the boundary where the mercury wetted the strip. There was a small hole in the anode opposite the anchoring mechanism through which the light flux from the cathode spot in the parts of the discharge closest to it passed to a glass prism and from there through the diaphragm D to the photomultiplier P. By focusing the image of the spot in the plane of the diaphragm and moving the diaphragm up and down by means of a micrometer screw, we could observe the different parts of the light flux coming from the spot itself or from the neighboring parts of the discharge. The dual-trace oscillograph O permitted the oscil-

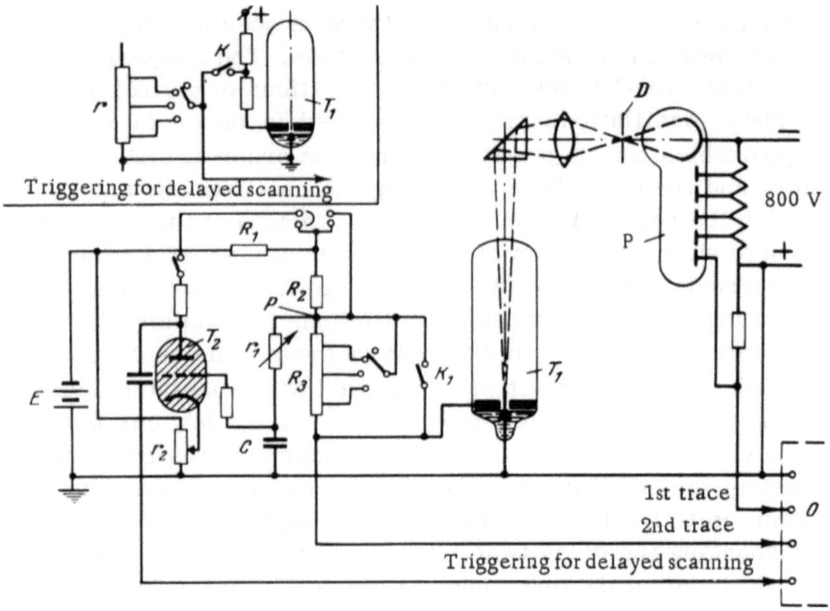

Fig. 32. Wiring diagram of tube T_1 for the study of oscillation processes in the arc.

lations in voltage and light flux to be observed simultaneously. To investigate the voltage oscillations in the arc with the spot moving freely over the cathode, it was only necessary to raise the mercury level in the tube slightly, thus covering up the anchor. Under these conditions observations on the radiation from the arc could only be made through the side walls of the tube.

In making observations at low currents, the current in the discharge circuit was suddenly reduced by switching in the additional set of noninductive resistances R_3 with a relay having an arc-extinguishing device. The delay circuit consisted of the variable resistance r_1 and the capacity C. When the switch K_1 is opened, the potential of the point P suddenly increases and at the same time the condenser begins to charge. When the potential difference on the plates reaches a definite value, the thyratron T_2 fires, triggering the delayed scanning on the oscillograph. In order to be sure that the phenomena observed when the current is reduced do not depend on the tube connections, and in particular do not come from an arc produced in the external circuit when the switch K_1 is opened, the second method shown inthe upper left-hand corner of Fig. 32 was also used to control the current. In this

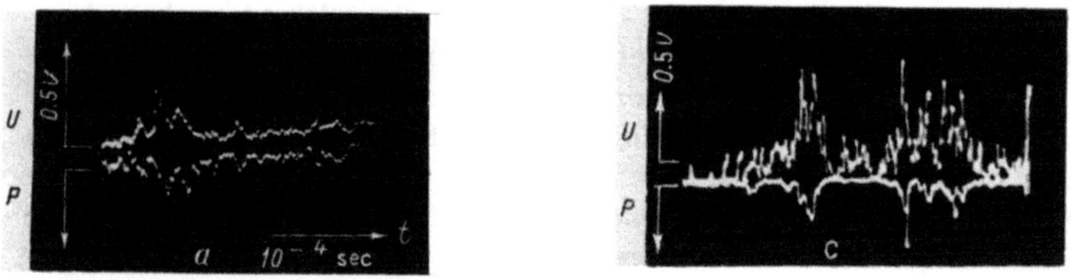

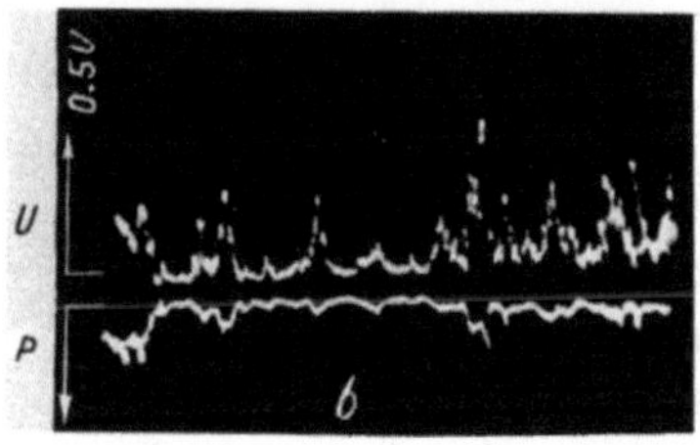

Fig. 33. Oscillograms of oscillations in the voltage U of a short arc and the light flux P of the cathode spot, with the spot anchored at the boundary where the molybdenum is wetted by mercury. a) I = 1.4 A; b) I = 0.4 A; c) I = 0.27 A.

form of the circuit, the current was reduced by shunting the tube with the resistance r_1, the value of r_1 being chosen so that the discharge current was reduced by the desired amount. Consistent results were obtained from all these forms of the experiment. We shall begin our presentation of the results with a short description of the oscillation processes occurring in an arc with an anchored spot.

b. Oscillations in an Arc with an Anchored Spot. The oscillation processes occurring in an arc with the spot anchored are of interest because oscillations occurring under the special conditions in which the discharge runs quietly without any secondary phenomena, such as chaotic spot motion or mercury boiling, can only come from inherent instability in the arc itself.

A typical dual-trace oscillogram for a short arc with an anchored spot at a current of 1.4 A and aperiodic conditions in the discharge circuit is shown in Fig. 33a. The upper trace shows the variation in anode voltage, and the lower shows the change in light flux from the cathode spot itself; the voltage increases in the upward direction, the light flux in the downward direction. Both the voltage and the light flux undergo continuous synchronous oscillations. The amplitude of the oscillations in U is of the order of 0.1 V for a mean arc-electrode voltage of about 7.5 V. The oscillations cannot be identified in either amplitude or form with the statistical fluctuations inherent in every discharge. This indicates that there are continued changes in the arc in the cathode spot. At the same time, at a current of 1.4 A, an arc with an anchored spot can maintain itself practically indefinitely. If we judge the stability of the discharge from this one fact, we would have to say that when the arc is operating this way it is completely stable. But, as a matter of fact, the discharge is kept running as long as it does by oscillation processes. Oscillations of this sort occur in a mercury arc no matter how large the currents are. But a process that can only maintain itself by means of continuous oscillations cannot be considered stable.

One of the characteristic features of the oscillograms under discussion is that they are unsymmetrical. They are obviously made up principally of pulses corresponding to an increase in voltage and light flux. This suggest the idea that the oscillations are the result of a superposition of positive pulses in rapid sequence. In this case we would have to try to resolve them into separate pulses by reducing the current, thus going to simpler experimental conditions. Just how correct this assumption proved to be is shown by the oscillograms (Figs. 33b and c) for currents of 0.4 and 0.27 A respectively. The nature of the oscillations changes appreciably as the current is reduced. At low currents, the coordinated oscillations in voltage and light flux actually take on the form of single or grouped pulses. Sometimes something of the nature of a smooth spot occurs between pulses. The amplitude of the pulses varies approximately from 0.01 to 1 V, increasing in general with decrease in current. The duration of the individual pulses, including the amplifier distortion correction, is not greater than 10^{-7} sec. Most often, extinction of the arc is preceded by a group of high-amplitude pulses. At currents below 0.07 A, an arc with an anchored spot can no longer exist even for a short time, from which it may be concluded that the threshold current value for this form of arc discharge lies somewhere near 0.07 A.

c. Oscillation Processes in an Arc with the Cathode Spot Moving Freely Over the Mercury. From the standpoint of interpreting the laws governing arc extinction as found in the previous section, special interest attaches to the oscillation processes occurring in a normal arc with the spot moving freely over the cathode. For this type of discharge, we shall first consider the voltage oscillations occurring in the range of discharge currents corresponding to the low-current exponential segment of the curve

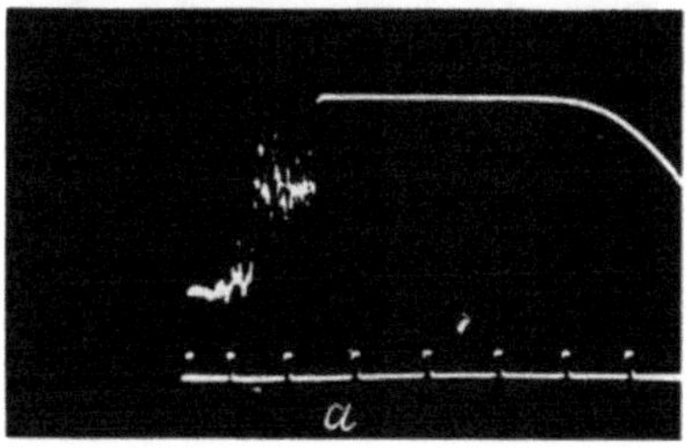

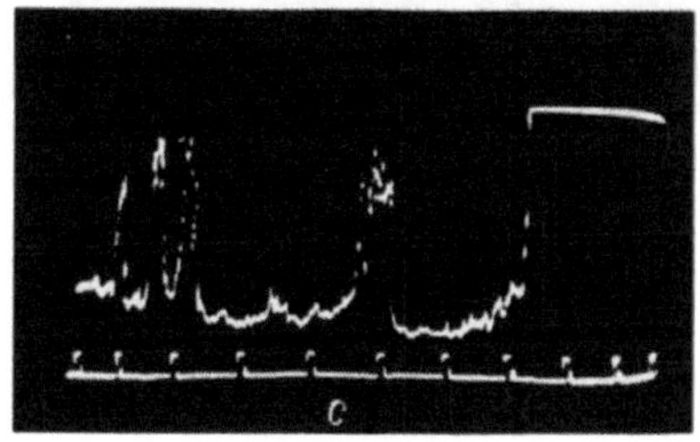

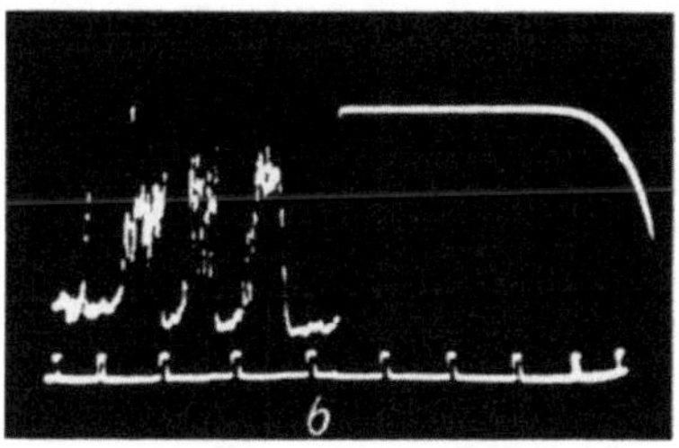

Fig. 34. Oscillograms of voltage oscillations in a short arc (d_a = 0.5 cm) with a homogeneous mercury cathode in the current range 0.07-0.5 A. Amplitude of markers 1 V, frequency 10 kc. a) I = 0.08 A; b) I = 0.3 A; c) I = 0.5 A

ϑ(I), where the simplest experimental conditions occur. Oscillograms of the arc voltage taken for a number of increasing currents in the range 0.07-0.4 A are given in Fig. 34. These show that the voltage on the electrodes of a short arc with a moving spot undergoes cyclic changes of a complex sort. They may be thought of as a superposition of two types of oscillations with periods of different lengths. One type is in the form of pulses following one another with an interval of about $4 \cdot 10^{-6}$ sec, and varying in amplitude. These pulses are a constant accompaniment to the arc and constitute its basic rhythm or pulse. Longer cyclic voltage changes occur also. There is a more or less gradual increase in voltage during the greater part of each of these changes. This cannot help but give the impression that some sort of process with thermal inertia is occurring in the cathode spot. As a rule each of the long periods ends with a group of relatively high-voltage pulses with an amplitude of the order of 10 V. On the oscillograms, these groups usually have the form of a monolithic pulse with a broad base and a modulated top, which may be partly due to insufficient resolving power in the oscillograph. The final rise in voltage is replaced by a sudden drop-off to its original value, which corresponds to the start of a new period. The length of these periods is on the average about 10^{-4} sec, with a considerable spread on both sides.

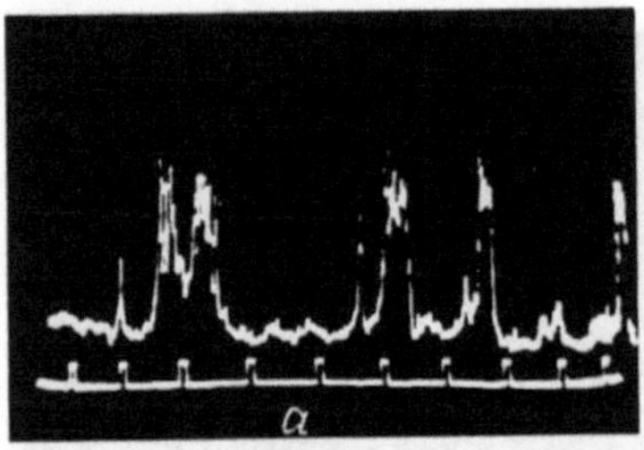

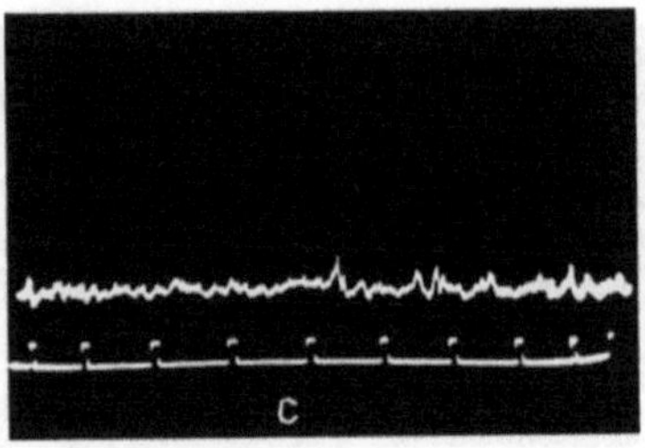

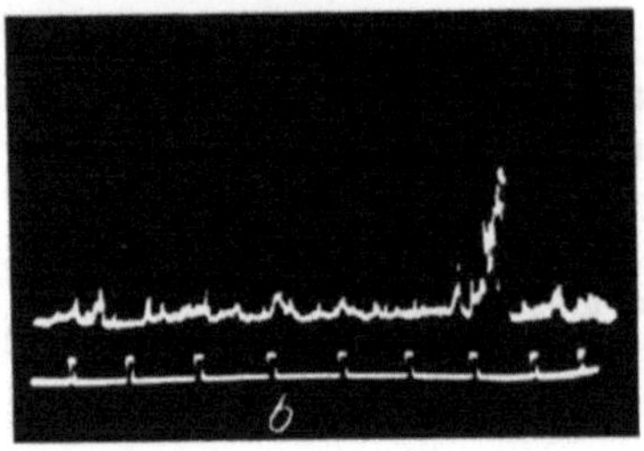

Fig. 35. Oscillograms of voltage oscillations in a short arc (d_a = 0.5 cm) with a homogeneous mercury cathode in the current range 0.5-2 A. The voltage and time scales are the same as in Fig. 34.

These features in the oscillograms of the arc voltage for currents up to 0.5 A lead one to the idea that, under these experimental conditions, and as long as the arc lasts, it undergoes the same development cycle time after time, passing through a number of repeating states. In describing these states, it is natural to take as the measure of instability the amplitude of the oscillations making up the fundamental rhythm of the arc. They always increase in amplitude at the end of the period, from which it follows that the most stable part of the period is the beginning and the least stable is the end. Thus, there is a steady increase in instability of the discharge within each period. The lógical end of the process should be extinction of the arc. Actually, in the current range under discussion and under normal conditions, the discharge is extinguished in the overwhelming majority of cases at the end of the period, when the electrode voltage has risen. These considerations allow us to consider each of the initial cycles in the voltage change as the normal arc development cycle under these conditions, while extinction of the discharge is its natural end. This provides

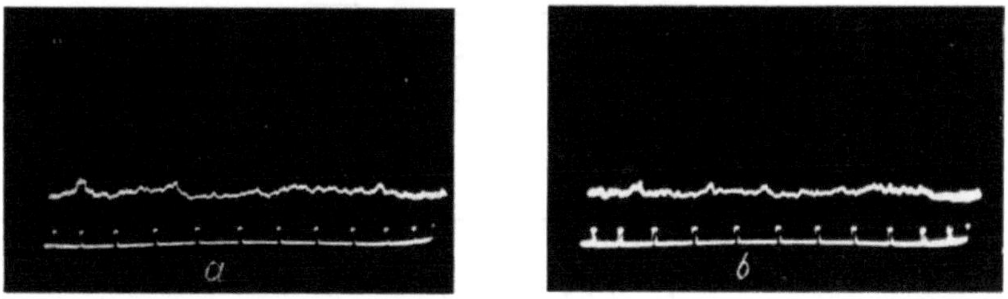

Fig. 36. Oscillograms of voltage oscillations in an arc at discharge currents of 5 A (a) and 10 A (b).

a simple and obvious connection between the voltage oscillations and spontaneous extinction of the arc.

However, these simple laws governing the oscillations are only found in a range of discharge currents up to approximately 0.5 A. Even in this range, when the current is increased to 0.4-0.5 A we note an appreciable reduction in the frequency of occurrence of pulses belonging to the high-voltage group. With further increase in current, large voltage rises become less and less frequent in the oscillograms. A careful study of the oscillograms shows that the reason for this is that there are more and more periods that only end with a small voltage rise of the order of 1 V. This change in form of the oscillations with increase in current is clearly visible in oscillograms (Fig. 35) arranged in order of increasing current from 0.5 to 2 A. Along with the change in the oscillations, another change is noted in the conditions under which arc extinctions occur. For currents from 0.5 to 2 A, in addition to extinctions of the arc occurring with considerable rise in voltage, sudden cessations of the discharge, without any preceding voltage rise, appear; this would indicate that there is a gradual increase in the instability phenomena in the discharge. As low as 5 A, the cyclic changes in voltage on the arc electrodes are almost exclusively represented by oscillations belonging to a low-voltage group with an amplitude of the order of 1 V, as shown by the oscillograms (Fig. 36) taken for currents of 5 and 10 A. Further increase in current has no appreciable effect on the form of the oscillations. They have this form for any larger discharge current and are thus an inalienable attribute of the arc.

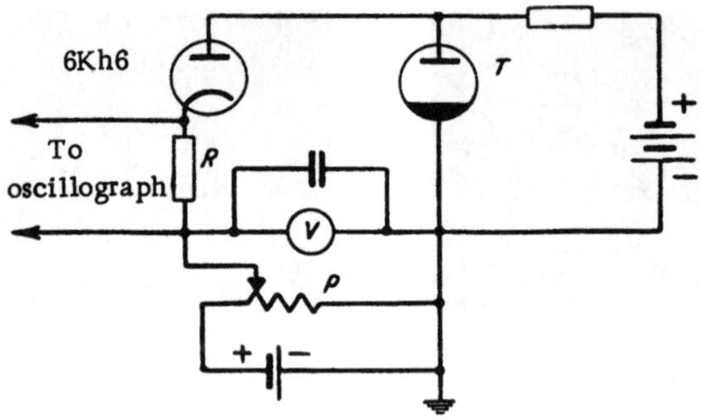

Fig. 37. Null method of measuring the voltage on the arc electrodes.

28. The Transitional Form of the Arc Discharge

We cannot fail to note, in studying the oscillograms in Fig. 34, another important feature of the oscillations in arc voltage which enables us to get at the origin of the cyclic changes from a somewhat different point of view. In this current range, of all the states that the discharge passes through, it stays a particularly long time in two extreme states which are markedly different from one another in respect to both voltage and stability. It is natural to interpret these fundamental or characteristic states as two different forms of arc discharge. An approximate idea of the cathode drop in the two forms may be obtained from the values of the electrode potential difference corresponding to them. The null method of voltage measurement used for this purpose is illustrated by the circuit of Fig. 37 and the oscillogram of Fig. 38. A type 6Kh6 diode was connected in parallel with the tube T containing the arc. The circuit formed by the diode and the tube contained an auxiliary source of emf, opposite in polarity to the voltage to be measured, in the form of a semiconductor

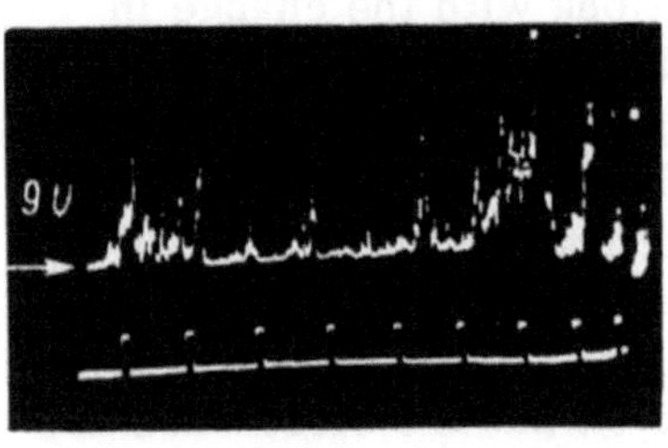

Fig. 38. Oscillogram of oscillations in arc voltage at a current of 0.3 A, illustrating the null method of measuring the voltage. The arrow was drawn at a level of 9 V.

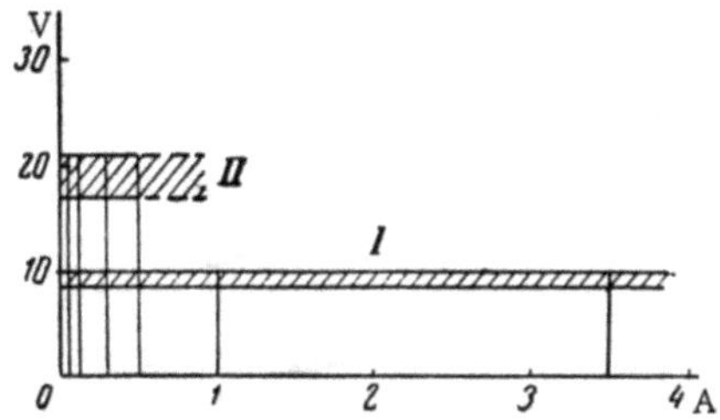

Fig. 39. Static volt-ampere characteristics of the fundamental (I) and transitional (II) form of the arc.

rectifier loaded by the potentiometer P. The balancing voltage could be varied and read with an ordinary voltmeter. Equality between the measured and balancing voltages was indicated by an electronic oscillograph with its input terminals connected to the ends of the resistance R, which is in series with the diode. With the oscillograph connected in this way, only those parts of the oscillogram remain undistorted which correspond to a potential difference greater than the balancing voltage. The rest of the oscillogram is smoothed out to form a horizontal line as shown in Fig. 38. At the points where the peaks intersect the line, the potential difference on the tube electrodes is equal to the balancing voltage to the accuracy of the contact potential difference of the diode.

The results of the measurements for currents from 0.07-3.5 A with a short arc (d_a = 0.3 cm) are shown in Fig. 39 by two volt-ampere discharge characteristics. The currents at which the measurements were made are shown by vertical lines. Each of the characteristics is a horizontal cross-hatched band in which the voltage readings lie. The fact that the characteristics are not shown as lines but as continuous bands reflects the uncertainty in the measured value. The uncertainty is due to the fact that the voltage is changing both within each period and on passing from one period to another. The lower of the cross-hatched bands shows the range of voltages corresponding to the start of a period. It gives the cathode drop of the more stable form of arc, which we shall in future call the fundamental form and designate by the Roman numeral I. The upper cross-hatched band is the range of voltages at the end of the period. It corresponds to the less stable state of the arc which we shall call the transitional form and denote by the Roman numeral II. The voltage is independent of the current in both forms of the arc. For form I, the operating voltage lies between 8.5 and 10 V, while for form II under these conditions the voltage varies from 17 to 21 V.

These voltages are approximately in the ratio 1:2, and they

are close to the first and second ionization potentials of mercury. We are also struck by how close they are to the two lower characteristic electron loss levels in metals [168], which may suggest the idea that there is some connection between these potentials in the arc and the quantum structure of the energy bands in the metal. However, the data in the next section enable us to assume rather that the characteristic arc states under discussion are associated with processes taking place in the cathode region of the discharge.

What we have said here about the two forms of arc throws new light upon the changes in form of the voltage oscillations observed with increase in current. In the low-current portion of Eq. (13), which extends up to 0.5 A under normal conditions, it is obvious that an alternating type of discharge is taking place, changing systematically from one form of arc to the other. This type of discharge does not have any single, clearly defined value of cathode drop. Under conditions in which there is an alternating type of discharge we can only speak of a mean value of the cathode drop during some interval of time. For intervals greater than 10^{-4} sec, the mean value will always lie between the limiting values given above for the two forms of arc. A special feature of the second part of Eq. (13), i.e., in the approximate current range 0.5-2 A, is that in this part the transitional form of the arc is gradually replaced by the fundamental, i.e., more stable, form (see Fig. 35). One of the results of this process is the break in the curve at the point where the two parts come together. Another result is the reduction in the mean voltage at which the discharge operates as the current is increased. Our observations throw some light on the origin of the so-called falling volt-ampere characteristic of the arc at small electrode separations.

29. The Relation Between the Voltage Oscillations and the Cathode Processes in a Short Arc. Instability of the Cathode Drop

The preceding sections have been devoted primarily to a phenomenological description and determination of the laws governing the voltage oscillations in a short arc. It was stated that the oscillations come from the cathode, and that there is a transitional form of the arc which has a higher cathode drop. For these statements to have a firm foundation, it must be shown by direct experiment that a connection exists between the oscillations and the cathode processes, and at the same time one must be sure that the periodically observed voltage pulses are accompanied by an increase in cathode drop.

We hardly need state that this problem cannot be solved by the usual probe method used in the study of a gas-discharge plasma under the simpler conditions obtaining in the positive column. In view of the exceedingly small dimensions of the cathode spot and its constant random motion over the cathode, a probe in this case is too coarse and immobile an instrument. Much better possibilities are offered by the optical method described below for investigating the cathode processes in the arc.

Any discharge region with large field gradients can serve as a source of high-energy electrons. This applies equally to the cathode-drop region. In passing through the cathode-drop region, the electrons acquire an energy corresponding to its instantaneous value $U_c^!$. If this value is subject to oscillations of some sort, similar oscillations will be observed in the mean energy of the electrons reaching the negative luminescence region. It is possible to establish the presence of these oscillations by observing the changes in relative brightness of different parts or individual lines in the spectrum of the discharge luminescence. Such an optical method can serve as a basis for investigating the cathode processes in an arc discharge. Particularly clear results with extreme simplicity of the experimental equipment are given by this method in a discharge taking place in a mixed atmosphere of mercury vapor and neon. The use of precisely this inert gas as the second component is dictated by two considerations. First, neon cannot be excited by electron impact as long as the cathode drop stays between 8 and 10 V, i.e., as long as it corresponds to the normal or fundamental form of the mercury arc, and thus it cannot play any active role in the mercury discharge. The first appearance of neon luminescence serves as a sign that the value of $U_c^!$ is getting close to or exceeds the first neon excitation potential of 16.8 V. Thus, the neon luminescence provides a sensitive indicator for the transitional form of the arc. Second, the spectrum of neon is greatly different from that of mercury in the visible region. It contains many bright lines in the red region, which is extremely poorly represented in the mercury spectrum. Thus there is a simple way of isolating the neon luminescence in pure form, either by visual observation or by photographing the discharge through a red filter. By using this

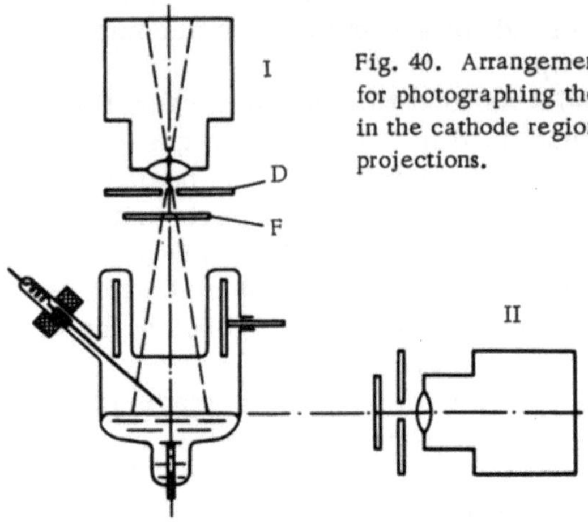

Fig. 40. Arrangement of an experiment for photographing the neon luminescence in the cathode region of an arc in two projections.

method, it is not only easy to show that electrons with increased energy are present in the discharge, but also where they come from, as is illustrated below.

The arrangement of an experiment using the optical method of investigating the oscillation processes in the cathode region of the arc is shown in Fig. 40. The tube shown was made so that the luminescence of the discharge could be observed in two projections. The discharge space was bounded below by the mercury cathode and above by the plane wall of the tube, through which the luminescence was photographed as projected on the cathode (projection I). The program of study included oscillographic observations of the neon luminescence using a photomultiplier with a cesium–oxygen cathode, taking the spectrum of the luminescence, and finally, photographing it through the red light filter F. The photographs were taken on isopanchromatic film with a sensitivity of 250 All-Union State Standard units. The diaphragm D, with an opening of less than 1 mm, cut off all but the narrow central beam of rays, so that the depth of focus exceeded the dimensions of the space being photographed. The discharge was excited with a contact needle. When neon was introduced into the tube, the cathode region of the arc showed a flickering red luminescence which decreased in intensity as the current was raised. Oscillograms of the photomultiplier current showed that the neon luminescence was discontinuous. The oscillograms showed sudden bursts of photocurrent in the form of isolated pulses. An especially important

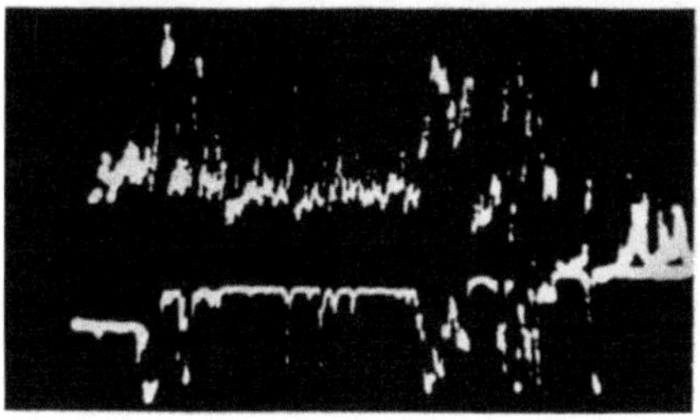

Fig. 41. Oscillogram illustrating the pulsed form of the neon luminescence (lower trace) and the synchronims of the flashes of light and the voltage surges (upper trace).

fact is that these neon bursts always occurred at the same time as the voltage pulses corresponding to the transitional form of the arc. This coincidence is illustrated by the oscillogram of Fig. 41, which is a simultaneous dual-trace oscillograph recording of the voltage oscillations on the arc electrodes and the intensity changes in the neon radiation. In order to get as much information as possible on the oscillogram, the time scale in Fig. 41 is greatly compressed in comparison to the previous oscillograms. Hence it is especially clear that the neon luminescence is only produced by the high-voltage pulses associated with the transitional form of the arc. As the current was increased, the frequency at which the pulses appeared decreased, which explains the apparent reduction in intensity of the neon luminescence at high currents.

Demonstrating that the neon luminescence is pulsating and that it coincides in time with the voltage pulses is only the first part of the problem. The second part consists in investigating the form and location of the regions where the neon luminescence occurs. A clear idea of both of these regions is given by the positive prints of the discharge, taken in two projections; the prints are given in order of increasing inert gas pressure in Fig. 42a and b. In taking these pictures, a red filter was used, which, for practical purposes, absorbed the mercury radiation completely. Accordingly, the light spots on the pictures can be ascribed only to neon luminescence. It is obvious from a comparison of the form of the spot in the two projections given that the regions where the neon luminescence occurs are hemispheres resting directly on the cathode. It is easily seen that the center of each hemisphere is the cathode spot in one of its instantaneous positions on the cathode. Actually, the trace of the spot visible on some of the pictures is a line joining the centers of the luminescent regions. As the gas pressure is increased, the radii of the hemispheres decrease

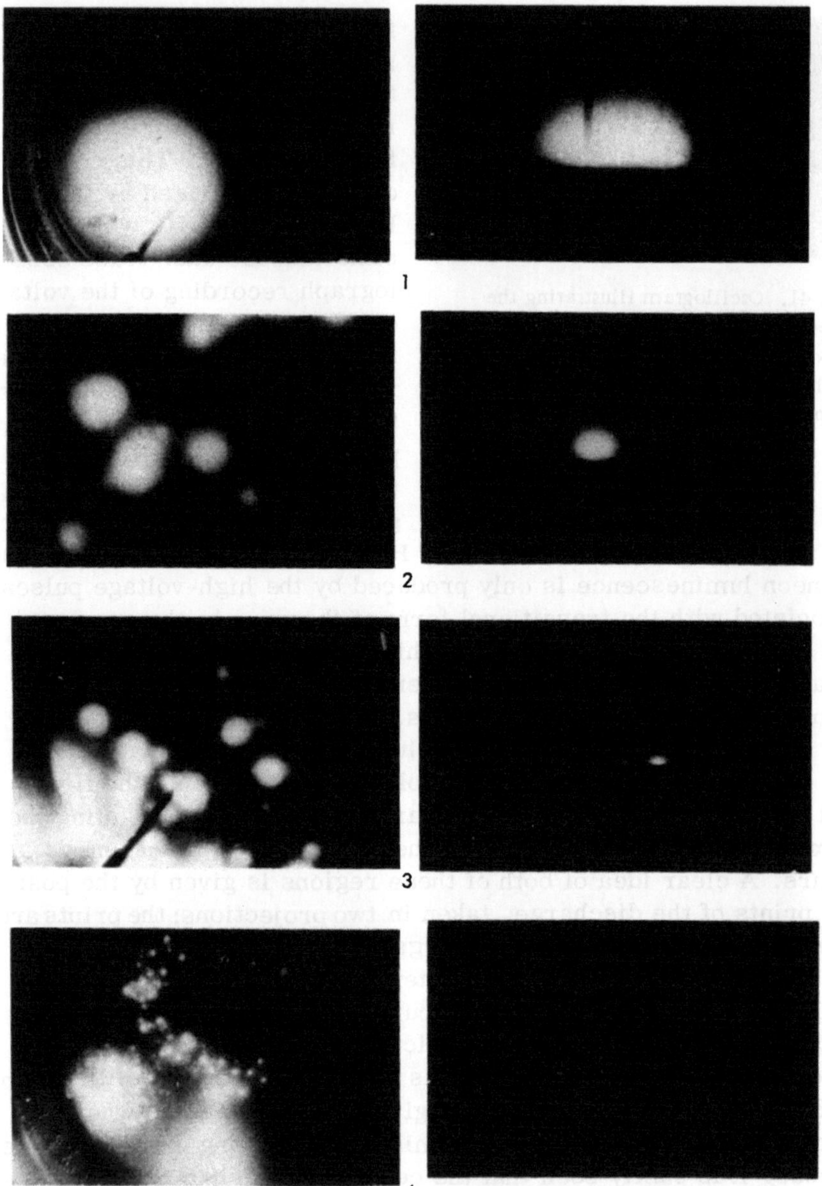

Fig. 42. Pictures of the neon luminescence regions taken through a red filter at successively increasing pressures. a) Projection on the cathode plane; b) side view. Magnification 0.75. 1) $p_{Ne} = 0.6$ mm Hg; 2) $p_{Ne} = 2$ mm Hg; 3) $p_{Ne} = 7.5$ mm Hg; 4) $p_{Ne} = 30$ mm Hg.

TABLE IX

Ne pressure, mm Hg	ρ_S, cm	
	Measured values	Calculated values
0.7	2.8	7.15
2	0.65	2.5
6	0.42	0.83
15	0.12	0.33
20	0.1	0.25

regularly until they contract into a bright spot which is always on the path described by the cathode spot in its random motion over the cathode. This reduction in the dimensions of the hemispherical regions where the neon is excited is clearly visible in the pictures of Fig. 42. The decrease is shown quantitatively by the figures of Table IX, which gives the results of measuring the mean radii of the luminescent regions (column 2) for the range of pressures used in the experiments.

In making the above observations, we could not help but notice that the extinction of the arc was always preceded by a bright neon flash, easily observable visually and in the pictures. It, too, consists of a luminous hemisphere. In the pictures which show the trace of the cathode spot, the trace always stops at one of the hemispherical regions where the neon is excited. This observation clearly emphasizes the relation between the critical states and the extinction of the arc. Extinction of the arc is, in essence, the end of a natural course of development of one of the critical states in a direction unfavorable to the discharge. In the neon experiments, the unfavorable development of events is marked by a beautiful effect worth special mention. In this case, the usual bright neon flash in the hemispherical region around the cathode spot suddenly changes into a less bright neon luminescence in the form of a red film which quickly covers the whole cathode surface, and just as quickly decays. The neon luminescence surrounding it shows that there is a relatively weak electron emission from the whole cathode surface. This phenomenon must arise from the formation of a residual space-charge plasma at the boundary with the cathode; in the resulting field the electrons acquire the energy needed to excite the neon. The development of the enveloping luminescence is a sign that the decay of the cathode spot is complete.

That the luminescence actually comes from the neon was shown by photographing the spectrum of the discharge. For these spectrograms, the cathode spot was excited with a semiconducting igniter on a part of the mercury cathode bounded by three quartz plates and the glass wall of the tube, as shown in plan in Fig. 43. The image of the arc was focused in the plane of the spectrograph slit in such a way that the image of the cathode spot with its motion limited by the quartz shields frequently intersected the slit. The resulting spectra of the discharge in a mixture of mercury vapor and neon are shown for three different neon pressures in Fig. 44. The right-hand end of each of the spectrograms shows the two bright yellow mercury lines occupying the total height of the image of the spectrograph slit. To the left are a number of bright neon

Fig. 43. Plan and elevation of tube used to photograph the neon spectrum. 1) Mercury cathode; 2) anode; 3) semiconducting igniter; 4) quartz plates, limiting the motion of the cathode spot. A — A: direction of observation.

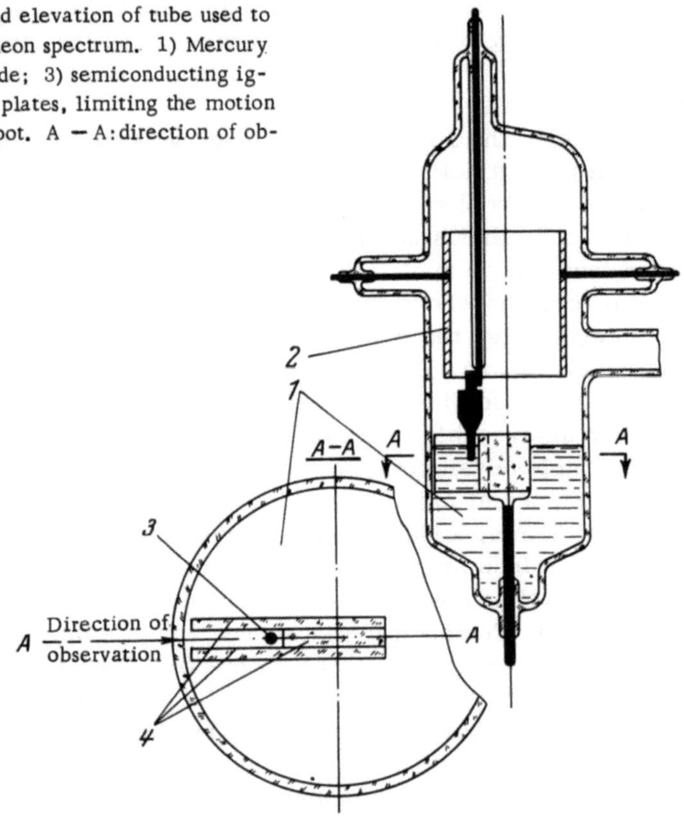

lines filling only a part of the slit. The neon lines decrease in intensity the further their origin is from the cathode surface; this effect increases as the neon pressure increases. The level of the mercury surface is marked on the top spectrogram by the band of continuous spectrum from the cathode spot. The shortening of the neon lines observed with increase in pressure corresponds to the contraction of the neon luminescence into the cathode spot, as has already been demonstrated by the pictures (see Fig. 42).

Since the regions where the neon is excited are hemispherical in form and are symmetrically located around the cathode spot, we can conclude that each of the hemispheres is a region where the energy of the fast electrons emitted by the spot is being dissipated, decreasing from its original value of about 20 down to 16.8 eV. But in this case the conclusion must be verified by calculating the relaxation radii ρ_s for the above energy losses and comparing them with the radii of the hemispheres as measured at the appropriate gas pressure p. Under the assumption that the value of ρ_s is determined principally by elastic losses in electron energy, the following equation may easily be found:

$$\rho_s = \sqrt{\eta \frac{m_g}{m_e} \cdot \frac{\lambda'_e}{p}}, \tag{18}$$

where η is the fraction of energy lost by the electrons in each hemisphere, m_g/m_e is the ratio of the atomic mass of the atoms of the gaseous medium to the mass of an electron, and λ'_e is the mean free path of the electrons at a gas pressure of 1 mm Hg.

Taking $\eta = 0.15$[1] and $\lambda'_e = 6.7 \cdot 10^{-2}$ cm, we may find numerical values of ρ_s for all the pressures at which the flashes were photographed. The results of these calculations are given in Table IX together with the measured radii of the luminous regions. The calculated and measured values of the

Fig. 44. Spectrogram of the luminescence in the cathode region of a mercury arc with neon present. a) p_{Ne} = 0.5 mm Hg; b) p_{Ne} = 7 mm Hg; c) p_{Ne} = 20 mm Hg.

[1] This figure is the ratio of the difference between the energies of the electrons in the center and at the periphery of the hemisphere to the total initial energy, which is assumed to have the rounded-off value of 20 eV.

radius are quantities of the same order of magnitude; they are inversely proportional to the pressure. Thus, there is every reason to believe that the luminous hemispheres are actually regions where the energy of the fast electrons emitted periodically by the cathode spot is being dissipated. The fact that the measured values of ρ_s are approximately 0.4 of the calculated values may indicate that a considerable role is being played by inelastic losses in dissipating the electron energy. The validity of this explanation for the difference between the calculated and measured values of ρ_s is supported by the bright neon luminescence in the regions in question.

Thus the inhomogeneities observed in the luminescence of the medium, consisting of discrete hemispherical regions with a special spectral composition, show that sudden transient changes occur in the mean energy of the electrons emitted directly by the cathode spot. This means that from time to time the cathode drop in the arc suddenly increases, reaching values in excess of 16.8 V. In view of the synchronism between the flashes of neon luminescence and the voltage pulses, it must be concluded that there is a direct causal relation between the sudden changes in cathode drop and the voltage surges at the arc electrodes that we have already investigated. Thus, the material presented shows that it is correct to assert that there is a relation between the oscillations in voltage at the electrodes of a short arc and the processes taking place at the cathode, as well as that there is a transitional form of the arc which has a high cathode drop.

The transient phenomena that we have investigated in the cathode region of the mercury arc, consisting of instability in the cathode drop and the periodic appearance of groups of electrons with high energy, in all probability occur in all cold metallic arcs. They may be looked on in the most general way as an indication that systematic disturbances occur in the equilibrium between the various processes occurring in the arc cycle. For that matter, the very occurrence of groups of fast electrons in the cathode region is a serious disturbance of the equilibrium, which can in turn produce a chain of successive changes in the discharge. Since they have greater ionizing power, the electrons will suddenly activate the ionization process not only in the cathode spot but in comparatively wide regions of the discharge, which in turn will increase the space charge and the positive-ion bombardment of the cathode. However, this by no means exhausts the role played by groups of fast electrons. In the normal long arc they can penetrate the lower boundary of the positive column, thus making the column unstable, resulting in peculiar flickering of the luminescence and sudden oscillations in voltage on the arc electrodes in synchronism with the oscillations in the cathode drop.

30. Arc Re-ignition and "Relay-Race"[1] Types of Arc Restoration

The information so far obtained on the oscillations and the conditions accompanying decay of the arc enable us to approach a physical explanation of the statistical laws already found to govern the extinctions. At the same time, it is possible to draw a number of conclusions concerning the mechanism by which a closed operating cycle is maintained in the mercury cathode arc.

As pointed out in Sec. 27c, for currents up to 0.5 A, extinctions of the arc occur in the majority of cases with sudden surges in voltage up to values corresponding to the transitional form of the arc. This has provided the basis for considering spontaneous extinction of the arc to be the natural result of steadily increasing instability in the discharge. But then a legitimate question arises: how do we explain the fact that there are a large number of voltage pulses on the oscillograms which do not end by interrupting the arc? With a little rephrasing, this question may be formulated as follows: why is the arc not interrupted in all cases at the very first rise in voltage? A careful study of the oscillograms shows that this is precisely what does in fact happen at currents of about 0.1 A or less. A case of this sort is recorded on the oscillogram of Fig. 34a. However, as the current is increased, decay of the arc occurs later and later, until it occurs at one of the subsequent voltage rises. The oscillograms show that at the same time, as the current is increased, a larger and larger number of periods end by restoring the fundamental form of the arc discharge, which is easily seen by comparing oscillograms taken at different currents. Obviously the reason for this phenomenon is to be found in the intervention of a restoring mechanism, which we can be sure exists since the duration of the arc is independent of

[1] The author's metaphor is a little faulty in that it is not a "new runner" that takes the old one's place; rather it is a redistribution of current among the remaining spots that occurs – Ed.

the external circuit constants such as emf or inductance (see Secs. 24b and 25). The general principle of action of the restoring mechanism is entirely clear from the fact that the start of a new period is always preceded by a more or less large rise in voltage. This can only mean that the rise in voltage is precisely what causes restoration of the arc.

On the basis of the observations made above, it is a simple matter to lay out the whole sequence of arc restoration which occurs with the aid of the transitional form of arc discharge. Judging from a number of indications, the cyclic changes in the arc come from alternate gradual cooling and reheating of the metal in the cathode spot. Then the critical state occurring at the end of the cycle can be accounted for by reduction of the mercury vapor density at the cathode spot to a level which is too low to maintain the normal form of high current density arc, in which the ionization in the glow luminescence region is maintained principally by cumulative or stepwise processes. If the partical concentration is reduced, the efficiency of the cumulative process is also quickly reduced, and this starts a sudden reduction in discharge current. However, as a result of the ballast resistance and inductance in the circuit, any reduction in current causes an automatic rise in the voltage on the arc electrodes, which also increases the potential of the plasma with respect to the cathode. Here the increased voltage is added across the thin layer of space charge in the cathode-drop region of the arc, and this suddenly increases the drop. From this, the relation between the instability of the cathode drop demonstrated in the preceding section and the internal instability of the arc is clearly visible.

The next stages of the restoring process may be arrived at with confidence on the basis of the results of the investigation made on the hemispherical regions containing excited neon atoms. The natural reaction of the arc to an increase in cathode drop will be to increase the cathode emission current, at the same time increasing the mean energy of the electrons reaching the plasma from the cathode-drop region. These streams of fast electrons can increase the ionization of the mercury vapor even when the concentration of the vapor has begun to decrease, the result being to start activating the arc. The final stage, upon which the result of the critical state depends, is an increase in positive-ion bombardment of the cathode. The result of this increase in energy flux to the cathode will, with a favorable combination of circumstances, be a new and violent outburst from the cathode, followed by restoration of the high vapor density in the cathode region of the arc.

The process of arc restoration involving the transitional form is, from all indications, a type of repeated arc breakdown of the discharge gap. Its distinctive feature must be considered to be that for a short length of time there is a sudden change in the nature of everything that goes on in the cathode region of the arc, including the ionization of the mercury vapor. It is highly probable that single ionization becomes important. In this stage, the cathode drop is maintained on the average at a level such that the increased ionization of the vapor compensates for the reduction in density. This state, which we call the transitional form of the arc, is extremely unstable and must end either by restoring the fundamental form of the arc or by going out. The oscillograms show that the first result depends on random coincidence of circumstances. Thus restoration of the arc through the agency of the transitional form is of the nature of a probability process. For given experimental conditions, each value of current corresponds to a completely definite probability of extinguishing the arc, P(I), and the associated probability $\overline{P}(I)$ that the arc will be restored out of the critical state. Since restoration and extinction of the discharge are mutually exclusive events, the probabilities will, under any experimental conditions, satisfy the relation

$$\overline{P}(I) = 1 - P(I). \tag{19}$$

As a result of having a definite time-independent probability of restoring the arc, we get the purely random distribution (12) in Part A of this chapter for the individual readings of arc lifetime at any given current. The numerical extinction probability may be expressed as the ratio of the number of times the arc decays to the total number of cyclic voltage changes in a large number of readings. If $\Delta\vartheta$ is the mean duration of a single voltage change cycle, and ϑ is the mean arc duration for a given current, then $P(I) = \Delta\vartheta/\vartheta$. Using the empirically determined relation between ϑ and I in the form (13), the probability of extinction or restoration may easily be expressed as a function of current. For currents up to 0.5 A, the probability of extinguishing the arc will be given by the relation

$$P(I)=\frac{\Delta\vartheta}{\vartheta_0}\,e^{-\psi_0(I-I_0)} \tag{20}$$

According to our earlier observations (see Sec. 24a) the threshold arc current I_0 for a mercury cathode is about 0.07 A. If the current is reduced to this value, the probability of extinguishing the discharge will approach 1, so that the quantities $\Delta\vartheta$ and ϑ_0 will be identically equal. These conclusions are supported by direct observations.

First, in the oscillograms of the type shown in Fig. 34, at a current of 0.07 A the arc always breaks off right at the beginning of the very first voltage change cycle. Second, the mean duration $\Delta\vartheta$ of the cycle is, within the accuracy of observation, the same as the mean arc duration ϑ_0 at the threshold current $I_0 = 0.07$ A.

Equation (20), in which the additional substitution $\Delta\vartheta = \vartheta_0$ should be made, gives the probability of extinguishing the arc in the current range 0.07-0.5 A, for which restoration occurs predominantly as a type of repeated arc breakdown. The probability of restoration increases with increase in current, as follows from (19) and (20). With the value of the parameter $\psi_0 = 4.5\ A^{-1}$ found for normal conditions in the mercury arc, it is easily calculated that increasing the current from 0.07 to 0.5 A corresponds to increasing the probability of restoring the discharge from approximately 0 to 0.86. Thus, in Eq. (13), ϑ_0 must be considered as essentially the normal lifetime of the cathode spot. The factor multiplying the exponential is a probability coefficient expressing the increase in arc duration that results from repeated restoration.

By using the results of measuring the mean arc duration in the current range 0.5-2 A, we can, in a similar way, express the extinction probability of the arc in this range as

$$P(I)=e^{-\psi_1(I-I_0)} \tag{21}$$

A special property of this part of the range is that, at currents above 0.5 A, division of the spot begins to play an appreciable role. Here, for some portion of the observation time, the current is distributed between two spots. The probability of two or more spots existing simultaneously on the cathode increases with increase in current according to the relation

$$P_{>1}(I)=1-e^{-\psi\,(I-I_0)}, \tag{22}$$

where ψ is a constant depending on the physical properties of the cathode and above all on its boiling point. However, in a situation where there is more than one spot, there will be a great change in the conditions under which the spots themselves exist and decay, as well as in the arc-restoring mechanism. First, if one of the spots decays under these conditions, it cannot be accompanied by formation of the transitional form of the arc since the transitional form cannot compete with the more stable fundamental form of the arc which is maintained by the remaining spot. This also eliminates the possibility of restoring the spot that has gone out. Second, if one of the spots goes out, the fraction of the current that it was carrying will be taken on by the other spots, and, accordingly, this particular decay will not, as a rule, cause extinction of the arc, but only redistribution of current among the different parts of the spot.

Thus we arrive at the idea that there is a new type of arc restoration that takes place by redistributing the current among the independent cathode spots. In contrast to restoration by repeated arc breakdown, this process will maintain a closed arc cycle as long as there are two or more spots on the cathode in the same time. The principle of action is simply that, if one of the spots goes out, the automatic increase in voltage on the arc electrodes stimulates the activity of the other spots, producing a redistribution in current, which temporarily restores the disturbed equilibrium. There is some reason to assume that this type of arc-restoring mechanism, which may be called the "relay-race" mechanism, does not require as large voltage rises to bring it about as the process in which the arc is restored by arc breakdown. That an arc-restoring mechanism of the "relay-race" type exists will be shown later in discussing the structure of the cathode spot, which forms the subject of the next chapter. A thorough discussion will be given there of all the details.

The material given above provides the basis for considering the exponential increase in arc duration with increasing current to be under all conditions the result of increasing the probability of restoring the arc from a systematically occurring state of decay. The break in the ϑ (I) curve comes from the change in the arc-restoration mechanism. For currents up to 0.5 A the discharge is predominantly restored from its critical states by repeated arc breakdown. With increase in current, beginning at 0.5 A, redistribution of current among the various decaying parts of the cathode spot gradually gains predominance in restoration of the arc. This latter process will occur for all large discharge currents and this is why it is possible to maintain this form of discharge for a long time.

C. THE PROBLEM OF ARC STABILITY AND HOW IT IS AFFECTED BY VARIOUS INTERNAL FACTORS

31. Physical Features of Possible Ways of Stabilizing the Arc

Having made a study of the phenomena accompanying spontaneous arc extinction as well as of the actual mechanism by which this form of discharge is maintained, we must again, but this time from a somewhat different point of view, consider the question of what effect various internal factors have on the stability of the arc. Here we shall be talking about possible reasons for the large increase in arc duration that we have observed under the following conditions:

a. when the cathode is heated to the boiling point,
b. when the cathode is cooled to solidification,
c. when the cathode spot is anchored at a thin film of mercury wetting another metal,
d. when a gaseous medium is present,
e. when a longitudinal magnetic field is applied.

a. Mechanism of Arc Stabilization by Cathode Heating. Turning first to the boiling mercury cathode, we recall that a fundamental property of the curves giving the arc duration ϑ as a function of current under these conditions is that the values of ϑ are high at low currents. The oscillograms for these conditions (Fig. 45) show a marked increase in the period between the voltage pulses in the high-voltage group over that found under normal conditions (from approximately 10^{-4} to $3 \cdot 10^{3}$ sec). This must be viewed as a direct indication that there is a relation between the voltage rises and the periodic cathode cooling process in the spot, which naturally will be greatly retarded in an arc where the cathode is boiling. There is little doubt that this increase in length of the period to a considerable extent explains the relatively long duration of an arc with a boiling cathode at low currents, as reflected in the increase in the parameter ϑ_0 in Eq. (13). Along with the increase in period of the oscillations, there is an appreciable increase in probability of a repeated arc-breakdown type of arc restoration. This increase is shown by the fact that it is only comparatively rarely that the oscillograms show a case where the arc is broken off in its transitional form, but one such case is shown in Fig. 45a. The greater part of the extinctions occur under these conditions as a result of a sudden break-off in current without an appreciable decrease in the current beforehand, as may be seen in Fig. 45b. The result of the reduction in frequency of the first type is a marked upward shift of the high-current portion of the ϑ (I) curve.

Thus, the increase in arc duration produced by heating the cathode to the boiling point is the result of the superposition of two effects. One of the effects is the reduction in the rate at which heat energy flows out of the cathode spot, and the other is the increased probability of restoring the arc from its transitional form. As is shown by the similar increase in arc-restoration probability when a gaseous medium is present, the increase must come from the increase in mercury vapor pressure at the boiling cathode surface and the resulting increase in the intensity of ionization in the cathode region of the spot, an indication of which is also given by the smooth form of the voltage pulses.

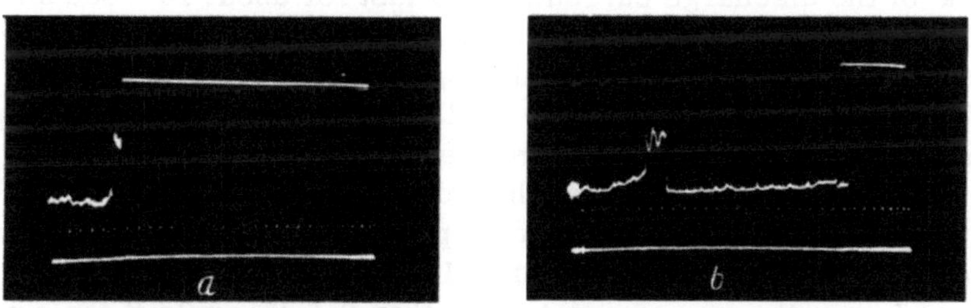

Fig. 45. Oscillograms of voltage oscillations in an arc with a boiling mercury cathode (d_a = 0.3 cm, I = 0.2 A). Marker amplitude 10 V, frequency 10 kc. a) Arc broken off in the transitional form; b) direct breakoff of the fundamental form of the arc without the transitional form.

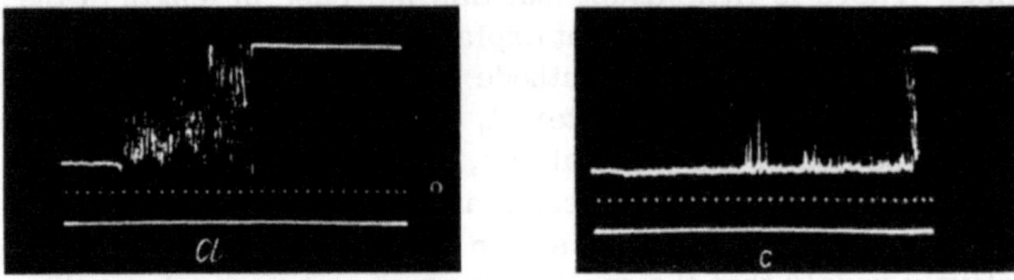

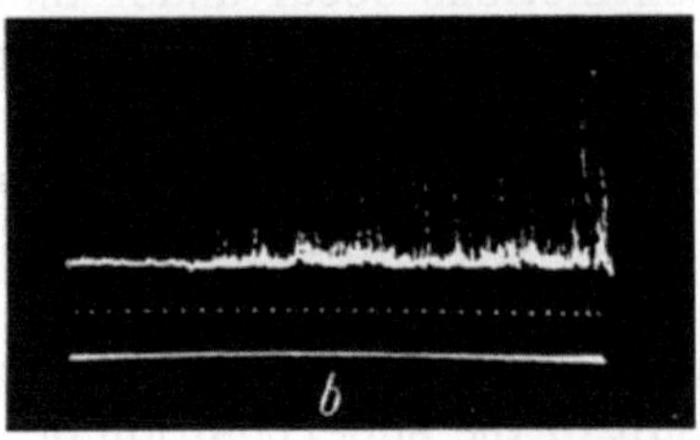

Fig. 46. Oscillograms of voltage oscillations in an arc with a solid mercury cathode (T = −48°C, d_a = 0.3 cm). Marker amplitude 5 V, frequency 10 kc. a) I = 0.075 A; b) I = 0.2 A; c) I = 0.33 A.

b. The Nature of the Phenomena Occurring in a Solid-Cathode Arc. Changes of a completely different sort are observed when the cathode is frozen. The most typical features of a solid cathode are the large reduction in the amplitude and the change in form of the voltage oscillations. The change in form shows up in the fact that the voltage rises on the oscillograms for these conditions take on the form of narrow isolated pulses of all possible values from hundredths and thousandths of a volt to 10 V, depending on the magnitude of the discharge current. They last for about 10^{-7} sec and nothing definitely cyclic can be seen in the way they occur, although at times the amplitude of the pulses suddenly increases and they occur very close to one another. All these features of the oscillations in the solid-cathode arc are clearly seen in Fig. 46, where the oscillograms are arranged in order of increasing currents. As the current is increased, the amplitude of the oscillations greatly decreases. At a current of 1 A, the amplitude of the oscillations is, on the average, not more than 0.5 V, while if the current is increased to 5 A, the amplitude decreases to about 0.2 V, without any substantial change with further increase in current.

The oscillograms show further that a solid-cathode arc can

exist at currents as low as 0.04-0.05 A, but here practically nothing is observed but the transitional form of the discharge, which shows continual sharp voltage oscillations with an amplitude of about 10 V (Fig. 46a). This fact is of interest in two respects. First, we may conclude from this that with a solid cathode there is an appreciable shift in the threshold arc current toward the low side (approximately from 0.07 to 0.04 A).

Second, the increase in amplitude and frequency of the pulses observed as the current is reduced right up to the point where a continuous band of oscillations is formed, as in the transitional form of the arc, throws some light on where the transitional form comes from. Apparently what we have been calling the transitional form of the arc does not in fact represent any definite state of equilibrium between the various processes in the arc cycle, but rather a series of sudden departures from the equilibrium state where some processes are successively enhanced at the expense of others. But then the oscillations, consisting of isolated pulses about 10^{-7} sec long, occurring at higher currents should represent the same kind of random disturbance of the equilibrium. The only thing to show that there is any difference between them is that under the existing conditions the arc is so easily restored that a single pulse is sufficient to bring it back to its original state. In this case the pulses can supply two kinds of information on the restoring mechanism in the arc. First of all, the pulse length will be exactly such that the original state can be restored during one pulse, from which the restoration time is found to be 10^{-7} sec. Another quantity which can serve as a quantitative measure of the restoration process in the arc is the pulse amplitude. In a solid-cathode arc, the amplitude of the pulses is almost an order of magnitude less than the amplitude in a normal liquid cathode arc at the same current. All this indicates that when the cathode is solidified it is much easier to restore the arc. The reduction in the threshold current, accompanied by displacement to the left of the break in the ϑ (I) curve, means that in addition to making it easier to restore the arc there is, when the cathode is solidified, an increase in arc stability in the narrow sense of the word. The diagnostic symptoms of the changes noted are, however, quite different from those found with a boiling cathode, and accordingly they must have different causes.

A possible source of the special features observed in the solid-cathode arc is first of all the crystal structure of the solid metal surface itself. It has been found by direct observation of various metals [169-172] that the cold emission is quite unequally distributed over the metal surface, being concentrated predominantly at isolated crystal faces, and in particular at the sharp points on the joints between faces. The unequal distribution of the emission current shows up clearly in the so-called patch structure of the emission image in picture tubes. It is attributed to the different values of the electron work function on the different faces of a crystal [173-177]. The work functions for a number of tungsten crystal faces have been measured by two different methods [175, 176], which gave results in surprisingly good agreement. According to these data, the work function changes from one face to another approximately from 4.3 to 5.9 eV. It was also suggested [178] that the unequal distribution of the emission over the surface of crystals comes from unequal distribution of the electric field, which is concentrated principally at the various faces and vertices of the crystal.

In view of this data it is to be expected that when the cathode changes from the liquid to the solid state there will be sudden change in the conditions under which electrons are emitted as a result of crystallization of the mercury, and that this will make it much easier to restore the arc. Above all, when the change occurs, the homogeneous conditions existing on the cathode surface will be destroyed, since there will be parts of the surface where the conditions for forming a cathode spot are greatly improved. It is true that the cathode spot will quickly destroy the crystals by melting, which may be one of the reasons the spot decays rapidly. However, here the spot may be formed again by the restoring mechanism acting on neighboring intact crystals, while the parts of the cathode that were destroyed are automatically restored as a result of cooling. The process by which the arc is maintained on a solid cathode may be thought of as continuous destruction by the discharge of its own foundation, which is automatically restored by rapid cooling and crystallization of the metal. Obviously, this continuous destruction of the crystals by melting is one of the possible reasons for the sudden disturbances of the equilibrium between the various processes in the arc cycle and, along with this, of the random motion of the cathode spot over the solid cathode.

If it were possible to explain the increase in arc stability observed when the cathode solidifies solely in terms of the effects just considered, this would be a new and powerful argument in favor of the cold-emission theory of the arc discharge. Unfortunately, the question of possible reasons for the improvement is far from being this simple. Actually, solidification of the cathode is accompanied by a number of other effects which are capable of either directly or indirectly affecting both the stability of the arc and the efficiency of the restoring mechanism. An improvement in arc stability when the cathode is solidified will be effected by the elimination under these conditions of the possibility of sudden deformations occurring in the mercury surface resulting from the violent convection and turbulent phenomena in the cathode spot. An important factor here may be the reduction and regulation of the heat transfer from the cathode spot. That these factors can actually be responsible for the increased duration of the solid-cathode arc is shown by the reduction in threshold current that occurs, along with the great reduction in restoration time, when the cathode solidifies.

c. The Increase in Arc Stability When the Cathode Spot is Anchored on a Thin Liquid Mercury Film. It must be assumed to have been proven that there is a definite resemblance between the phenomena observed in the solid-cathode arc and those observed when the spot is anchored on the thin liquid mercury film wetting other metals. Under both sets of conditions there is a relatively high value for the initial arc duration ϑ_0 at currents near the threshold value, as well as an exceedingly large increase in ϑ with current at small values of the current. There is much in common in the voltage oscillations shown by the two types of discharge, including the fact that they are of small amplitude and in both cases are in the form of isolated narrow pulses lasting about 10^{-7} sec, and no cycle is followed. These similarities between an arc with a solid cathode and one with a cathode consisting of a thin film of liquid mercury wetting a solid metal must mean that there is some common feature in the way in which the stability of the discharge is increased in the two cases. But the conditions under which the cathode spot exists on a thin mercury film differ from those under which the cathode spot is on a homogeneous liquid cathode principally in the way in which the heat is conducted away, the mercury is evaporated, and new mercury is supplied. In a thin mercury film there can be neither violent convective motion of the mercury nor explosive bursts of greatly overheated liquid, which is apparently what causes the cyclic voltage changes in an arc with a homogeneous liquid cathode. It must be assumed that a particular characteristic of an arc with the spot anchored in place is that the heat transfer, evaporation and supply of mercury occur in the ordered way appropriate to laminar liquid flow. In view of this an analogy may be drawn between the way in which the cathode spot is maintained on the mercury film and the way in which a liquid fuel is burned in a wick, where the fuel is supplied from a suitable reservoir as it is consumed. There is no doubt that this sort of "wick" system for supplying and evaporating the mercury is one of the fundamental reasons for the large increase in stability of the arc when the cathode spot is anchored in place. It is obvious that a large role will be played here by the regularity introduced into the heat transfer from the spot. It is also not impossible that the increased stability of an arc with a film-type cathode is to some degree due to a change in electron work function under the complex conditions existing in the bimetallic cathode.

Phenomena of a similar sort will occur in a solid cathode at low currents when the metal surface melts. However, if the discharge current is increased beyond a certain limit, substantial changes will occur in the processes as a result of broadening and deepening of the zone around the spot where the cathode is melted. But then the effect of deep-melting of the metal around the cathode spot will be precisely what causes the sudden break in the ϑ (I) curves when the current is increased, a proof of which is that the slopes are equal, and the high-current parts of the curves are the same in solid and liquid cathodes.

The increase in stability of the cathode spot on a thin mercury film gives a good explanation for the tendency of the spot to anchor itself to the boundary where the mercury wets the metal in contact with it. Actually, as soon as a portion of the spot leaves this part of the cathode it decays automatically in favor of the portion that it left, which is under more favorable conditions.

d. The Mechanism by Which a Gaseous Medium Affects the Arc. The question of what increases the duration of the arc when a gaseous medium is present is considerably more definite. At relatively low pressures, the effect of the medium on the arc reduces purely and simply to increasing the efficiency of the restoring mechanism without any obvious signs of increasing the stability of the arc in the exact sense of the word, as may be seen from the nature of the change in the ϑ (I) curve. This is easily understood. At low pressures the gas does not produce any substantial changes in the conditions existing in the actual functioning cathode spot, where the concentration of neutral and charged particles is quite high and is controlled by the discharge itself. The gas can only play an appreciable role in the critical arc states which are accompanied by a sudden reduction in the concentration of mercury atoms in the cathode region of the discharge and an increase in cathode drop. In these critical circumstances, the gas atoms excited by fast electrons can help to increase appreciably the intensity of ionization through ionizing the mercury vapor by collisions of the second kind, thus making a contribution to the arc-restoration process. That it becomes easier to restore the discharge from its transitional form when a gas is present is shown not only by the increase that we have found in the probability of a positive result, as seen from the increase in the exponent of ψ_0 in Eq. (16), but also by the appreciable smoothing out of the voltage pulses in the oscillograms taken with a gas present.

If the density of the gas is high enough, in addition to the effect on the restoring mechanism, there is a stabilizing effect, which is expressed by the shift of the break in the ϑ (I) curve toward low currents, and then by an increase in the slope of the high-current part of the curve (see Fig. 24). The change in the curve indicates that at these pressures the concentration of gas atoms becomes comparable with the concentration of mercury atoms at the cathode spot. Hence it follows that it is in principle possible to find the mercury vapor pressure in the cathode region of the arc from the change in form of the ϑ (I) curve produced by gas.

e. The Mechanism by Which the Arc is Stabilized by a Magnetic Field. If we are going to understand a number of the properties and peculiarities of behavior of the cathode spot on a mercury cathode, the mechanism by which a magnetic field acts to stabilize the arc is of major importance. It has already been pointed out (Sec. 26) that, when a magnetic field is applied, the ϑ (I) curve undergoes changes consisting in an increase in slope of both segments and a displacement of the break in the curve toward low currents. Further investigation has shown that the displacement of the break in the curve toward lower currents is a sign that the point at which a single cathode spot is no longer formed occurs at lower current. This type of change means that

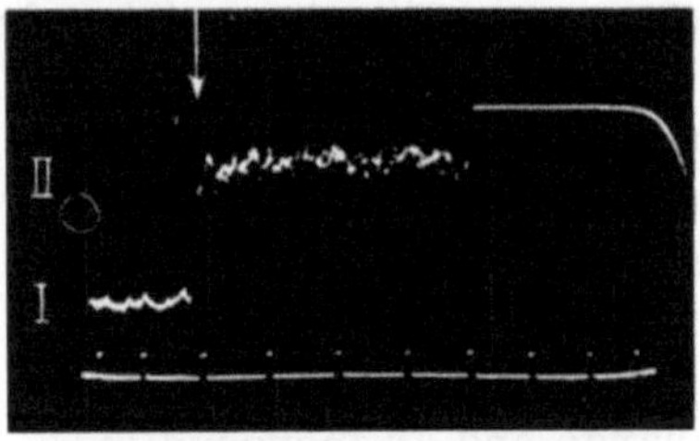

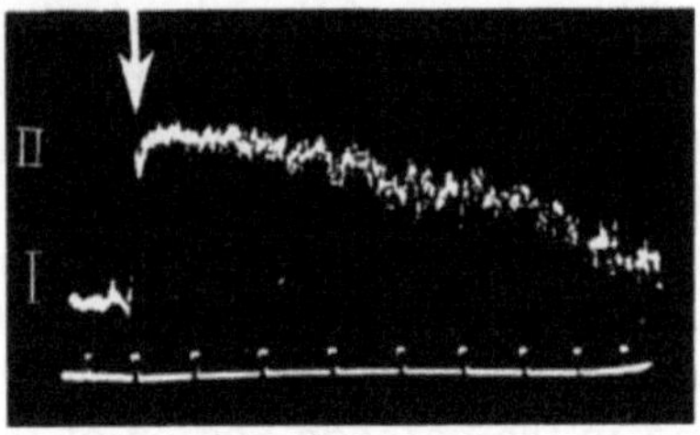

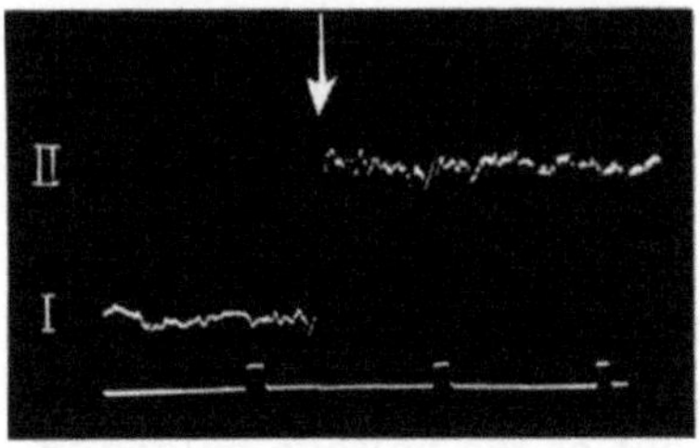

Fig. 47. Oscillograms of voltage oscillations in an arc with a longitudinal magnetic field of about 7 kOe applied. I = 0.08 A. The instant at which the current drops to this value is shown by the vertical arrow. I) Fundamental form of arc; II) transitional form of arc.

it is easier to form a spot, so that it may be interpreted as a sign that the field produces a true increase in stability of the arc. The magnitude of the change may be gauged by observing that the break point is shifted by the magnetic field from 0.5 A to approximately 0.1 A, which is equivalent to a reduction to less than one fifth the current required to form a full-blown single spot.

Additional information on the action of a magnetic field may be extracted from the oscillographic observations. In their most essential features the results of these observations are as follows. First of all, applying a longitudinal field of up to 7 kOe reduces the threshold arc current from 0.07 to 0.04 A or less. Here, for currents from the threshold value to 0.1 A, what is observed is exclusively the transitional form of the arc. In this range, the duration of the arc is increased in the presence of a field, since the field increases the stability of the transitional form, so that it is particularly convenient to observe the transitional form under these conditions. The oscillograms are shown in Fig. 47. Here the transitional form of the arc occurs immediately after the current is reduced to 0.08 A from its initially high value. The instant at which the current is reduced is shown by an arrow; II is the transitional form of discharge itself. It looks like a step, but it differs

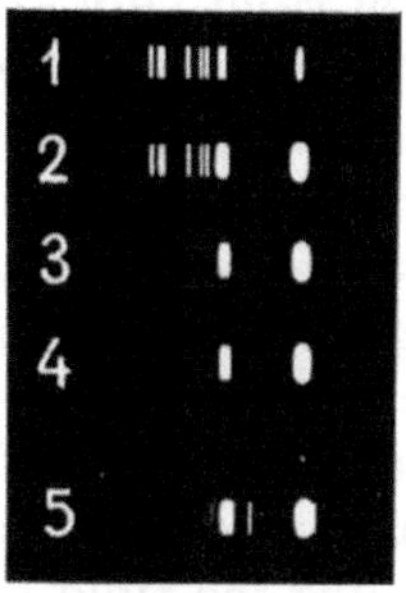

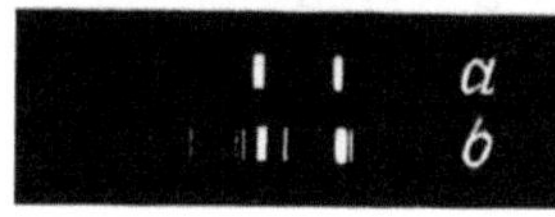

Fig. 49. Effect of a longitudinal magnetic field on the spectrum of a mercury arc at I = 3.5 A. a) H = 0; b) H = 7 kOe.

Fig. 48. Spectrogram of luminescence of arc in mercury vapor and neon, illustrating extinction of the neon spectrum by a magnetic field. I = 1.5 A, p_{Ne} = 1 mm Hg. 1) H = 0; 2) H = 0.4 kOe; 3) H = 1 kOe; 4) H = 2 kOe; 5) H = 4 kOe.

from the fundamental form I in that the amplitude of the oscillations is greater. Usually it ends by breaking off the arc. With this type of discharge, the luminescence occupies a rather large region on the cathode surface (of the order of 1 cm^2), with bright luminous points constantly appearing and disappearing on the periphery. This may mean that this form of discharge is basically of low current density with isolated, rapidly decaying emission centers with high current density appearing in the background. Sometimes this form of arc changes to the fundamental form by gradual reduction of the amplitude of the oscillations and, at the same time, reduction of the overall voltage level. The start of a change of this type has been caught on one of the oscillograms.

In a magnetic field, the transitional form of arc is completely replaced by the fundamental form even at currents less than 0.2 A, and this is a reliable indication that under these conditions formation of the cathode spot is complete at considerably lower currents. Thus, the changes observed in the voltage oscillations in the presence of a field support the conclusion that the magnetic field exerts a stabilizing effect on the arc.

The stabilizing effect of a magnetic field may be demonstrated by means of a quite vivid experiment. It has already been pointed out that the instability of the arc in a mixture of mercury vapor and neon shows up as a flickering neon luminescence coming from the cathode spot, easily observable on the background of the mercury luminescence. If the discharge current is greater than 0.15-0.2 A, applying a longitudinal magnetic field of even relatively small intensity eliminates the neon luminescence. The disappearance of the neon luminescence in a magnetic field is illustrated in Fig. 48, which shows a series of spectrograms of the luminescence from a discharge in mercury vapor and neon at 1.5 A; the spectrograms are given in order of increasing field strength. The top spectrogram is for the discharge without magnetic field. Here the neon is represented by a series of bright lines in the red end of the spectrum, and the mercury by the yellow and green lines at the right. Applying a magnetic field and increasing the field strength greatly reduces the brightness of the neon lines in the spectrograms, which shows that the cathode drop is being stabilized. Our attention is also attracted by some new lines that appear with strong fields. These same lines are observed in the luminescence of a pure mercury discharge in a strong longitudinal magnetic field, as illustrated by the spectrograms of Fig. 49. Analysis has shown that these lines belong to the spark spectrum of mercury. They come from singly ionized atoms of the metal. The brightest are the lines at 5425.25 A, 5677.17 A, 5871.73 A, and 5888.94 A. The excitation of the spark lines of mercury under these conditions is a sign that there is a large increase in the degree of ionization of the mercury vapor in the magnetic field.

In discussing the reasons for the increase in mercury arc stability observed in a magnetic field, a number of effects must be taken into consideration which in principle can exert an effect on the stability of the arc. One of the most probable ways that a field can act on the stability of a discharge is to reduce electron diffusion losses from the cathode. As a result of the exceedingly small dimensions of the cathode portions of the arc and the large particle concentration gradients at the boundary of the cathode, one of the most serious sources of particle loss will be diffusion into the surrounding space. A magnetic field, by bending the trajectories of the electrons during motion between collisions with the atoms in the gas, will greatly reduce these losses. Electron diffusion in a magnetic field acquires some peculiar features. First, the field only changes the diffusion rate of the electrons in the transverse direction, which makes the medium anisotropic with respect to diffusion. Second, as the density of the medium increases, the transverse diffusion rate at first increases, and only after it reaches a definite density does it start to decrease. This inversion of the relation is the result of limiting the freedom of electron motion in one direction normal to the lines of magnetic force. The electrons only move in the transverse direction as a result of collisions with gas atoms. In extending the concept of diffusion to the peculiar conditions of electron motion in a magnetic field, Townsend [179] derived a formula[1] relating the transverse diffusion coefficient D_H in the magnetic field to the ordinary diffusion coefficient D in the absence of a field:

$$D_H = \frac{D}{1+\left(\frac{\lambda_e}{R_L}\right)^2}, \tag{23}$$

where R_L is the mean value of the Larmor radius, and λ_e is the mean free path of the electrons measured along the curved path. It is assumed that the value of λ_e defined in this way will be independent of the field and equal to the usual linear mean free path of electrons in the same gas medium but without the field. The quantity R_L is related to the field strength H and the rate of thermal motion of the electrons, w, by the familiar formula

$$R_L = \left(\frac{m_e}{q}\right)\frac{w}{H}. \tag{24}$$

Here m_e/q is the mass to charge ratio of the electron, expressed in electromagnetic units. A check of Eq. (23) under practical conditions has shown that it usually gives values of the coefficient D_H which are much too low, the possible reasons for which have already been discussed in the literature [180,181], and Eq. (23) cannot yet be considered finally established. Neverteleless, the way in which D_H depends on λ_e and H is correctly described by Eqs. (23) and (24).

It may be concluded from the above equations that for any given density in the medium corresponding to some value of λ_e, there will exist a completely definite minimum or threshold value of the field strength H_p at which the field can still reduce the electron diffusion rate. This fact points the way to a direct quantitative check of our assumption that there is a relation between the stabilizing effect of a magnetic field on an arc and the reduction in diffusion losses at the cathode spot.

[1]The formula is given in somewhat modified form.

The threshold value of field strength H_p may be found for each value of λ_e from the condition

$$\frac{\lambda_e}{R_L} = 1, \tag{25}$$

where R_L may be found from Eq. (24). No substantial error is introduced in Eq. (24) if w is replaced by the mean square velocity and set equal, for two degrees of freedom, to $\sqrt{2kT_e/m}$. Since the electron temperature T_e is under the radical sign, we can also without great error in the results take the rounded-off value $T_e = 10^4$ °K, which gives the approximate formula

$$R_L = \frac{3.1}{H}. \tag{26}$$

Using this formula and Eq. (25), we can easily find the relation between the threshold value of the field strength H and the pressure p in the gas:

$$H_p = \frac{3.1}{\lambda_1}\left(\frac{273}{T_g}\right)p, \tag{27}$$

where λ_1 is the mean free electron path for a gas pressure of 1 mm Hg with the magnetic field reduced to zero, and T_g is the temperature of the gas in the cathode region of the arc. Using the value of λ_e from the kinetic theory of gases, and setting T_g equal to 1000°K, we can obtain numerical threshold values of H_p for helium and xenon in the range of pressures to which the experimental and calculated values of the threshold field strengths are placed side by side in Table VIII. It may be seen that for helium there is practically complete agreement between the two values. For xenon, the calculated values are one half the measured values. It should be noted that in the latter case it is impossible to expect any better agreement between experimental data and calculations based on values of the mean free path taken from the kinetic theory of gases, since as a result of the Ramsauer effect the actual collision cross sections for electrons with Xe atoms will be considerably different from those used in kinetic theory.

A confirmation of the fact that diffusion loss of charge from the cathode spot plays an essential role in the instability of the arc, and along with it a confirmation of the above mechanism for the action of a magnetic field on the arc, may be found directly in the form of ψ_0 as a function of H in the empirical equation (17). It is only necessary to compare Eq. (17) with Eq. (23). If we substitute in Eq. (23) the value of R_L from (24), it is easily seen that the value of ψ_0 is inversely proportional to D_H. Accordingly, the arc duration in the first exponential segment of the curve may be expressed by the relation

$$\vartheta = \vartheta_0 e^{\alpha \frac{I - I_0}{D_H}}, \tag{28}$$

where α is some constant quantity.

In writing Eq. (13) in this way, the role played in the arc-stabilization mechanism by the field's reduction of the diffusion losses stands out with extreme clarity. Comparing the coefficients of H^2 in (17) and (23) enables us to find the mean free electron path λ_e in the cathode region of the arc and the corresponding reduced mercury vapor pressure. The values of the order of 10 mm Hg found from the calculation are in agreement with the data of other observations, and in any case do not seem absurd.

Another possibility of explaining the stabilizing action of a magnetic field on an arc is provided by the fact that the effect of a field is smaller when the arc has a solid cathode. It forces us to admit that the field stabilizes the arc because it exerts a damping effect on the mercury. That such an effect necessarily exists follows from the laws of magnetohydrodynamics, according to which a magnetic field prevents the formation of streams of liquid flowing in a direction transverse to the field vector. This results in a marked reduction in heat transfer from the cathode spot by convection.

From the point of view of magnetohydrodynamics there is in principle still another possibility of explaining why the arc is unstable and how it is stabilized by the magnetic field. It must be kept in mind that the cathode region of the arc may possess electrodynamic instability of one of the types shown by a plasma thread when it is compressed by its own magnetic field H_i [182,183]. That a suggestion of this sort must not be considered absurd is shown by calculating the magnetic pressure at the boundaries of the cathode spot:

$$p_H = \frac{H_i^2}{8\pi}. \tag{29}$$

As a result of the large curvature of the magnetic lines of force in the arc field near the cathode spot, this expression gives pressure values that are too high. However, in the present case, this is not particularly important, since the data on the concentration n_e and the temperature T_e of the electrons in the cathode region of the arc are even less reliable, and these are what determine the plasma pressure kept in equilibrium by the magnetic field. Neglecting the ion pressure in the plasma, the equilibrium condition may be written in the form

$$\frac{H_i}{8\pi} = n_e k T_e. \tag{30}$$

The strength of the arc field at the boundaries of the cathode spot may be expressed approximately by the relation

$$H_i = 0.1\pi a j, \tag{31}$$

where a is the radius of the spot and j is the current density. Substituting (31) in (30) and eliminating the radius of the spot from the equation $\pi a^2 = i/j$, we find an expression for the maximum electron concentration in the cathode plasma at which the kinetic electron pressure can still be held in equilibrium by the magnetic pressure of the inherent field of the arc:

$$n_{e\,\max} = 1.25 \cdot 10^{-3} \frac{ij}{kT_e}. \tag{32}$$

It is easily found from this formula that for a current of 1 A, and assuming $j = 10^7$ A/cm^2 and $T_e = 10^4$ °K, the electron concentration that can be held in equilibrium is about 10^{16} cm^{-3}.

It follows from the above calculation that the cathode region of the arc is actually capable of being a plasma thread, compressed by its own field. But then serious attention must be given to the electrodynamic version of the theory of arc stability, especially in view of the stabilizing action that a magnetic field is found to exert. In our case it is easy to lay out a way of verifying this possiblity. It consists in comparing the effects of longitudinal and transverse magnetic fields on the arc. The duration of the arc in a transverse field was measured with the cathode design described in detail in one of the subsequent chapters (see Fig. 90). In this cathode, the cathode spot in the transverse field executed an ordered motion in a closed trajectory. It was also found here that the duration of the arc increases with increase in field strength. For a comparison of the stabilizing effects of longitudinal and transverse fields, curves are constructed in Fig. 50 showing the currents corresponding to a discharge duration of about 10 sec as a function of the magnetic field strength for horizontal and transverse orientation. The constant-duration curve shown by the solid line gives the results for an arc in a longitudinal field, while the dashed curve gives those for a transverse field. Both curves show monotonic and equal reduction of the values of discharge current corresponding to the arc duration used. The fact that the one curve is as a whole somewhat above the other must be due to some difference in the experimental conditions between longitudinal and transverse orientation of the field. For these differences not to be reflected in the comparison of the effects of the two types of fields, there is some point in comparing the relative values of the currents for the two curves at H = 1000 Oe, calculated as the ratio i_H/i_0, where i_0 is the current at H = 0. It is easily seen that for either orientation of the field this ratio is about 0.6.

Thus the stabilizing action of the magnetic field is independent of its orientation with respect to the cathode, as long as we exclude secondary effects such as the arc breaking off in a transverse field as a result of sudden increase in operating voltage at the discharge. In view of this, it seems highly improbable that the increase in arc stability observed in a magnetic field is electrodynamic in nature. On the other hand, observations made recently by the author have shown that in some arcs with metallic cathodes, such as solid and liquid indium, applying a magnetic field does not in general produce any increase in duration of the arc. This fact forces us to reject the idea that there is a connection between the effect and any damping action of the field on the liquid cathode. As against this, it may be thought of

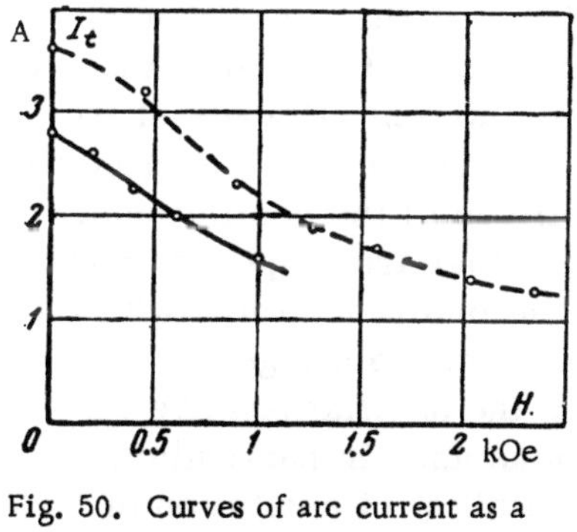

Fig. 50. Curves of arc current as a function of magnetic field strength for a constant discharge duration of about 10 sec. ——longitudinal field; -----transverse field.

as an additional serious indication that the reason for the increase in stability of the mercury arc in a magnetic field is the reduction in electron diffusion losses from the cathode spot and the resulting increase in probability of stepwise ionization of the mercury vapor. Actually an effect of this sort is only observed in metals having a set of atomic energy states favorable to stepwise ionization. An especially effective set of states in this sense is possessed by metals of the II group of the Periodic Table since they have triplet levels of the type 6^3P_{012}, including the metastable levels 6^3P_0 and 6^3P_2. The level systems in the I and III group metals are considerably less favorable to stepwise ionization. In view of this, it may be foreseen that on metals belonging to these groups (with the exception of the Sc subgroup) it will in general be impossible under ordinary conditions to observe a form of discharge with a cathode drop approximating the ionization potential, as occurs in the fundamental form of the mercury cathode arc.

32. Internal Instability of the Mercury Cathode Arc

In this chapter, a detailed discussion is given of two different ways in which a mercury cathode arc can depart from the conditions required to maintain a stable discharge, namely, by extinguishing spontaneously, and by undergoing various types of oscillations. A characteristic of spontaneous extinction proved to be that this form of arc instability occurs over a rather wide but limited range of discharge currents, the range changing with the experimental conditions. On the other hand, oscillations are observed in the arc under all conditions, thus constituting an inherent attribute of the arc. There are various stabilizing factors that can produce a more or less substantial reduction in the amplitude of the oscillations. However, even under these conditions, neither the amplitude nor the frequency characteristic of the oscillations is identical with the usual statistical fluctuations inherent in every discharge.

Another very important fact is that the oscillations are observed under all external circuit conditions and, in particular, under conditions which eliminate the possibility of excitation of small current fluctuations and thus generation of self-oscillations in the circuit containing the so-called "negative" resistance of the discharge. It must be especially emphasized that, in all the experiments described above, since the conductivity of the arc is high the discharge circuit is aperiodic and incapable of maintaining oscillations.

That the oscillations are not the result of the falling characteristics of the discharge is also shown by the following considerations. First of all, in the short arc under discussion here, which has no positive column, there is no falling characteristic in the strict sense of the word. As was shown above, the operating voltage of the normal and transitional forms of the arc is independent of the current. Under these conditions the only thing that can be thought of as a falling characteristic is the reduction in the mean static voltages with increase in current; this is exclusively the result of reducing the frequency of occurrence of the voltage pulses corresponding to the transitional form of the arc. But this reduction in mean static voltage cannot possibly be a source of instability in the arc for the simple reason that it is one of the consequences of the instability in the arc. The true dynamic characteristic of the discharge, as we shall see below, continually increases.

Thus the voltage oscillations must be attributed to processes occurring inside the mercury arc, and they are thus a sign of internal instability in this type of arc cycle. The correctness of this deduction is supported by the relation established above between the voltage oscillations in the short arc and the processes occurring at the cathode. There are a number of indications that the oscillations are the result of systematic disturbances of the equilibrium between the various processes in the arc cycle. Apparently, the mercury cathode arc does not represent any definite state of equilibrium between these processes, characterized by some value of cathode drop, but is rather a state in which all the processes in the arc cycle are constantly varying in intensity within some range of values determined by the experimental conditions. This is equally true of both the transitional and fundamental forms of the arc. More than this, since the oscillations are observed in the mercury cathode arc under all experimental conditions, and this is one of the most stable varieties of metallic arc, it may be concluded with a high degree of confidence that internal stability is a general property of all cold arcs.

All the oscillographic observations show that the conception formed of a perfectly definite cathode drop and cathode current density for every arc is an idealization admissible only as a first approximation to the facts, and then only in solving a definite limited group of problems which does not include problems in arc stability.

The nature of the voltage changes observed in the arc is such that they may be classed with relaxation oscillations. This fact forces us to assume that nonlinear elements of some sort occur in conjunction with cumulative processes and large energy losses during each period. The nonlinear elements in this case are evaporation of the cathode metal and electron emission, along with ionization of the gas. The associated cumulative processes may be local heating of the cathode by ion bombardment, or formation of the plasma and the space charge next to the cathode. The energy losses will come predominantly from charge diffusion losses in the cathode region of the arc, and from dissipation of the heat liberated at the cathode. It would be too early at the present time to try to develop a quantitative theory of the oscillations in the arc cycle in view of the large degree of uncertainty in the basic data. Nevertheless, the general picture of how the oscillations arise as a result of disturbances in the equilibrium between the processes in the arc cycle can already be drawn rather clearly on the qualitative level.

A number of conditions must be satisfied if the arc is to be maintained in its relatively stable fundamental form. One of these is violent boiling of the metal in some small region of the cathode surface, since only boiling of the metal can supply the high concentration of atoms in the cathode spot required for the discharge. Another very important process in the arc cycle is electron emission by the cathode. These processes are related to one another through other processes such as ionization of the metallic vapor by electrons and formation of a space charge consisting of positive ions moving toward the cathode. These ions are the agent which keeps the cathode boiling and which at the same time ensures the electron emission required for the discharge. The first striking aspect is the complex overall reciprocity between all these processes, the second is the impossibility of maintaining equilibrium between them for any length of time.

One of the inevitable sources of disturbance to the equilibrium will be that a single agent, namely the positive ions, must simultaneously maintain two such entirely different processes as liberating electrons from the cathode and boiling the cathode itself. The first of these processes is practically inertialess and, according to cold-emission theory, depends in intensity only on the instantaneous values of the collective ion charge density in the thin layer next to the cathode. In contrast to this, the second process is cumulative in nature, being related to the need for the cathode to accumulate a definite amount of heat energy ahead of time, and it depends exclusively on the power supplied by the ions. It is entirely improbable that the conditions nominally necessary to maintain both processes will always be reached at the same ion current densities and cathode drops in the arc. But in a situation of this sort we can only imagine the following two ways of maintaining the discharge:

a. maintaining some sort of equilibrium state in which one of the processes in the arc cycle is enhanced;
b. alternate enhancement of some processes at the expense of others.

The choice between these two possibilities is already decided by the violent nature of all the processes in the cathode spot, coupled with its exceedingly small dimensions, as well as by the difference in nature of the various processes in the arc cycle pointed out above. It is especially typical of the cold-cathode arc that relatively enormous powers have to be concentrated in microscopic volumes of materials. This situation eliminates the possibility of enhancing any of the processes and at the same time maintaining an equilibrium state. This leaves only the second of the two possibilities of maintaining the cold-cathode arc, namely, continuous oscillations in intensity of all the processes with alternate enhancement of some at the expense of others. But with the strong dependence of the cold-emission current on the electrostatic field intensity, even small changes in the field will produce very large disturbances in equilibrium. This means that a serious review must be made of the question of whether or not it is permissible to treat the conditions in the cold arc with a theory and method of calculation based on the idea that this form of discharge is some kind of equilibrium state.

It may be concluded from everything that has been said above that the principal sources of the internal instability that we observe in the mercury arc or, moreover, in any other cold arc are:

1. the great complexity of the conditions required to maintain the arc cycle, resulting from the fact

that it has the additional function of forming the medium required for the discharge by evaporating the cathode;

2. combining in one cycle practically inertialess processes such as electron emission or gas ionization, and the cumulative process of heating the cathode;
3. the highly nonlinear character of electron emission, ionization, and evaporation;
4. concentration, because of their very nature, of the processes of the arc cycle in a microscopic region of the discharge, which necessarily leads to developing large powers in small volumes of material, with large losses of energy and charge.

It is perfectly obvious that in view of the instability of the arc cycle there are only the following two ways out under any imaginable experimental conditions: either the continuity of the cycle will be broken at some instant of time, which is what we encounter in spontaneous extinction, or, under more favorable conditions for coordinating the various processes in the cycle, the cycle will be maintained in a continuous state for an indefinitely long time by means of oscillations and alternate enhancement of processes. In the latter case, the instability of the arc will only show up in the continuous voltage oscillations which our investigations have revealed. It will be clear from this that, in itself, having no arc extinctions at a suitably high current is no criterion for saying that this particular form of discharge has acquired absolute stability under the conditions at hand. It follows from all the material given that the increase in arc duration with increase in current is a result of increasing the probability that it will be restored out of the critical states corresponding to sudden disturbance of the equilibrium between the various processes in the arc cycle. This fact gives a good explanation for the two most general laws governing spontaneous extinction, namely, the purely random distribution of the intervals between extinctions, and the exponential form of the two segments of the curve giving the mean arc duration as a function of current. The first of these laws is a direct consequence of the fact that for every current there is a definite time-independent probability of restoring the arc, expressable as an exponential coefficient, while the second of the laws is the result of increasing this probability. Thus, the fact that spontaneous extinctions stop at high currents means only that under these conditions, instead of a single manifestation of internal instability in the arc, its other manifestations predominate. These include voltage oscillations in the arc and the instability phenomena in the cathode spot itself, discussed below.

Chapter 3

THE FOCAL STRUCTURE AND THE CONTINUOUS DECAY AND RE-FORMATION OF THE CATHODE SPOT ON A MERCURY CATHODE

33. Fundamental Problems and Methods of Investigating the Cathode Spot

In our investigation of arc stability in the preceding chapter we discussed the properties of the mercury cathode discharge as a whole, without going into the detailed behavior of the cathode spot itself, which is the most important part of the discharge, since its properties are essentially what determine the specific features of the arc discharge. At the same time, the phenomena occurring at the cathode are precisely what cause the instability of the arc. It is accordingly to be expected that an investigation of the behavior of the cathode spot will provide much interesting information on the arc cycle. This will substantially enlarge our ideas on the cold arc as an unstable form of discharge. In proceeding to the solution of this problem we need to set up a program of study that will permit us to obtain answers to the following questions:

1. How does the internal instability of the arc cycle show up in the behavior of the cathode spot or its various parts?
2. How does the cathode spot manage to coordinate the processes in the arc cycle so that the current is on the average maintained at the level fixed by the external circuit constants?
3. By what operations on the cathode spot is the transition effected from one steady current state to another when the external circuit constants are changed?
4. Is it possible to establish a relation between the internal instability of the arc cycle and the vari-

ability of the cathode spot and its tendency to be in constant motion over the cathode?

Before carrying out this program, we must first make a detailed study of the structure of the cathode spot under the various conditions in which it exists. Some particularly useful ways of penetrating the structural details of the spot and the changes occurring there are provided when the spot is anchored at the thin film of mercury at the boundary of the region where the mercury wets the metal in contact with it. As we shall see below, an exceedingly revealing picture of the fine structure of the spot is obtained under these conditions. On the other hand, since the random motion is absent, and the spot can be given a forced orientation on the cathode, it is in this case possible to follow in most detail all the continuous changes which the spot as a whole and its various parts undergo with the passage of time. In view of these facts, special attention will be given below to investigating the structure of an anchored cathode spot. The conclusions arrived at in the course of these investigations may then be extended to the normal arc with a homogeneous liquid cathode by simply taking into consideration the special features of structure and behavior of the arc peculiar to these circumstances.

In carrying out our program, of all the available methods of observation, including high-speed and instantaneous photography of the spot with modern electronic and optical equipment, the most satisfactory proved to be the primitive method using continuous mechanical or mirror photographic scanning of an enlarged image of the spot. In the first case, the image was projected onto a photographic film fastened to a uniformly rotating drum as shown schematically in Fig. 51. The spot was anchored by a small strip of sheet molybdenum or nickel, located in such a way on the mercury surface that the line where it was wet by the mercury was parallel to the axis of the cylinder. The image of the cathode spot was scanned in a direction normal to the strip and the line where it was wet by the mercury; the cathode spot was drawn out along this line and was a narrow, brightly luminescent band. The anchor was set at an angle to the cathode surface so that the upper edge protruding out of the mercury did not partially mask the spot. The side of the anchor facing down into the cathode was covered tightly with mica to keep the cathode spot from being anchored there. In making the photographs, the length of exposure was set approxi-

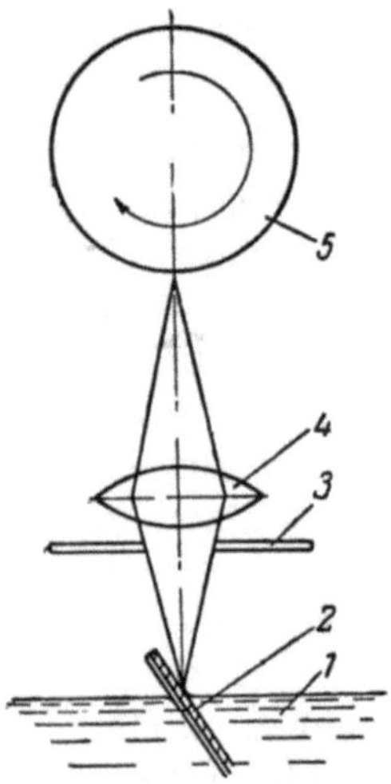

Fig. 51. Mechanical method for continuous photographic scanning of the image of the spot on a film moving at constant speed. 1) Mercury; 2) molybdenum anchor; 3) diaphragm; 4) objective; 5) rotating drum with film fastened to it.

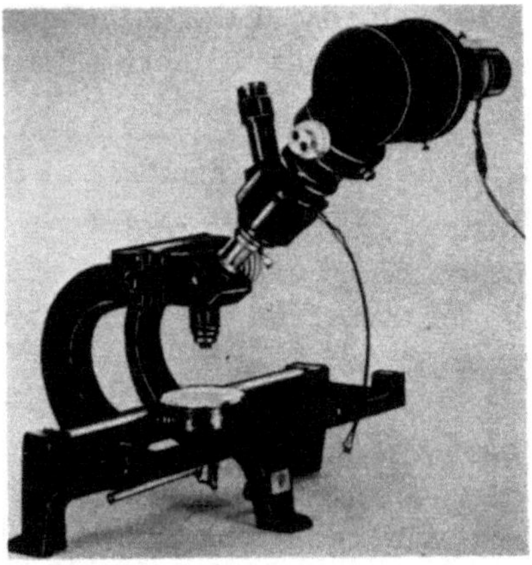

Fig. 52. Exterior view of rotating film photographic chamber, combined with an MIR-12 microscope.

mately equal to one rotation of the drum, so that the film only showed one trace of the image of the spot.

The camera and the rotating drum were construced in such a way that if necessary they could be attached to a microscope by means of the universal adapting head shown in the photograph of Fig. 52. In this case it was possible to get a greatly enlarged image of the cathode spot, which considerably extended the possibilities of investigating the fine structure. A special discharge tube was constructed for these observations; it consisted of a shallow metal reservoir for the mercury cathode and a ring-shaped anode close to it, with a ground joint around the top of the reservoir. Its general construction is shown in Fig. 53. The upper wall of the tube, through which the photographs were taken, was a piece of ordinary optical glass, fitting tightly on the ground joint. Since the distance between the glass and the cathode spot was small, the cathode spot could be photographed with short-focus objectives, which gave the required magnification of the image.

The second method of observation used was ordinary mirror scanning with photographic reproduction of the image of the cathode spot. This method gave pictures with relatively small magnification, but they could be taken at high scanning rates. In both forms of the apparatus, means were provided for photographing the cathode spot starting at low-discharge currents and going up to the threshold value. Since the duration of the arc is extremely small under these conditions, the observations could only be made by synchronizing the shutter with the instant at which the discharge current was reduced to the desired value. The current was reduced by suddenly connecting an additional noninductive resistance of proper value into the discharge circuit. A special relay was used to synchronize this operation with tripping the shutter. Provision was also made for taking a simultaneous oscillogram of the voltage on the arc electrodes by means of delayed scanning of the oscillograph. Thus photographic scannings of the cathode spot and oscillograms were taken for the same period. The rate of rotation of the drum and the mirror were determined with high accuracy by a device consisting of a miniature magnet, fastened to the shaft of the rotating part, and a fixed winding in which an emf was induced. By measuring the frequency of the emf with an oscillator and Lissajous figures on an oscillograph, the rate at which the image of the cathode spot was scanned could always be found to the desired degree of accuracy.

These methods for continuously scanning the image were used to investigate the cathode spot when it was moving at random over a homogeneous liquid cathode surface. Under these conditions, the observations were quite difficult, partly because of the random motion of the object itself and the disturbance of the mercury associated with it, and to no less an extent because of the intense luminescence of the discharge, which gives a continuous background in the pictures and numerous spots of light reflected from the waves on the mercury. There are times when, for some reason, the trace of the cathode spot on the pictures disappears completely. In all probability this is caused by the formation of a rather deep depression in the mercury as a result of the forces developed in the cathode spot. If the depression is at an oblique angle, the rays may be masked by the walls of the depression and thus not reach the objective. In spite of all these complications, with the aid of observations made on an anchored spot it is possible to draw some perfectly definite conclusions even from the unclear pictures taken in this way. A considerably more convenient method of photographic

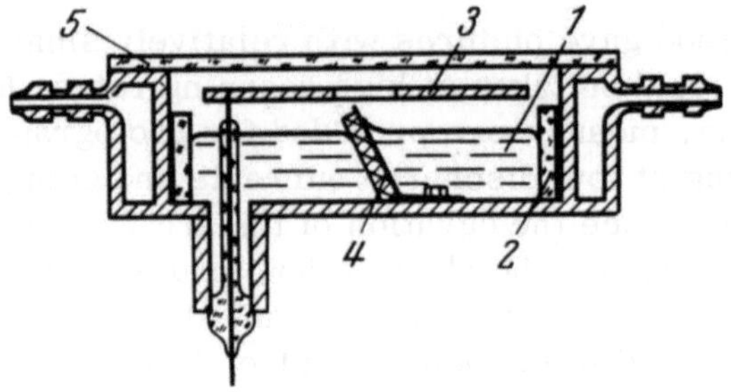

Fig. 53. Tube used to investigate the structure of an anchored spot. 1) Mercury; 2) protective glass; 3) anode; 4) anchor; 5) glass window for photographing the spot.

scanning of the image of the spot in an arc with a homogeneous liquid cathode is to photograph the spot when it is moving rapidly in a definite direction in a magnetic field tangential to the cathode surface. In this method, in which the observations are made with an ordinary Zenith S film camera, the effect of continuous image scanning is achieved from the motion of the object itself. As we shall see later, there are a number of advantages to this method of scanning the image of the cathode spot. It must, however, be kept in mind that in this method the spot is at all times under somewhat artificial conditions, being subject to the effects of a strong magnetic field. This fact obliges us to treat the results of the observations with a certain amount of circumspection, always checking them against the methods described previously. In what follows, we shall have to turn more than once to this method of observation under conditions where it becomes the basic method, and where there is some point in stopping to consider it in more detail.

A. STRUCTURE AND BEHAVIOR OF THE CATHODE SPOT ON MERCURY UNDER STEADY-STATE CURRENT CONDITIONS

34. Investigation of the Structure of the Spot Where It Is Anchored at the Boundary of the Region Where a Metal Is Wet by Mercury

We cannot orient ourselves in considering the extremely complex phenomena occurring in the cathode spot without comparing photographic scannings of the image of the spot made at differ-

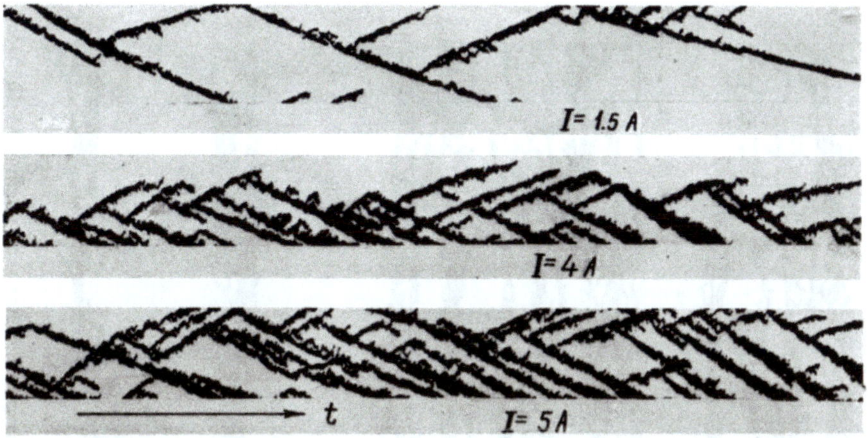

Fig. 54. Photographic scanning of the image of the cathode spot when anchored at the boundary of the region where the mercury wets a molybdenum strip. Magnification 2.5; time scale: 1 cm = $7 \cdot 10^{-3}$ sec.

ent magnifications and scanning rates. Several typical pictures of the spot, taken at low magnification and low scanning rate when the spot is anchored by a strip of sheet molybdenum, are given in Fig. 54. In the pictures, direction of the time axis is from left to right, and it makes a right angle with the line where the mercury wets the molybdenum. The trace of the spot moving over the emulsion looks like a regular sort of fringe consisting of straight segments extending out to the side like branches. Each of the branches corresponds to an autonomous region of cathode emission or autonomous cathode spot. Two or more branches on the same vertical section of the picture means that there are two or more autonomous spots in existence on the cathode at the same time. The number of spots increases with increase in current. Here the current going through one branch is on the average about 0.7 A. The impression of regularity in the figure is due to the fact that the branches are straight and all have the same angle. Both of these observations show that the autonomous spots move along the mercury meniscus at a constant rate. The rate of the motion is equal to about 24 cm/sec, which is close to the velocity of propagation of capillary surface waves on the mercury. This agreement indicates that the motion of the autonomous spots at a uniform rate along the line where the molybdenum is wet by the

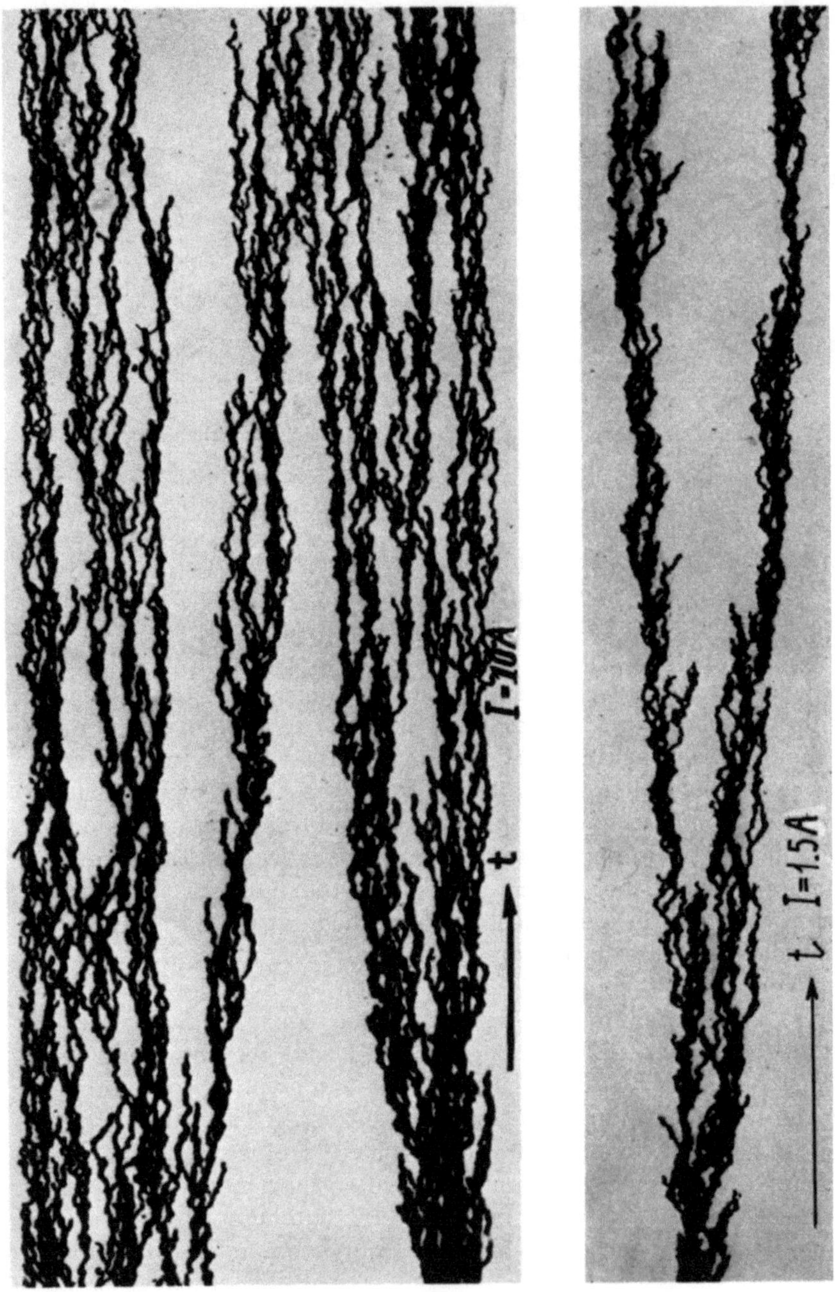

Fig. 55. Pictures of an anchored cathode spot with a magnification of 19 and increased scanning rate. Time scale; 1cm = $0.7 \cdot 10^{-4}$ sec.

mercury is caused by a capillary wave which keeps the thickness of the film of mercury that wets the molybdenum at the meniscus from being uniform. It is obvious that each spot is located on a part of the mercury film that has some optimum thickness, and that it moves along with this part of the film during the undulating change in thickness of the film.

It must be noted that observations of this sort were described earlier by Tonks [8], as has already been pointed out in Chapter 1. However, with the low resolving power apparatus that he used, he could not observe the fine structure of the various parts of the cathode spot, and he missed a number of the interesting properties of the finer constituent parts of the spot.

Indications of the fine structure of the branches are already to be found in the pictures of Fig. 54. For further consideration of the details, we must turn to the pictures in Fig. 55, which show the cathode spot with a magnification of 12 and with a greatly increased image scanning rate. A comparison of both sets of pictures will convince us, first of all, that each of the branches in the image of the spot, shown in thepictures of Fig. 54 as straight segments, consists in reality of a number of interlaced wavy lines. This inhomogeneity in the image shows that there is a great deal of inhomogeneity in the current distribution in each of the autonomous spots. The current is seen to be concentrated at the separate, very small portions of the surface occupied by an autonomous cathode spot. But this means essentially that each of the autonomous spots making up the cathode region of the arc in turn consists of finer cells which carry practically the whole current. A detailed analysis of the pictures makes it possible to determine a number of the individual properties of these cells, as well as to find how they interact. The most important results of an analysis of this sort are given below.

35. Data on the Elementary Cells in the Cathode Spot

In looking at enlarged pictures of the type shown in Fig. 55 we are struck first of all by the fact that for a certain period of their existence the spots are in intimate contact with one another, breaking up into smaller groups of two, three, or more spots each, which rotate around one another. That they rotate around one another is shown by the wavelike way in which their traces are interlaced and by the characteristic way in which they change their shape when the scanning rate is decreased so as to give the one-sided cusps characteristic of a cycloid. From the size of the period between neighboring loops in the traces it may be calculated that the period of rotation of the cells is of the order of $2 \cdot 10^{-5}$ sec.

With their obvious preference for intimate contact with one another, the cells also show that they can exist individually, from time to time separating by distances many times greater than their dimensions. It is noteworthy that the traces of separate cells often split up and then come back together again or interlace with the traces of other cells. Both the traces of these waltzing pairs of cells and the single traces can be followed through repeated approaches and withdrawals. We can conclude from this that the cells do not lose their autonomy even when they are close together.

A very important fact is that in the steady state of the arc the number of cells in the cathode spot remains constant in time. Here, independently of the magnitude of the discharge current, one cell always takes, on the average, the same current i_1, lying between 0.08 and 0.1 A, which is very close to the threshold value, I_0, of the current in the mercury arc cathode which we found in Secs. 24 and 27. That these last statements are correct for the case in which the cathode spot is anchored on molybdenum is shown by the values given in Table X for a number of currents, found by counting the individual traces in different cross sections of the pictures. The first column of the table gives the values of the discharge current to which the counting results apply. The second and third columns show respectively the mean number of cells in existence simultaneously on the cathode, and the mean current carried by each cell. The small variations observed in the figures in the third column must be ascribed to difficulty in separating the individual cells when their traces are closely inter-

TABLE X

I, A	$\overline{n}$	$i_1 = \frac{I}{\overline{n}}$, A	$\frac{\Delta i_1}{i_1}$
0.3	4.0	0.07	—0.3
1.15	8.4	0.14	+0.4
1.5	16.7	0.09	—0.1
2.1	17.6	0.12	+0.2
3.0	29.0	0.10	0.0
5.0	50.4	0.10	0.0
	Mean	0.10	

laced. It is perfectly obvious, if we neglect these variations, that the mean current carried by one cell of the cathode spot is independent of the discharge current and lies in the vicinity of 0.1 A. The inevitable failures to record some of the spots oblige us to assume that the figures in the third column are somewhat too high, approximately within the limits of the deviations of the individual figures from the mean, shown in the last column of the table. In view of this, it must be assumed that in fact the mean value of current i_1 carried by one cell of the arc lies within the limits 0.08-0.1 A. Similar results were found by counting the cells in pictures taken with the cathode spot anchored on nickel.

The properties observed in the cells, namely, that they are autonomous, that they carry the same average current, and that this current is very nearly the threshold current of the mercury cathode arc, enable us to regard the cells as the simplest constituent parts of the cathode spot or as elementary cells of the arc. If this is really true, we can try to isolate a single cell by suddenly reducing the discharge current to the value $I = i_1$, and at the same time photographing the cathode spot. With a pure mercury cathode, attempts of this sort lead to a negative result in the majority of cases, since the single cell that is left decays too rapidly. It was found, however, that it becomes much easier for the cells to exist individually if the mercury surface is covered with thin layers of powdered metal such as iron and nickel. The results of experiments on isolating single cells under these more favorable conditions are shown in the pictures of Fig. 56a-c. The first of the pictures is for a sudden current reduction from 1.5 to 0.16 A, i.e., to twice the current carried by the single arc cell. The second and third pictures were taken with the current reduced from the same initial value to 0.086 and 0.06 A, respectively. In the first case, the trace remaining on the picture shows that two cells were in existence on the cathode at the same time. The trace remaining on the second picture obviously belongs to a single cell. Finally, reducing the current to 0.06 A always produces instantaneous breakoff of the arc.

The experiment just described clearly shows that it is correct to conclude that the cells are capable of individual existence, as well as that the value for the mean current in a single cell, i_1, is correct and that it is related to the threshold current I_0 in the mercury cathode arc. In addition, the picture of Fig. 56a contains

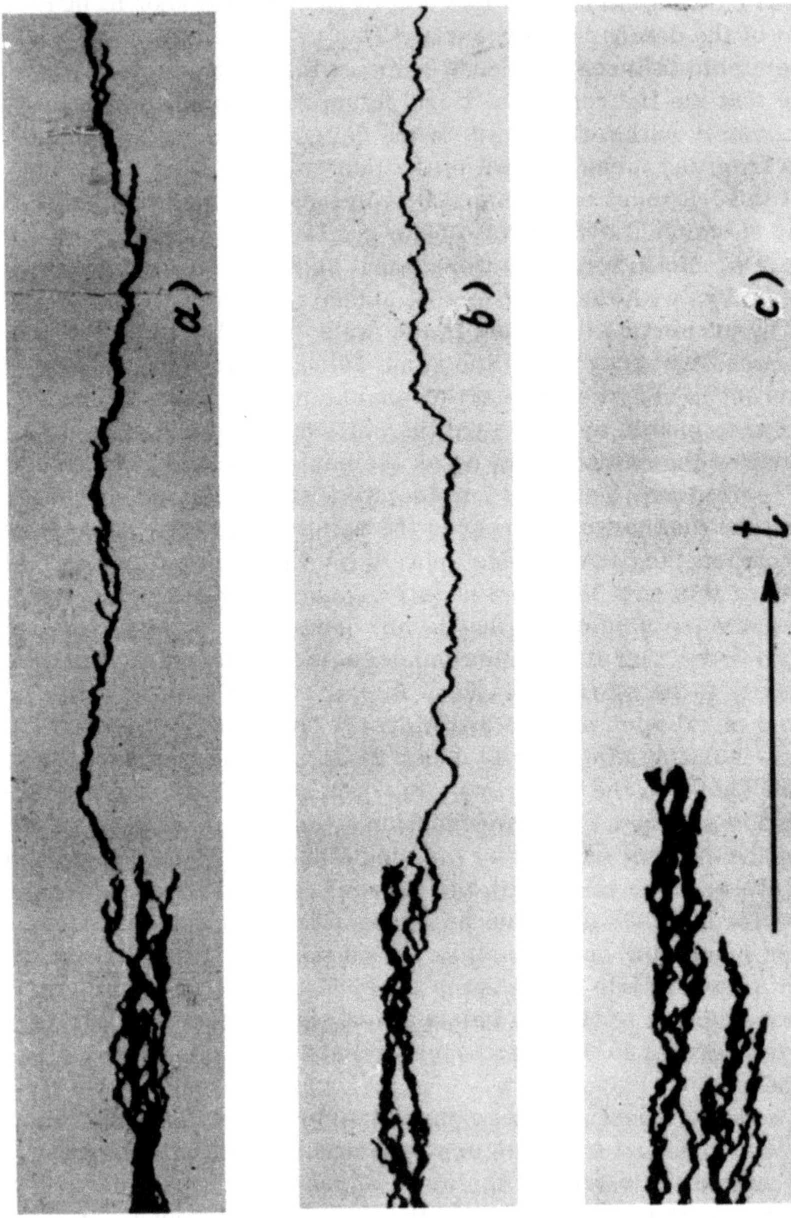

Fig. 56. Results of experiments on isolating single elementary arc cells by suddenly reducing the current from 1.5 A to 0.16 A (a), 0.086 A (b), and 0.06 A (c), with the spot anchored at the boundary where nickel is wet by the mercury. Magnification 24; time scale: 1 cm = $5 \cdot 10^{-5}$ sec.

definite indications that at different instants of time the current in a given cell can depart appreciably from the norm. This occurs in particular if one of the cells suddenly decays. It is obvious that at such instants the whole of the discharge current is distributed among the remaining cells. In the present case one single cell takes on the whole current, and so it has twice as much current concentrated in it. It may be seen from the picture that this state of affairs cannot exist for a very long time, and the result of this sudden departure of the current from the norm is the formation of a new cell on the cathode. We shall give special attention to this latter fact in the next section. It is now only important to emphasize in this connection that the fact established, that the mean current in a single cathode-spot cell is constant, in no way means that the spots are incapable of carrying currents for a short time that are appreciably different from the norm. Obviously, the current i_1 must be considered simply as some optimum value of current which provides the greatest stability for an autonomous cathode-spot cell, which is simply the smallest structural unit of the arc that can maintain an arc cycle by itself.

The picture of Fig. 56b, showing a single cathode-spot cell, was convenient to use in finding the mean current density in the cathode spot. To this end, the image of the cell was magnified several hundred times, and its cross-sectional dimension δ was found for different sections of the picture. The mean value of δ found from the measurements proved to be $1.56 \cdot 10^{-3}$ cm. If the cross section of the cell is a perfect circle, this diameter corresponds to a current density of the order of $0.5 \cdot 10^5$ A/cm^2, i.e., a factor of 5 greater than the value found by Tonks under similar conditions. It must be kept in mind that the figure given is only for some effective value of the current density in the cell for its visible surface, which is probably greatly exaggerated.

36. Decay and Division of Elementary Cathode-Spot Cells Under Steady-State Conditions

In the preceding section we intentionally disregarded all those questions of stability and interaction between the cells which go beyond the range of purely structural problems. Nevertheless they are of the first order of importance for an understanding of the nature of the internal instability of the arc and of the peculiarities in behavior of the cathode spot in a cold-cathode arc.

TABLE XI

I, A	Δt, sec$\cdot 10^3$	Δn	$\frac{\Delta n}{\Delta t}\cdot 10^{-4}$	$\nu = \left(\overline{\frac{\Delta n}{\Delta t}}\right)$	ν/I
0.3	1.08 0.41 1.53 1.27 1.34	21 11 35 24 27	1.95 2.68 2.29 1.89 2.02	$2.17\cdot 10^4$	$7.23\cdot 10^4$
1.5	0.38 1.02 0.32 0.50 0.54 1.52 0.60 0.15 0.13 0.17	42 100 33 55 48 130 57 14 15 17	11.1 9.8 10.3 11.0 8.9 8.5 9.5 9.3 11.5 10.0	$9.99\cdot 10^4$	$6.66\cdot 10^4$
2.1	0.21 0.26 0.08 0.23 0.28	25 54 12 45 50	11.9 20.8 15.0 19.5 17.9	$17\cdot 10^4$	$8.10\cdot 10^4$
3	0.35 0.14 0.09 0.33	65 30 18 70	18.6 21.4 20.0 21.2	$20.3\cdot 10^4$ Mean	$6.78\cdot 10^4$ $7.2\cdot 10^4$

A source of information on stability and interaction between the cells may readily be found in the pictures that we have already discussed (Fig. 55) and those like them. Returning to them we find first of all that single cells and whole groups of cells suddenly disappear from time to time, as may be seen from the broken-off traces in the pictures. Cell decay of this sort occurs systematically in every steady state of the arc, as may be seen from the results, given in Table XI, of counting the broken-off traces in various parts of the pictures. The first column of the table gives the currents at which the counts were made. The second and third columns give the length of the time intervals observed and the corresponding number of cell decays recorded. The fourth and fifth columns contain the frequencies of cell decay found from measurements at various currents, along with the mean values. Finally, the figures in the last column show the mean cell-decay frequency at 1 A. The number of cells that decay, Δn, is proportional to the time of observation t and the magnitude of the discharge current I, as follows from the constant value obtained when the mean cell-decay frequencies, $\nu = (\overline{\Delta n/\Delta t})$, are divided by the current. As in the preceding table, the small variations in the figures in the last column are in this case due to some break-offs being inevitably missed in the observations, which gives some reason to assume that these figures as a whole represent low values for the cell-decay frequencies.

Thus, by all indications, cell decay is a phenomenon that follows a definite law, even under steady-state current conditions. Since the number of cells is constant in time for any given stationary state, we must conclude that, along with cell decay, the opposite process of formation of new cells is taking place, and that equilibrium is maintained between the two processes. Nevertheless the pictures do not show any traces of new cells that suddenly start on the free parts of the cathode surface. The reason for this is that, in the steady states under discussion here, new cells are formed exclusively by division of cells already in existence, with the result that the traces of new cells always start at one of the wavy lines that belongs to a cell that has already been formed. This makes it very difficult to find cells that are being formed. However, there is no particular need to do this, since in any steady current state the number of cells formed by division over any long period of time will be exactly equal to the number of cells that decay, the data on which have already been given in Table XI.

If we know the mean number $\overline{n}$ of cells in the spot at some current I, or the mean current in a single cell, $i_1 = I/\overline{n}$, and if we have found the cell-decay frequency ν, then the mean lifetime ν_1 of an elementary cell may be easily determined. Actually, the value of $\overline{n}$ is found by counting the number of cells n_k in existence simultaneously for successive small time intervals Δt_k, and then averaging:

$$\overline{n} = \frac{1}{t} \sum n_k \Delta t_k.$$

But $\Sigma n_k \Delta t_k$ is the total lifetime of all the cells during the time t covered by the observations. This quantity may be otherwise defined as the product of the mean cell lifetime τ_1 and the total number of divisions νt, which leads to the simple relations

$$\bar{n} = \nu\tau_1; \tag{33}$$

$$\tau_1 = \frac{1}{(\nu/I)\, i_1}. \tag{34}$$

It follows from Tables X and XI that the values of ν/I and i_1 are independent of the discharge current. Accordingly, the mean cell lifetime τ_1 is also independent of current. By making use of the values of ν/I and i_1 given in the tables, we can easily find the numerical value of τ_1. We find in this way that for the arc with an anchored spot that we have been discussing, the value of τ_1 is about $1.3 \cdot 10^{-4}$ sec.

It is easily seen that the lifetime τ_1 of the elementary cells in the cathode spot is a quantity of the same order of magnitude as the values that we have already found (see Secs. 25, 27 and 30) for the arc duration ϑ_0 near the threshold current I_0, and the period $\Delta\vartheta$ of the cyclic changes in the cathode drop, which is equal to ϑ_0. The fact that these values are the same is quite important. We now have a direct indication that there is a relation between the internal instability in the arc already discussed and the instability of the elementary cells, i.e., the instability of the arc cycle itself on each of the small parts of the cathode surface where the cells occur. These observations can only mean that in each of the autonomous cells the arc cycle can only maintain itself for a limited time. which is completely definite for any fixed set of experimental conditions. For the relatively favorable conditions under discussion here, where the cathode spot is fixed, the normal duration of the cycle is only about 10^{-4} sec. The fact that the arc discharge as a whole can maintain itself for a longer or even unlimited time is the result of the arc cycle reoccurring on other parts of the cathode surface. It is the continuous replacement of one set of active centers by others that enables the arc cycle as a whole to maintain itself for a long time. But this is achieved at the expense of the cycle jumping continuously from one part of the cathode to the other. Thus, the so-called steady state of the arc is in essence nothing more than a certain state of equilibrium between cell decay and cell formation.

37. Interaction Between Cells and Why They Move Over the Cathode

Before we pass to a discussion of the cooperative mechanism

between cell decay and cell division, we must try to explain what it is that forces the cells to move over the cathode, and what the phenomena are that lie at the basis of their tendency to collect into groups.

The fact that the cells remain in intimate contact with one another for a relatively long time shows that a definite interaction exists between them. The interaction must not be thought of in terms of the usual types of ponderomotive interaction which control material bodies and currents, since the cells cannot be identified with either the one or the other. Each cell is a region of the cathode, active in the sense of electron emission, maintained in this state by a complicated concatenation of processes. The concept of interaction between these regions must include any change in stability of every cell as it is affected by the rest of the cells. The tendency of the cells to come into intimate contact with one another must be thought of from this point of view as a direct proof that they are more stable when they exist collectively. The increase in stability noted when the cells are in intimate contact with their own kind serves to define something which may be called their "bond strength." A quantitative measure of the bond strength of the cells, and along with it of the increase in stability that forms the basis of the bond strength when the cells exist collectively, is provided in a natural way by the relative time ζ that each cell remains in intimate contact with the other cells. This quantity is comparatively easily determined in the simplest case where there are not more than two cells on the cathode at one time, as shown in Fig. 56a. Analysis of pictures of this sort gave $\zeta \approx 0.7$, which is apparently a measure of the degree of attachment of the cells on a thin mercury film. It is completely logical to assume that anything that makes it more difficult for the cells to exist on the cathode will increase the bond strength and cause the cells to join up into larger and larger groups.

The tendency of the cells in the cathode spot to distribute themselves in groups cannot be entirely explained simply in terms of an interaction of the type considered above. It is beyond doubt that to some degree this tendency is a natural manifestation of the instability of the genealogically related cells, which cannot, as a result of their limited capability for individual existence, get very far from the point where they were split off. Every such group of cells, bound together by their common origin, is an independent genealogical branch of the cathode spot. It maintains its existence by replacing defunct cells with new ones through division, as occurs in the field of biological phenomena. If a branch dies out, it is the result of gradual degeneration, in that cell decays predominate over cell divisions. Thus, each of the branches observed (see pictures, Fig. 54) corresponds to a decadent group of genealogically related cells. The degeneration of the various branches of the cathode spot is the result of competition between them, and shows that there are relatively unfavorable conditions for the existence and division of cells in the branch in question, which depends on how many there are, how compact they are, and the concurrence of random factors. Just how correct it is to interpret the branches in the image of the cathode spot in this way is shown by the picture which the branches form, which reminds us of an ordinary family tree (see Fig. 54). The picture is so characteristic that it is in itself sufficient to show the focal structure of the cathode spot and that the individual parts of the spot are unstable.

When we come to the question of why the cells are observed to move over the cathode, we must at the very start admit that there are various reasons, as is shown by the complex nature of the motion. The tendency of the cells to cover more of the cathode surface when the discharge current is increased, as is clearly shown by the pictures in the next section, shows beyond any doubt that an interaction of repulsive type exists between them. The source of the repulsion may be simply the magnetic field of the arc. However, under the conditions considered here where the cathode spot is anchored at the thin film where the mercury wets the metal, the interaction only shows up in an obvious way as some sort of collective effect of the mutual repulsion between the cells as the current is increased. A repulsion of this sort is not observed explicitly in the behavior of any two neighboring cells. The paths that they follow in a number of cases converge repeatedly and diverge again. Behavior of this sort cannot in general be the result of interaction between cells. It can be caused bv various sorts of hydrodynamic effects. As was already noted in Sec. 34 in connection with the pictures of Fig. 54, there is some reason to assume that the uniform motion of individual groups of cells along the mercury meniscus comes from capillary-type waves propagating over the mercury surface. These waves, as it were, drive the groups of cells from place to place, continually dragging them with them. The mutual location of the cells on the cathode will inevitably change during this change in position. The cells may move on the cathode as a result of such trivial phenomena as the removal of mercury immediately under them by evaporation. The same effect can produce rotational motion of two or more related cells around their common center. As a matter of fact, if there is some binding force between the cells, due to it being easier for them to exist when they are in close contact with one another, and one of the cells moves away from a part of the cathode that has been depleted of mercury, this will produce a coordinated motion of a second cell or of the rest of the cells. But under such conditions, the freedom of motion of the cells is limited primarily to rotational motion alone. Once it has started, the rotation cannot stop until the binding force between the cells is broken. This is not due to any inertia in the cells, but simply to the fact that they leave behind them a part of the cathode that has been depleted of mercury. The reason for the decay of a group of cells may be that more of the mercury is used up in the region where the cells are turning around.

We do not pretend that the reasons for the cell motion given above are correct or complete. The one thing that is beyond doubt is that the general reason the cells continually change their position on the cathode and are so mobile is that they are unstable. In view of this instability and the continuous changes in the conditions under which a cell exists at any part of the cathode, each one of the cells will always be moving about over the cathode, shifting itself automatically to regions where the conditions of existence are simplest. It is precisely at this level that we must consider the tendency of the cells to remain in motion at all times on the thin mercury film, and principally in intimate contact with other cells, when they have maximum stability.

38. Coordination Between Cell Decay and Cell Division. The "Relay-Race" Type of Arc Restoration

It was found in the preceding sections that at steady-state current the number of elementary cells in the cathode spot remains constant, and the number increases in proportion to the current. The fact that the number of arc cells existing on the cathode at one time is constant, and that the number depends on the current, shows that together with the instability in the cells there is some coordinating mechanism which maintains the relation between the cells and controls the process of cell division through the external circuit of the arc. The coordinating action of this mechanism is clearly shown in the pictures (Fig. 54): the decay of one of the dying branches of the cathode spot is as a rule accompanied by the appearance of new branches proceeding out from the main trunk or from the largest cell groups. A careful analysis of the pictures taken at high magnification and a high rate of scanning of the cathode-spot image is sufficient to show that, at moderate currents, coordination of decay and formation of cells by division is a process which occurs in detail; i.e., each decay produces rapid replacement of the missing cell by division of one of the other cells. This form of current redistribution between the cells suggests a "relay race," and is clearly visible on the picture discussed above (Fig. 56a). The formation of new cells can be observed by the sudden thickening and increase in optical density of the line representing the trace of a single cell. This thickening of the line occurs every time another trace breaks off. It is easily seen that each such act of passing along the signal is in essence nothing other than the arc-restoration process already discussed in Sec. 30, and consisting of redistribution of current between the cells, which we concluded must occur from an investigation of the oscillation processes in the arc. For brevity we called this type of restoration of the discharge the "relay-race" type.

More detailed information on the coordination mechanism of the arc, and at the same time on relay-race type restoration, may be obtained by comparing the photographic scannings of the cathode spot with the oscillograms of the voltage on the arc electrodes taken at the same time. One pair of synchronized pictures and oscillograms is given in Fig. 57 along with their common time scale. As usual, the direction of the time axis is from left to right. The discharge current used in the experiments was about 0.3 A, so that there were four elementary arc cells on the cathode at one time. It may be seen from the picture that for the greater part of their existence the cells were in intimate contact with one another, joining up in two's, three's or four's, as shown by the close interlacing of the traces. Sometimes the traces diverge, showing that the cells were single. As a rule, this corresponds to a general lowering of voltage. It may be concluded from this that the cells

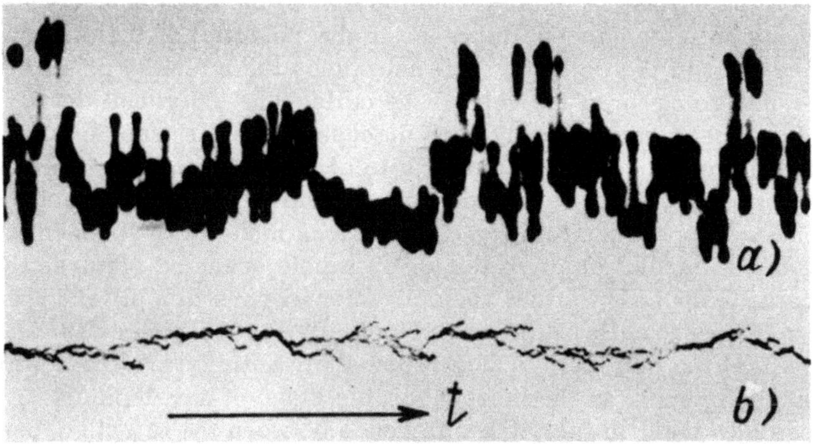

Fig. 57. Oscillogram of voltage oscillations in an arc (a) and synchronized photographic scanning picture of the image of the spot (b) at a current of 0.3 A. Time scale: 1 cm = $0.7 \cdot 10^{-4}$ sec.

diverge automatically as soon as the conditions for their existence for some reason get better, and that they come together again as conditions become more difficult. Usually cell decay is accompanied by a narrow voltage pulse with an amplitude from several hundredths to a tenth of a volt. There is every reason to believe that this pulse is precisely what produces the division of the remaining cells, which restores the equilibrium from its disturbed state. It is of course evident that it is impossible to establish exact correspondence between every pulse and a definite cell decay on the pictures, since we are not in a position to record all the cell decays. All the material discussed, however, enables us to assume with a very definite degree of credibility that every pulse observed on the oscillogram is actually caused by the decay of one or more of the four spots, and that the pulse produces division of one of the remaining cells. But then the results of counting the cell decays given in Table XI must be considered very much too low.

From an analysis of all the material given, the following simple scheme may be set forth for the way in which the arc coordination mechanism works. If a cell decays, the voltage on the electrodes rises abruptly as a result of the reduction in voltage drop on the ballast resistance and the emf in the inductance, which produces an instantaneous increase in the potential of the plasma with respect to the cathode. This increases the intensity of all the processes going on in the rest of the cells. The individual details of this process have already been discussed earlier (see Sec. 29), so that there is no point in stopping to consider them now. Here we shall simply point out in addition several essential new factors which have appeared from the observations made in the present section. A careful study of the photographic scanning pictures of the cathode spot shows that the activation occurs to a different extent in different cells, affecting principally the isolated cells. It seems that the only cells which renew their activity are those which are at the moment in the most favorable stage of development. It is possible that, in fact, the difference between the effects depends simply on the concurrence of random factors. However this may be, the result of the phenomenon is that, when the steady state is disturbed, the current is unequally distributed among the rest of the cells, and is concentrated principally in the isolated cells. It has already been emphasized in Sec. 35 that in a steady-state discharge the elementary cells of the cathode spot can only exist as autonomous regions with a closed arc cycle for very definite values of the current from approximately 0.08 to 0.1 A. Reducing the current below 0.08 A produces immediate decay of the spot; i.e., it ceases to exist. On the other hand, if the current is raised above 0.1 A, the cell divides into two cells. In view of this instability of the cells, unequal distribution of current among them will automatically produce division of those cells which are carrying more than the normal current as a result of the redistribution. If this forms an excessive number of cells in the cathode, one of the cells will automatically decay as a result of the reduction in current. Thus the steady state of the arc, disturbed by the decay of one of the cells, is again restored.

From the scheme for the coordination mechanism of the arc set forth here in its general features, it follows that in maintaining the current at a fixed level, a considerable role is played by the same instability of the elementary cells in the cathode spot which causes the instability of the arc. The instability of the cells is reflected not only by the way in which they decay systematically, but also by the way in which they divide systematically. While the decay of cells is caused by the reduction of the current in them to below normal, the cells divide because the current is raised above normal. In all probability, the current is continually being redistributed among the cells as long as they exist. When cells decay, the redistribution takes on the nature of a discontinuous process which causes the most active cells to divide. The time required to do this may be found from the duration of the voltage pulses which produce the redistribution of current. The duration of any such pulse gives the length of time from the instant a cell decays, thus stimulating division of one of the other cells, to the instant at which the steady state is restored. Thus, from the duration of the coordinating pulses we can find how long it takes the relay-race mechanism to restore the arc, and along with this the time required to form a new cell. This time interval we shall designate below as t_d. It follows from the oscillograms that, in order of magnitude, t_d is about 10^{-7} sec. In view of the fact that the slowest process in the arc cycle is heating the cathode to boiling, it appears obvious that heating the cathode at a new place is precisely what limits the time required to form a new cell.

39. Special Features of the Structure of the Cathode Spot on a Homogeneous Liquid Cathode

The conception of the spot as an assembly of unstable cells, arrived at from investigating the structure of an anchored spot, may, with slight changes, also be extended to the conditions in which the spot moves freely over the cathode. One of the reasons for this generalization is that in going from the first set of conditions to the second there are no fundamental changes introduced into the basic physical processes of the arc cycle. The possible changes are of a purely quantitative sort. That this conclusion is correct is shown by the fact that, when moving about on a homogeneous cathode, the spot continually undergoes the same characteristic changes as were observed when it was anchored on the thin mercury film. Figure 58 gives two photographic scanning pictures of a cathode spot, showing its behavior on a homogeneous liquid cathode under widely different experimental conditions. Figure 58a was taken with the spot moving at random over the cathode; i.e., it is for the normal conditions under which the spot exists on mercury. In contrast, Fig. 58b shows the spot when it is forced to move in a definite direction in a magnetic field tangential to the cathode. In spite of the fact that neither picture shows the regular

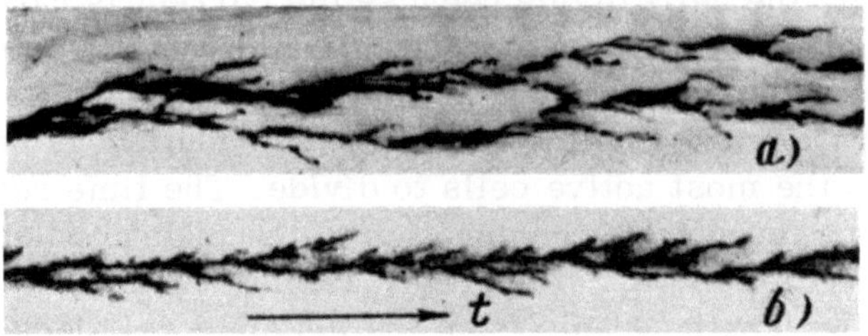

Fig. 58. Photographic scanning pictures of cathode-spot image. a) Moving at random over cathode; b) directed motion in a magnetic field.

figure characteristic of the image of an anchored spot, both cases clearly show our familiar "genealogical" scheme, consisting of a trunk with a series of decaying branches. The conclusion must be drawn from this that, in the liquid mercury cathode arc under any conditions, the cathode undergoes the same type of changes with time, which shows that it has a focal structure and is unstable.

The changes in structure of the spot shown in Fig. 58 for the condition where the spot is on a homogeneous liquid cathode are of second order of importance. Figure 59 helps to explain what they are; this picture caught the instant the spot became anchored on an amalgamated copper plate. The originally relatively monolithic spot splits up into a number of cells, a sudden drop-off in brightness occurring at the same time. Obviously, the conditions under which the spot exists on a homogeneous cathode are marked by a high degree of strength of the bond between cells, corresponding to values of $\zeta \approx 1$. The reasons for the increased bonding between the elementary cells in the spot on a homogeneous liquid cathode may easily be grasped if we compare the data given previously (see Secs. 25 and 31) on the stability of an arc with a homogeneous

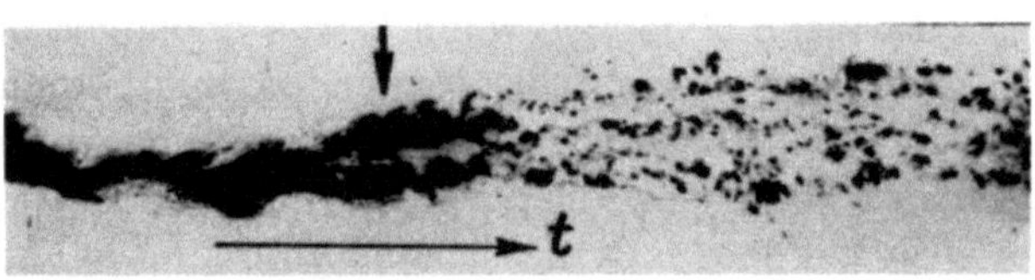

Fig. 59. Photographic scanning of the image of a cathode spot which has caught the instant at which the spot becomes anchored on a copper plate (shown by the arrow).

cathode with the condition where the spot is anchored on a thin mercury film. It may be concluded from this comparison that the conditions under which the cells exist on a homogeneous cathode are considerably less favorable, and this automatically causes the cells to stick together, which may be expressed quantitatively as an increase in the value of ξ. Without doubt, the complicated conditions under which cells exist on a homogeneous cathode must be held responsible for the comparatively high values of the cathode drop found in this type of arc. It is clear from this that the cathode drop of a mercury arc is not uniquely related to the ionization potential of mercury.

Thus, the first feature of the structure of the cathode spot on a homogeneous mercury cathode is the high bond strength between the elementary cells, which may be expressed by setting $\xi \approx 1$. What this means in fact is that under these conditions the individual properties of the cells are of second order of importance in comparison with their group properties. The group properties will be reflected first of all in the fact that in the homogeneous cathode arc the cells will not reveal their existence in quite as definite a way as in the anchored-spot arc. In this case we have to deal almost exclusively with closely bound groups of cells of the most diverse dimensions. Obviously, each of the decaying branches in Fig. 58 is just such a group. We shall call such a group simply an autonomous cathode spot.

The changes noted in the structure of the spot change the nature of the interaction between the various parts of the spot. A direct result of the intimate joining of the cells into associations or groups will be a considerable increase in the role played by the inherent magnetic field of the arc in the interaction between various parts of the arc. As was already pointed out in Sec. 37, the special feature of this problem is that interaction in this case must be thought of not as some sort of ponderomotive effect of one current on another, but rather as any change in the stability of one part of the cathode spot brought about by another part. This is the level at which we must consider the role played by the magnetic field of the arc in the interaction between the various parts of the cathode spot. With the large current densities existing in the cathode spot, the intensity of the inherent field of the arc can reach impressive values in this region. In view of the stabilizing action of a magnetic field on an arc as shown above, every part of the cathode spot can exert an appreciable effect on the other parts of the spot by means of the field of the current which it is carrying. One must keep in mind this important result of the compact structure of the cathode spot when investigating the behavior of the spot on a homogeneous liquid cathode.

B. HOW AN ARC DISCHARGE IS SET UP, AND HOW IT CHANGES FROM ONE STEADY STATE TO ANOTHER

40. Setting up the Steady State of the Arc at a Rate of Current Increase Less Than 10^8 A/sec

The results given above of the investigation of the structure and the properties of the various parts of the cathode spot were for a steady-state arc. In order to get some idea of the actual process by which the arc is set up and the mechanism controlling the cathode emission current, it is necessary to further investigate the phenomena that take place in the cathode when the current is increased.

Consider first the reaction of the arc to a change in the external circuit which causes the discharge current to rise at a rate not greater than 10^7 A/sec. The most common method of controlling the discharge is to change the discharge current by making more or less sudden changes in the ballast resistance connected in series with the arc. Some idea of the phenomena which this produces in the discharge is given by the oscillograms of current and operating voltage of the arc shown in Fig. 60. These show the reaction of an arc with an anchored spot to suddenly shorting out part of the ballast resistance, thus raising the current from 1 to 8 A. The oscillograms are arranged in such a way that increasing the current corresponds to a rise in the curve, while increasing the voltage corresponds to a drop in the curve. At the instant the resistance is

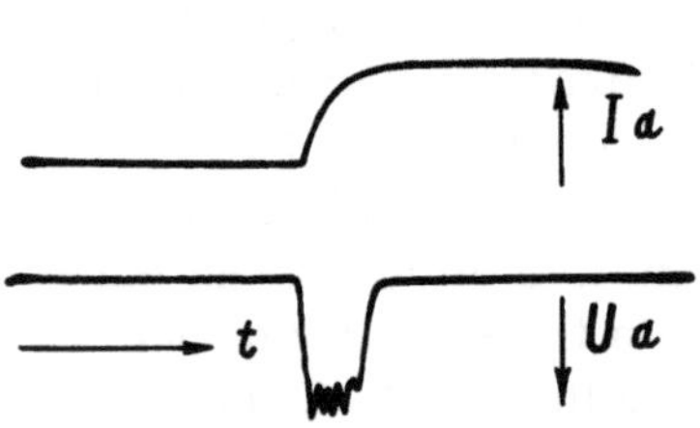

Fig. 60. Oscillograms of arc current and voltage showing the reaction of the discharge to suddenly shorting out part of the ballast resistance, thus raising the current from 1 to 8 A. Time scale: 1 cm = $1.3 \cdot 10^{-4}$ sec.

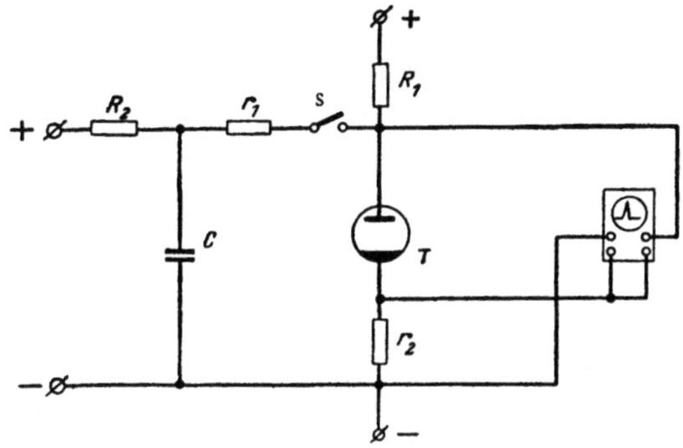

Fig. 61. Diagram of connections of the tube T used to investigate the reaction of the arc to a sudden increase in current produced by connecting the charged condenser C to the tube.

shorted out, the voltage on the arc electrodes increases by several volts, with an accompanying increase in current. After having reached its maximum value and gone through several sharp oscillations, the voltage again drops to normal. By this time, the discharge current has already reached the value corresponding to the new conditions imposed on the arc. According to our earlier steady-state conception of coordinating pulses (see Sec. 38), the temporary increase in voltage on the arc electrodes is in essence the coordinating impulse which serves to form the new arc cells at the cathode and make it possible for the increased current to flow. From this point of view, having a series of maxima on the voltage curve may mean that the process by which the required number of cathode-spot cells is formed by division of the existing cells goes through a series of stages associated with the characteristic features of the division process itself.

In studying the reaction of the arc to changes in the external circuit it is convenient to use another discharge-controlling method in which the rate of current rise is to a smaller degree limited by the inductance of the circuit, as this considerably simplifies the experimental conditions. In this method, shown schematically in Fig. 61, the current is increased by suddenly connecting the tube T through a small inductive resistance r_1 to the capacitance C, which has been charged from a constant-voltage source. By changing the resistance and the voltage to which C is charged it was possible to get a smaller or larger increase in current. The capacity of the condenser bank used in these experiments was about $2 \cdot 10^{-3}$ F and the smallest resistance was 3.5 Ω. The time constant of the discharge circuit was at least two orders of magnitude greater than the time t_s required to set up the steady state of the arc. Thus, the process of setting up the new steady state with increased current occurred with practically constant voltage on the plates of the condenser bank. The oscillograms and photographic scanning pictures of the cathode spot for these experimental conditions are shown in Fig. 62.

In Fig. 62a we must distinguish the region where the voltage rises to its first, usually highest, maximum (part AB) and the region of longer duration where the voltage drops to its normal value (part BC). The reason for the distinction is that these regions correspond to two perfectly definite stages in the process by which the arc adapts itself to a higher current, and the stages have clearly different physical characteristics. The steady increase in voltage observed in the first part of the oscillogram is essentially nothing more than the ordinary process of charging the interelectrode capacity through a resistance, complicated by leakage through the arc, which is comparable to the discharge current and variable in time. Under these conditions, the process of charging the capacity will depend at any instant on the ratio between the charging and leakage currents. The large changes in the leakage current will distort the process considerably. The time change in the relation between the charging current and the leakage current is, in the final count, precisely what causes the complicated form of the voltage pulse shown in Fig. 62. If we investigate the form of the pulse, we can determine how the conductivity of the discharge changes and then find the reasons for the change. Here we have to bear in mind that an increase in voltage on the arc electrodes always means that the charging current is greater than the leakage, while a reduction in voltage means that the ratio is the other way around. Any reduction in the steepness of rise or fall of the voltage curve is an indication that the ratio between the charging and leaking currents is leveling off, while every maximum or minimum is a sign that they are temporarily equal during large changes in leakage.

Starting with these ideas, we would conclude that the first maximum on the voltage oscillogram (point B) corresponds to a large increase in conductivity of the discharge, indicating an increase in emission from the cathode. The reasons for this are

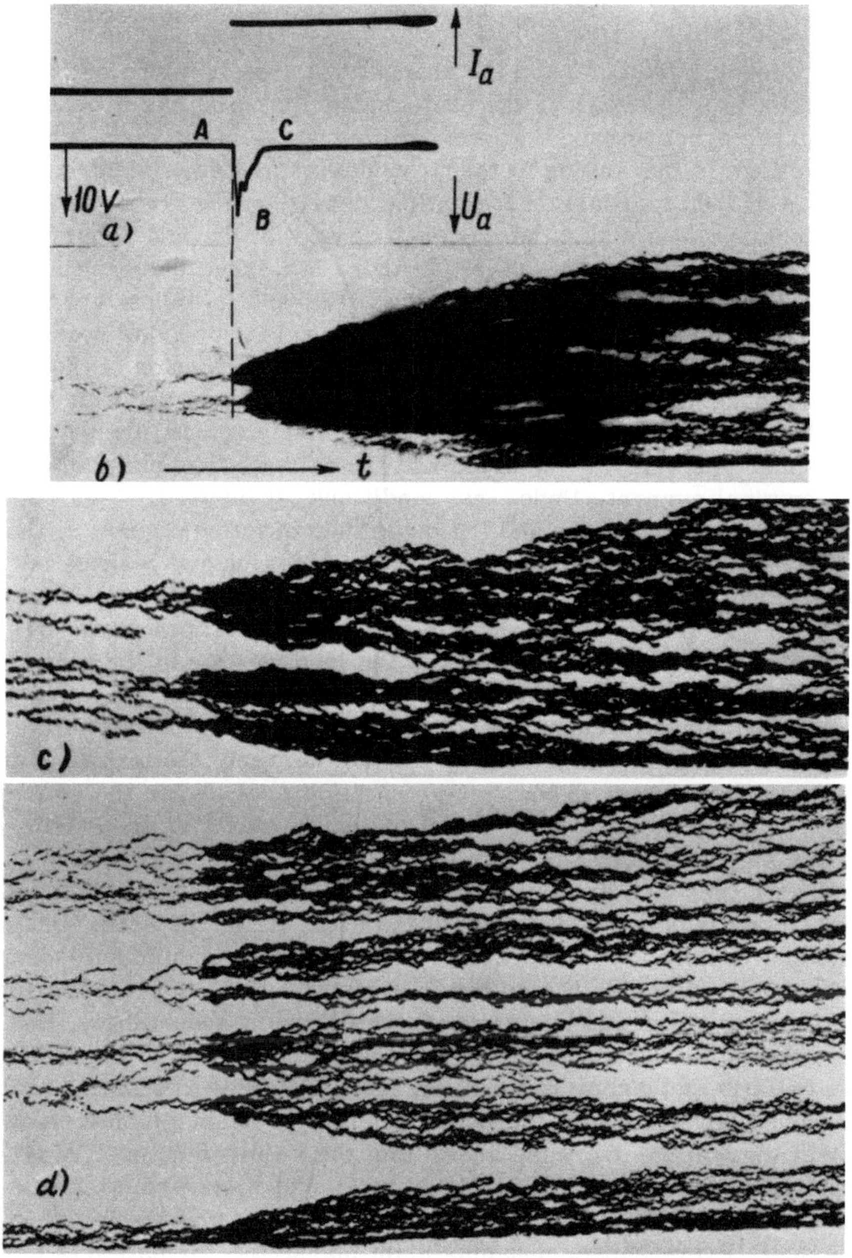

Fig. 62. Oscillogram of arc voltage (a) and synchronized photographic scanning picture of the image of the cathode (b), showing the reaction of the arc to suddenly connecting in a charged condenser which increased the current from 1 to 13 A; c) current increased from 2 to 10 A; d) current increased from 4 to 12 A. Time scale: 1 cm = $1.4 \cdot 10^{-4}$sec.

easily found if we compare the changes with the development of the cathode spot shown in the photographic scanning pictures of the image synchronized with the oscillograms. In the stage where the voltage is increasing to the first maximum, these pictures show a sudden increase in the optical density of the traces of the elementary cells making up the cathode spot at the initial current. The increase in density shows that there is a large increase in brilliance of the luminescence of the cells, and is a sign that there is a large increase in density of the arc-cycle processes in the cells in question. Thus, the initial stage of the reaction of the arc to the change in external circuit conditions, brought about by the sudden increase in voltage and current, may be called the stage of enhancement of the arc cycle and the cells as they existed with the original current. Under our conditions, it lasted 10^{-6}-10^{-7} sec. The rate of voltage rise and the magnitude of the maximum in the pulse are determined by the ratio of the charging and leakage currents. This explains why it is that, at high initial potential differences on the plates of the shunt capacity and small discharge resistances, we find that the slope of the leading edge of the pulse is greater, and that the maximum itself is higher.

This stage of enhancement of the arc cycle is preparatory to the second stage in setting up the new current, the turning point being the maximum on the voltage oscillograms. When the voltage rise suddenly stops at point B and the curve starts to drop steeply, it is a sign of sudden increase in the cathode emission current. Since the increase in current occurs while the voltage on the arc electrodes is decreasing and, at the same time, the cells are dividing rapidly, which serves to increase the emission surface, we can only consider the increase in current to be the result of an increase in the number of cells on the cathode. Accordingly, the essence of the second stage is physically that division of the existing cells is stepped up to form the additional cells required to set up the steady state with the new current. The nonmonotonic character of the voltage decrease shows that the required number of additional cells is not all produced at once, but in several stages, as is shown by the photographic scanning pictures of the image of the spot. Every subsequent maximum marks the formation of a new series of cells. We can tell from the number of maxima on the oscillogram how many stages of division are required under a given set of conditions to produce all the cells required for the

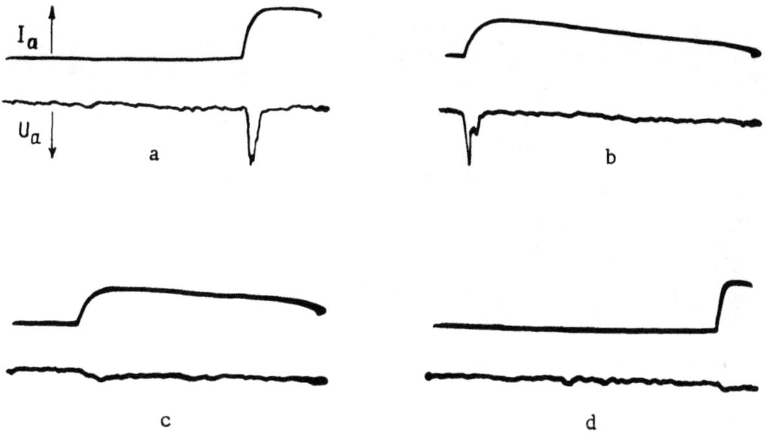

Fig. 63. Oscillograms of arc current and voltage showing two types of results from experiments on the reaction of a homogeneous mercury cathode arc to short circuiting part of the ballast resistance. a and b) At the instant the current increases, the voltage on the electrodes suddenly rises; c and d) increasing the current does not produce any appreciable voltage rise.

new current. The time intervals between the maxima are about 10^{-5} sec, and the whole process of setting up the steady state in the arc occupies a time approaching 10^{-4} sec. The pictures and oscillograms show clearly that one of the reasons the additional cells are formed in stages is that, as the current increases, it is unequally distributed among the cells. A more general reason for this phenomenon is to be found in a basic property of the division process itself, where the number of cells increases in geometric progression, and is only reached by successive doubling cycles.

The above experiments on the reaction of the arc to a change in the external circuit constants tending to increase the discharge current are actually for the condition where the spot is anchored. When experiments of the same kind were made on a normal arc with the spot moving freely over the cathode a new and interesting phenomenon was observed, namely, that there are two types of experimental results. In a certain percentage of the cases, the reaction of the normal arc to increasing the current was exactly the same as the reaction of the arc with the spot anchored, if we neglect the fact that there is some change in the form of the voltage pulse, in that it occurs in a shorter time. Normal oscillograms of this sort for the homogeneous liquid cathode arc are

Fig. 64. Picture illustrating spontaneous formation of new spots on the cathode at a rate of current rise of about $5 \cdot 10^8$ A/sec.

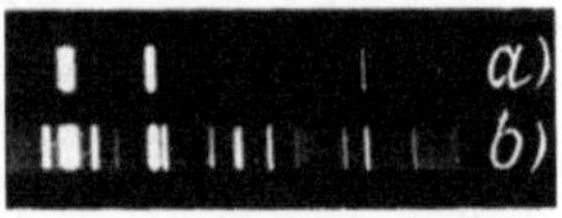

Fig. 65. Spectrograms of the luminescence from a mercury arc at small and large rates of current increase. a) $di/dt < 10^6$ A/sec; b) $di/dt \approx 10^8$ A/sec.

shown in Fig. 63a and b. In a number of cases, however, when the current was increased, there was no clearly defined increase in voltage, as is shown by the oscillograms of Fig. 63c and d. These two ways in which the homogeneous liquid cathode arc reacts to an increase in current must be interpreted as an indication that under these conditions the cathode spot can occur in two widely differing states. In one state the increase in current occurs in exactly the same way as in the arc with the spot anchored. In the other state it goes much more easily, without requiring any appreciable increase in voltage.

41. Nature of the Phenomena When the Current Is Increased at a Rate of 10^8 A/sec or Higher. Transition to a Spark Discharge

The observations in the preceding section are for the condition where the rate of current rise in the discharge circuit is comparatively low, not in excess of 10^7 A/sec. The nature of the phenomena occurring over the entire discharge and at the cathode changes greatly when the rate of increase in current reaches 10^8 A/sec and passes beyond this figure, with the result that this may be called the critical value.

One of the most important features of the arc reaction to sudden changes of this sort is that the cathode emission surface is increased not only by a step-up in cell division but also as a result of spontaneous formation of new cells on parts of the cathode that had been previously free of cells. The new cells show up in the form of numerous emission foci or cathode spots formed on the cathode without being connected to one another. This phenomenon has been described previously by Froome [12], who made his investigation by instantaneous photography with a Kerr shutter. In Fig. 64 we give a picture of spontaneously formed spots taken with an ordinary film camera. In these experiments, the discharge was produced in an aperiodic circuit, thus eliminating the possibility of an oscillatory discharge, which was further checked by the oscillograms. The formation of numerous emission foci on the cathode under these conditions cannot be explained in terms of periodic extinction and reignition of the arc. As the rate of current rise is increased, there is an increase in the number of spots formed spontaneously. The maximum dimensions finally reached by the spots are determined by the duration of the discharge current pulse. The form of the spots indicates that within each spot the emission surface increases with time, growing from the center to the periphery by continuous cell division, as may be seen from the large number of branches spreading out in different directions. It may be concluded from these pictures that at large rates of current increase, when the cells are subjected to particularly large effects from the external circuit, the cells can divide into not just two but a large number of cells at the same time. Here the division process can take on the form of continuous growth of the active cathode surface from the center to the periphery in regions having circular symmetry. The center of a formation of this sort, from which all the branches spread out in the form of rays, marks the region of the cathode where the emission first started.

The key to an understanding of the processes which lead to spontaneous formation of cathode spots on the mercury surface when the rate of increase in current is of the order of 10^8 A/sec is provided by a second feature of the reaction of the arc when it is subjected to a large effect from the external circuit. What we have in mind here is that we find an intense mercury spark spectrum. Two spectrograms of the mercury discharge luminescence for different current growth rates are shown in Fig. 65. In the upper spectrogram, the luminescence consists exclusively of the lines from neutral mercury atoms, as is usual in the arc spectrum. As the rate of current increase approaches the critical value of 10^8 A/sec, the spectrum begins to show the bright lines of singly ionized atoms and then the lines of doubly charged ions. At the same time the intensity of the neutral atom lines decreases appreciably. This is an unquestionable sign of almost 100% ionization of the mercury vapor in the discharge space. A characteristic thing is that the spark lines are absent in the spectrum of the luminescence from the cathode spot itself. This may be seen from the fact that if the image of one of the spots falls on the slit of the spectrograph the arc lines all show a section with greatly increased density, while the image of the spark lines is of constant density over its whole length. This difference between the spectrum of the cathode spot and the other parts of the discharge shows up in particular in the fact that if the cathode spot on the mercury is observed visually during large current pulses, it seems orange-colored in contrast to the luminescence from the discharge, and for very large rates of current rise it even seems red. This interesting visual effect may be used without any error to establish the fact that the rate of increase of current in a pulse is in any case in excess of the critical value of 10^8 A/sec.

The absence of spark lines in the spectrum of the cathode spots proves that the degree of ionization in the spot itself is very far from 100%, even when the current growth rate is above the critical value. The most probable reason for this is that this part of the discharge has a practically unlimited reservoir of neutral mercury atoms at its disposal in the continuously evaporating cathode. Nevertheless, under these conditions, the cathode spot is in a greatly enhanced state in which the current is being increased by continually increasing the intensity of the electrical processes in the arc cycle in the cathode spot, while the development of the emitting surface is lagging further and further behind. At subcritical rates of current rise, as we have seen, increase in the total cathode-spot surface is achieved exclusively by cell division. The rate of increase in current from this process of expanding the cathode spot and heating new parts of the cathode is limited by the time t_r required for a single division, and under normal conditions this time is about 10^{-7} sec. Hence it follows that cell division can only provide the necessary current growth rate up to a definite limit. The maximum rate of this process may easily be found for any value of discharge current i from the simple relation

$$\left(\frac{di}{dt}\right)_{\max}=\frac{i}{t_r}. \tag{35}$$

The maximum rate of current growth increases with increase in current as a result of the proportionate increase in the number of cells undergoing division. For a current initially of the order of 10 A, at which the current pulse is usually imposed, the steepness of rise will be of the order of magnitude of 10^8 A/sec. But this rate is the same as the critical rate at which new emission foci are found to appear spontaneously on the cathode with the spark spectrum showing up at the same time. Accordingly, the critical rate is the maximum value for the steepness of current rise at which division of cathode-spot cells can still occur. But what this means is precisely that higher values of the derivative di/dt can only be reached by a continuous increase in intensity of all the electrical processes in the arc cycle, and this requires a continuous increase in the voltage on the arc electrodes.

Everything that has been said above gives a good explanation of why the spark lines show up in the discharge, and why new emission centers are formed spontaneously on the cathode. Both these phenomena are to be regarded simply as a natural series of consequences of the greatly increased intensity of the arc-cycle processes in the cells, which cannot divide fast enough to provide the required rate of increase in cathode emission surface. Actually, the excitation of the short-wave spark spectrum owes its origin to the large increase in ionizing power of the electrons resulting from the rise in voltage on the arc electrodes. With the simultaneous operation of the strong electric field formed under these conditions, the short-wave radiation can in turn produce spontaneous formation of new emission foci on the cathode as a result of the photoelectric effect. This partly makes up for the insufficient rate at which emission surface is produced by the usual process of cell division. The spontaneous reduction of new cathode spots thus only serves to emphasize the fact that under conditions where the current is rising rapidly, the bottleneck in the discharge, which tends to hinder the process, is the insufficient rate at which cathode emission surface is formed by the division of cathode-spot cells.

If the mercury has dirt films on it visible to the eye, the new emission foci are formed principally on the films, as may be seen from the picture in Fig. 66. In contrast to Fig. 64, this picture illustrates a case in which the cathode is extremely dirty. Here the spot that was already on the cathode before the high-

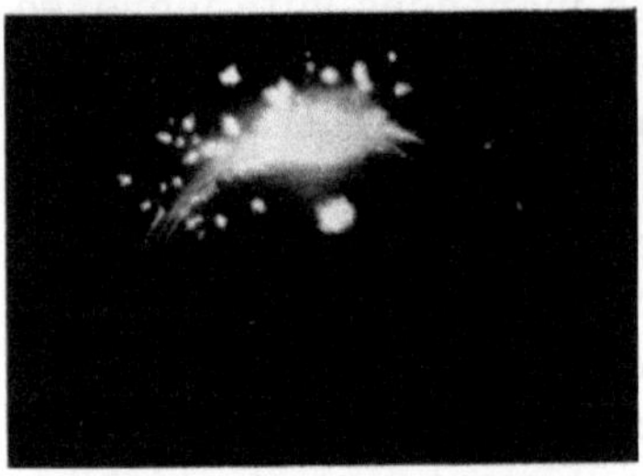

Fig. 66. Spontaneous cathode-spot formation at a current growth rate of about $5 \cdot 10^8$ A/sec. The picture also shows part of the circular trace of the spot formed before applying the current pulse.

current pulse was applied was executing continuous motion around a closed trajectory in a radial magnetic field. Part of the path of the cathode spot around its trajectory may be seen in the picture as a bright oval line. When the current pulse is applied, numerous new spots are formed on the cathode outside the region of motion of the original spot, where the mercury surface has been kept relatively clean. It may be seen that the spots are formed principally along the edge of the film on both sides of the clean strip of mercury. This experiment shows that for new emission centers to be formed in this case the important thing is not so much the high concentration of particles as a low cathode work function, and this emphasizes the probability of the photoelectric effect playing a role in the formation of autonomous spots.

The phenomena under discussion here that occur at high rates of current rise are very typical of the spark discharge, which gives some reason for assigning the form of discharge occurring under these conditions to the field of spark discharges, in spite of the extremely low-density medium, which is a somewhat unusual situation for this type of discharge. At the present time, all the spark discharge properties are assumed to come from the existence of a very narrow channel where the medium is 100% ionized, as a result of the large instantaneous currents concentrated in the channel and the large power supplied. Not enough attention is given here to the phenomena occurring in the immediate vicinity of the cathode. As is shown by the experiments described above on the low-pressure mercury arc, 100% ionization and the associated intense spark spectrum are also observed in a low-pressure discharge where the discharge channel has a large cross section. Obviously, it is not necessary to have a narrow channel to observe signs of a spark discharge, since the same result of 100% ionization of the medium with a spark spectrum may be achieved by supplying a high enough power to a wide discharge column, especially in discharges in a low-density medium. Judging from the results of the mercury arc experiments, it is the processes occurring at the cathode which are of first order of importance to an understanding of the special features of the spark discharge at low pressures. As a result of the restricted rate at which emission surface can be formed on the cathode, limited as it is by the slow process of heating the local surface regions to boiling, the necessary cathode emission can only be achieved at very high current growth rates by stepping up the electrical processes in the arc cycle. One of the results of stepping up the cathode processes is to get 100% ionization of the medium in the part of the discharge lying above the cathode, thus producing a spark spectrum. Another very characteristic accompaniment of the step-up in the arc processes is that new emission foci are formed on the cathode.

Thus, the clearest physical line of demarcation between the spheres of arc and spark discharges at low pressures is, in the opinion of the author, provided by finding the rates of current rise which can no longer be taken care of in the normal way, in which the cathode spot is enlarged by cell division, but are met principally by continuously increasing the intensity of the electrical processes occurring at the cathode.

42. The Essence of the Process Controlling the Current in the Cold-Cathode Arc. The Dynamic Volt-Ampere Characteristic of the Arc

The material given on the elementary cells in the cathode spot lends a new aspect to the question of how the discharge current is controlled in the cold-cathode arc and gives a more definite meaning to the dynamic characteristic.

The process by which the discharge is controlled includes in its essential features a number of effects on the cathode region of the arc which serve to set up the required value of cathode emission current. From the results of the preceding sections we can easily set up the whole sequence of changes in the discharge, caused by a change in external circuit conditions. The discharge is usually controlled by changing the ballast resistance. We have already in part traced out the sequence of phenomena that occur when the resistance is reduced. Here we need only sum up, illustrating the current control process for ease of comprehension by means of the diagram shown in Fig. 67. In the figure, the cathode spot in its original state, before the current is increased, is represented for simplicity by just two cells shown by the fine straight lines on the left. The time axis is directed from left to right. Cell divisions are indicated by doubling the lines, where, for clarity, only those divisions are shown which are needed to form the additional number of cells required for the new steady state of the arc with the current increased. The divisions which occur normally from time to time in the steady state have been omitted.

When the ballast resistance is reduced, the first thing that happens is an increase in the potential difference on the arc electrodes, as shown schematically by the voltage curve U. This automatically increases the cathode drop. The result of the increase is to raise the level of the electrical processes in the arc cycle in the existing cells, which produces a large increase in the discharge current i. This step-up in operation is shown in the diagram by thickening the lines. The fact that one of the lines starts to thicken somewhat after the voltage starts to rise emphasizes the observed phenomenon that the discharge current is unequally distributed among the cells in the transitional state. After the length of time t_d required for the cell to divide in two, and determined by the rate at which the new portion of the cathode is

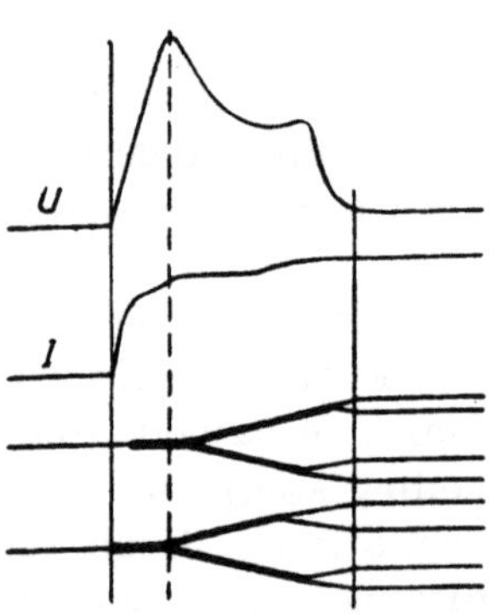

Fig. 67. Schematic representation of the process controlling the cathode emission current in the arc when the current is increased.

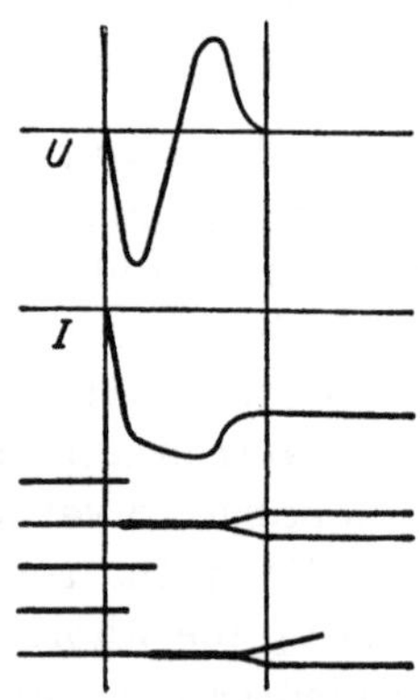

Fig. 68. Diagram of control of the emission current in the arc for the case where the current is reduced.

heated to boiling, the number of cells on the cathode has increased, which completes the first stage of adapting the arc to the new higher-current condition. At this instant, the voltage on the electrodes begins to decrease, accompanied by a further increase in current. Neither process is monotonic, but proceeds in separate steps following the step by step increase in the number of cells on the cathode and the way the load goes off them. The completion of the second stage is shown by the voltage dropping to normal, which serves to indicate that enough cells have been formed for the new steady arc state.

A somewhat different order of the processes is observed in the opposite transition from a higher current to a lower, as shown schematically in Fig. 68. If additional resistance is inserted in the circuit, the first thing that happens, as in any other operation that tends to reduce the discharge current, is that the voltage on the arc electrodes drops. By producing a temporary reduction in cathode drop, this operation paralyzes the activity of the arc cells, which as a result begin to decay. At a definite stage in the cell decay where the remaining cells are no longer in a position to maintain the new current in a normal way, the voltage on the electrodes, and with it the cathode drop, suddenly begins to increase again. Because of this the cell decay stops for a while and new divisions occur until the cells are brought up to the number required for the new steady-state current. If by chance the number of cells is too great, the unnecessary ones automatically decay, so that under these conditions the average current carried per cell is somewhat less than normal.

The results of the above observations enable us to draw a number of general conclusions as to the physical nature of the process controlling the discharge, what its characteristics are, and some of its special features, such as the tendency to extinction. From our observations, we would conclude first of all that both the ability to establish a steady state in the arc and the possibility of changing the operating conditions are based on the decay and restoration of the cells in the cathode spot. All known methods of controlling the discharge current are based on making use of this coordinating mechanism. It consists essentially in that any disturbance of the equilibrium number of cells required for the conditions existing in the external circuit produces a sequence of effects acting on the discharge, and primarily on the cathode region, the effects acting in such a way as to restore equilibrium.

One of the first links in the chain is a change in the voltage on the arc electrodes, which produces a temporary change in the cathode drop. The result of these changes is control of the number of cells active on the cathode. The process of setting up the equilibrium number of cells proceeds in two stages. The first stage is marked by a change in intensity of the electrical processes occurring in each cell, including a change in the emission current, in the intensity of ionization of the metallic vapor, and in the ion current to the cathode. In contrast to this, during the second stage the result of the above effects on the cathode spot is a change in the number of cells on the cathode. This is accomplished either by decay of the extra cells, those not needed for the existing conditions, or by cell division, depending on which side of equilibrium the arc is on. It is easily seen that the effect of the coordinating mechanism is based on the following properties of the elementary arc cells: they are unstable; they must carry a perfectly definite current in order to exist as autonomous regions; the cells can, for a short time, carry currents greatly different from normal; they are capable of division.

It follows from what has been said that for the very reason on which control of the discharge current is based, the current can only be increased by temporarily raising the voltage on the arc electrodes, while it can only be reduced by temporarily decreasing the voltage. However, after the new steady state has finally been reached, the voltage on the arc electrodes returns to its original value, since simply maintaining the steady state in a short arc of the type under discussion here requires the same mean voltage no matter how many cells there are on the cathode. The entire process of forming the equilibrium number of cells occupies a time of the order of 10^{-5}-10^{-4} sec. Hence it becomes obvious why there is a difference between the dynamic and static volt-ampere characteristics of the arc. If we measure the voltage on the arc electrodes while conditions are in continual change, without waiting for equilibrium to be established, which does not take more than 10^{-4} sec, the voltage will lag behind all the changes in current. This is a result of the fact that in the transitional states the change in discharge current is produced not only by changing the number of cells on the cathode, but to some degree also by changing the intensity of the arc-cycle processes in the individual cells, and this requires a corresponding change in voltage. If the rate of change in current is so small that the current only undergoes comparatively small changes during the time (10^{-4} sec) required to set up the steady state in the arc, the voltage changes will be small too. In this case we have very nearly the static characteristic of the arc. If, however, the current can undergo large changes in this length of time, we will be dealing with the positive dynamic characteristic, and the steepness will increase with increase in the rate of change of current. Thus, the dynamic characteristic of the arc is a functional relation between current and voltage taken under conditions where the change in current is produced not only by increasing the emission surface, but also by changing the intensity of the processes in each of the cells.

In the light of the above coordination phenomena between decay and formation of elementary cells on the cathode spot, a new aspect is also lent to the problem of the causes of arc extinction as observed under various experimental conditions (see Sec. 13). The most general way in which the arc is extinguished may be considered to be its natural reaction to producing conditions for the discharge to exist in, such that the coordinating mechanism is paralyzed. Two extreme cases may be cited, both quite typical of the arc, in which phenomena of this sort are observed. First, the coordination between cell decay and cell formation will be disturbed if the current is reduced to values such that isolated cells are left on the cathode, which is what is in fact observed. Second, a paralyzing effect is exerted on the coordinating mechanism of the arc by any factors which prevent the voltage on the arc electrodes or the cathode drop from increasing. Hindrances of this sort in the external circuit may be provided by connecting a very-large-capacity condenser in parallel with the arc electrodes (see Sec. 24). Of the internal factors, a strong paralyzing effect may be exerted on the arc by intense ionization of the mercury vapor in the discharge space, resulting from a powerful current pulse passed between auxiliary electrodes. Extinctions of this latter type are encountered in cases where the operating arc breaks off in mercury tubes under large changes in anode current.

C. RE-FORMATION OF THE CATHODE SPOT IN THE STEADY STATE AND THE ROLE PLAYED IN THE PROCESS BY A MAGNETIC FIELD

43. General Information on Spot Re-formation on a Cold Cathode

If we follow in detail the behavior of the cathode spot on mercury over the widest possible range of conditions, we are struck by one of its cardinal properties: the cathode spot never stands completely still. Actually, photographic scanning of an enlarged image of the spot under any conditions whatever shows that it is ceaselessly changing its shape by changing the relative positions of its constituent parts. In the majority of cases, the spot as a whole moves around over the cathode at the same time. The changes in shape have been noted in earlier sections of the book, even when the spot is anchored on metals in contact with the mercury, although in this case it will be difficult to observe the changes visually. The various types of motion of the spot as a whole may be observed in an arc with a homogeneous mercury cathode.

The spot exhibits the same variability on other metals under all conditions where it is impossible to heat even small portions of the metal to the temperatures required to produce appreciable thermionic emission, as in the case with metals belonging to the low boiling point group. It must be concluded from this that the variability is a universal property of the spot on metals, and is organically related to the very nature of the cold arc. But then it deserves the most serious consideration as being one of the important sources of information on this form of discharge. In order to get properly oriented in the consideration of the problems dealing with the nature of spot variability and the laws governing it, it is useful to turn again to the conclusions of previous sections, where a detailed discussion was given of the structure of the anchored spot and all its transformations were traced out.

As may be seen from the behavior of the cathode spot when it is anchored at the boundary of the region where the mercury wets molybdenum or nickel, the changes that it undergoes are of a complex sort. A detailed analysis of the changes leads to the conclusion that they may in principle be arranged in the following basic categories:

1. Systematic decay of elementary cells in the spot.
2. Redistribution of the current among the remaining cells.
3. Formation of new cells by division.
4. Motion of the cells over the cathode.

It is easily seen that the source of all the changes in the cathode spot under the above conditions is the instability of the elementary arc cells, as is shown with particular clarity by the systematic way in which they decay and the way in which they continually move over the cathode. But then a legitimate question arises: isn't it possible to give a quantitative description of all the changes undergone by the cathode spot on metals based on the idea that the elementary cells are unstable and including the data from observations showing how they are affected by various internal factors? The solution of this problem requires a detailed study of the behavior of the spot under the particular conditions that are most typical of the cold arc. It is most suitable in all respects to make an investigation of this sort on the normal low-pressure arc with a homogeneous mercury cathode of liquid type, and this will be the form to which the considerations presented below principally refer.

The systematic decay of the cells in the cathode spot, as traced out in detail in Sec. 36, has led us to the conclusion that the steady state of the arc is essentially a state of equilibrium between the simultaneous processes of cell decay and cell formation. This idea provides the starting point for a description of the changes in the cathode spot on liquid mercury. As a matter of fact, it is easily seen that continuous re-formation of the cathode spot occurs as the decaying cells are replaced by new ones. Replacing some cells by others is, however, only one side of the re-formation process. The re-formation process must be thought of on a wide scale, including, in addition to changing the cells making up the cathode spot, re-formation of the cells themselves, as shown by the way they move around over the metal. The following considerations will help to elucidate the mechanism.

In a cold arc with a homogeneous metal cathode, and particularly with a liquid cathode, there can be no continuous or stable emission centers, such as occur, for example, on a thermionic electron-emitting oxide cathode. Whatever the mechanism may be in this case by which electrons are liberated from the metal, the liberation is not due to any special properties of the metal in the cathode spot, but rather to a whole group of conditions occurring temporarily in the part of the discharge closest to the cathode, and maintained by the definite cycle of related processes which we call the arc cycle. Under these conditions, in view of the instability of the cycle, the very cycle of successive stages in each cell will be a source of cell re-formation, causing them to move over the cathode. It is not difficult to imagine how this occurs in fact.

For definiteness, let us assume that the electron emission from the cathode is produced by the electric field of the positive-ion space charge, as is assumed in the cold-emission theory of the cold arc. We shall not go into any further detail here on the question of how important a role is played by the tunnel effect, to which alone the term "cold electron emission" can be applied in the strict sense of the word. As we have always done before, in the absence of any convenient terminology, we shall assume cold electron emission to mean here any emission process which involves an electric field, including the Schottky effect and the tunnel effect. When the electrons are withdrawn from a metal by a mechanism of this sort, emission will occur automatically at the parts of the cathode surface where the positive space-charge density in the cathode-drop region reaches some critical value ρ_0. The space charge is produced mainly by the positive ions moving to the cathode from the region where the metal vapor is being strongly ionized by primary electrons, which in turn places definite requirements on the concentration of metal atoms above the cathode. The necessary conditions for emission can only be maintained in those parts of the cathode where the concentration of atoms reaches the threshold value n_0. The concentration of vapor must be at least such that when it is 100% ionized it can provide the critical space-charge value ρ_0 from the positive ions moving from the ionization region to the cathode. Even for the most moderate estimate of the value of ρ_0, the threshold value of the concentration n_0 proves to be so large that it could only be provided by the extremely intense evaporation of metal corresponding to a strongly superheated liquid. The necessary metal temperature is maintained in the cell by ion bombardment of the cathode.

Thus, each cell in the cathode spot must be made up of four different regions, related to one another through the processes in the arc cycle:

1. The cathode emission region S_e, with electrons flowing out of it.
2. The intense metal evaporation region S_T, with a high concentration of metallic vapor above it.
3. The intense ionization region V_q, which provides the positive ions needed to produce the space charge and heat the cathode.
4. The space-charge region V_ρ lying between V_q and the cathode.

It must be assumed from the start that the distribution of particle concentration and the distribution of the intensity of arc-cycle processes differ greatly over the cross sections of all these regions and that the distributions are continually changing with time. As a matter of fact, these four regions, which altogether occupy a microscopic region of the discharge, are kept in existence by two opposite tendencies. The usual processes of energy dissipation, such as particle diffusion from the high-concentration region into the surrounding space, as well as heat losses in the metal resulting from its thermal conductivity, tend to break up the regions. This is resisted by the factors which favor concentrating the particles and processes into small volumes. Such factors are, first of all, the extreme nonlinearity between electron emission and electric field strength, and between metal evaporation and metal temperature, as well as the inherent magnetic field of the arc, which limits the rate at which charge diffuses out of the cathode spot. If the arc-cycle processes in a cell were always in equilibrium, the simultaneous action of the dissipating and concentrating processes would result in establishing some equilibrium distribution of particle concentration and process intensity over the cross section of the cell. Since energy dissipation into the surrounding space plays a large part in establishing the distribution, the distribution would be given by a function decreasing monotonically from the center of the cell to the periphery, such as the normal distribution or Gauss error function. But a symmetric equilibrium distribution of this sort is an idealization, in no sense in accord with reality. Since equilibrium does not exist between the various arc-cycle processes in any cell, and since there are random fluctuations in particle distribution as well as various sorts of external forces acting on the cell, the symmetry and equilibrum at the four regions will always be disturbed.

In view of the coupling existing between the regions, these changes can shift the whole cell to a new location. Cold emission of electrons will be particularly sensitive to any changes in the distribution of particle concentration, because it responds rapidly and is strongly dependent on the field strength. The result of this will be an exceedingly high degree of variability and mobility of the emission region S_e. On the other hand, the cathode evaporation region S_T, the rate of motion of which is limited by the slow process of heating the metal, will have the lowest mobility, and will thus limit the rate of motion of the cell as a whole. While leaving unsolved the problem of tracing out all the unobservable changes which the emission region undergoes in course of time, we can nevertheless consider the final result, in the sense of how they effect the evaporation region. To this end we can imagine for simplicity that the center of some idealized emission region with uniform current density is constantly moving with respect to the center O of the evaporation region, describing a complicated random trajectory as shown schematically in Fig. 69. While this motion is taking place, the emission center is fixed at the point O until the center of the evaporation region moves away to some new point O'. The reason for this motion may in turn simply be that the symmetry of the motion of the emission center with respect to O is disturbed as a result of random or regular causes. It is natural to assume that the new evaporation center will be located at a part of the cathode where the relative duration of the emission center is a maximum. Actually, the distribution of the energy liberated by positive ions on the metal surface will be the same as the distribution of the probability of finding the emission center at a given point on the cathode. Shifting the metal evaporation center to a new point means that a single re-formation cycle of the cell has been completed.

The above scheme of the way in which an elementary cathode-spot cell is re-formed shows that the motion of the cells over the cathode may in fact be thought of as a result of the continual re-formation resulting from the instability of the arc cycle. Understanding the re-formation is a preliminary step toward a more extensive analysis of the factors which determine the behavior of the cathode spot on mercury. In accordance with the ideas presented, we shall from now on consider any changes in form or position of the spot on the cathode to be the result of the thoroughgoing re-formation which includes both the change in cells making

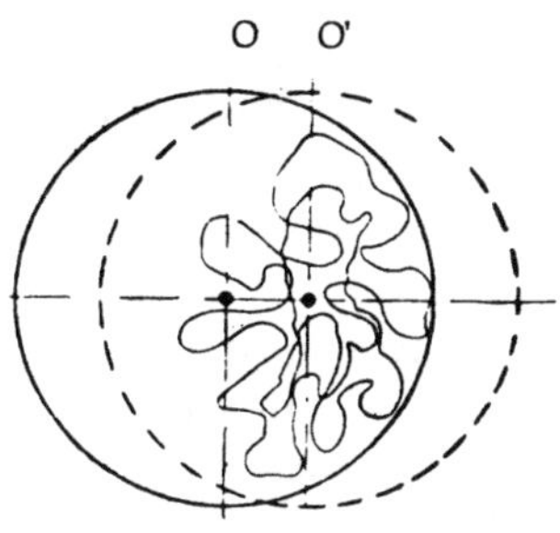

Fig. 69. Illustrating the process by which a single cathode-spot cell is re-formed, resulting in the metal evaporation center O moving to the new point O'.

up the spot and the re-formation of each of the cells individually. In using this idea to describe the behavior of an actual cathode spot on mercury, we have to determine at the very start precisely which changes in the spot are due to which type of re-formation, and how long the individual cycles normally last.

Turning first to the first type of spot re-formation, i.e., to the change in cells making up the spot, it is to be expected from the logic of the case that it will be the completely determining factor in the redistribution of current between the autonomous spots in the arc, and will accordingly determine the duration of the spots. As a matter of fact, as has already been found (see Secs. 36 and 37), an autonomous spot is nothing but a group of elementary cells related to one another by belonging to the same genealogical branch. The redistribution of current among individual branches is the result of the fact that cell division and cell decay are not in equilibrium in any of them. The current redistribution rate depends on the difference between the frequencies of cell division and cell decay in each of the branches, as well as on the mean duration τ_1 of a single cell. It is perfectly obvious that on the average the time required to change the cells making up the spot is equal to τ_1. Accordingly, the mean duration of a re-formation cycle which replaces all the cells in the spot will be equal to the mean duration of a single cell. From the data taken on the anchored spot arc, τ_1 is about 10^{-4} sec, which is probably too high. The same figure gives the duration of the first type of spot re-formation cycle.

The second type of spot re-formation, namely, re-formation of the cells themselves, is principally active in the redistribution of current between individual cells, and in the motion over the cathode of the cells and the autonomous spots which they form. In essence, this redistribution of the current from a decayed cell to a new one is precisely a single complete cell re-formation act. The time t_r required has already been found from the duration of the coordinating voltage impulse to be 10^{-7} sec, which may be taken as the duration of a single cell re-formation cycle. It may be seen that the second type of re-formation requires a time which is a factor of 1000 less than that required for the first type. This fact alone is sufficient reason to assume that the second type of re-formation is wholly responsible for the motion of the autonomous spots over the cathode. Apparently, t_r is the time required to heat the microscopic regions of the metal surface to violent evaporation. In this case, t_r will be inversely proportional to the energy which the ions acquire in the cathode-drop region. The energy of doubly or triply charged ions will be a factor of two or three greater than that of singly charged ions. This is a natural explanation for the large increase in the rate of motion of the cathode spot in strong magnetic fields, at the same time that spark lines appear in the arc spectrum [93], which will be discussed in more detail in the next chapter.

44. Model of the Cathode Spot on a Homogeneous Liquid Mercury Cathode

In order to complete the task of giving a description of the behavior of the cathode spot on mercury, using as a basis the idea that it is continuously undergoing re-formation, a definite model of the spot must be worked out, and in addition a general principle must be found which will permit us to predict in any special case what direction the re-formation process is going to take. Restricting the discussion of the problem to the special conditions of a low-pressure arc with a homogeneous mercury cathode in its normal liquid state, we shall base our development of the cathode spot model on the following considerations.

a. Structure and Fundamental Physical Processes of the Spot. According to the observations made above, the spot on a homogeneous cathode must be thought of as consisting of unstable cells, with the bond strength between the cells being close to 1. This means that, under the conditions we are interested in, the cells are practically unable to exist individually. Under these conditions, the individual properties of the cells become of second order of importance in comparison to their group properties. Accordingly the problem reduces to that of describing the behavior of tightly bound groups of cells. In view of this it may be shown that the idea of the cells being the basic structural unit of the cathode spot is in general superfluous. The following must, however, be kept in mind. First of all, the idea of elementary cells in the cathode spot is in complete agreement with the factual physical phenomena in the cathode region of the arc. That there is a threshold arc current for constant current density in the cathode spot means in fact that even at this current an independent arc cycle can be maintained for a finite length of time on some minimum possible cathode area. Increasing the current simply increases the number of such areas having an independent arc cycle. Whether they are tightly packed together or isolated depends on the increase in stability produced by collective existence. If we completely reject the idea of cells under some set of conditions, we would be limiting the generality of the model, which could then not be extended to other conditions. Accordingly, in the conditions discussed below, where the bond strength between cells is large, there is still some point in assuming that the cells actually exist.

By assuming the cells to be the basic structural unit of the cathode spot, we can apply to any autonomous spot the ideas worked out above, dealing with the physical properties in cells and the mechanism by which they are re-formed. Accordingly, any such spot must be represented as consisting of four different regions, namely: the surface cathode emission region S_e, the intense metal evaporation region S_T, the volume region V_q, where the mercury vapor is ionized by electrons, and the region V_ρ, separating them, which contains the positive space charge. The data on the mercury arc show that electron emission on the mercury is caused by the electric field from the space charge in the region V_ρ. After being accelerated in this region, the electrons produce ionization of the vapor in V_q by stepwise processes. The positive ions pass from V_q to the region V_ρ as a result of thermal motion. After gaining energy in this region, they transfer a certain portion of their energy to the cathode, which serves as a source of heat for the metal. It follows from the coupling between the processes in the arc cycle that all four of the regions will be inseparably bound up with one another. The coupling shows up in two ways. First, all four regions must be held together spatially, which limits the freedom of the regions to move separately over the cathode surface. Second, any change with time in the level of the processes in one of the regions will produce a chain of successive changes in the processes in the other regions.

b. Factors Determining the Stability and Dimensions of the Spot. Effect of the Inherent Magnetic Field of the Arc. Since high energy and particle concentrations are observed in the spot, and the spot itself is of microscopic dimensions, there will be considerable losses of both energy and particles to the surrounding space as a result of ordinary dissipation processes. In view of this the question arises as to what the factors are that stabilize a given state in the spot, and thus determine its microscopic dimensions. One of the principal factors of this sort is that, for an arc cycle to exist, an electric field of a definite high intensity must be maintained at the cathode. Since the field is produced by the positive-ion space charge, the necessary conditions can only be provided by having exceedingly high current densities at the cathode of the order of 10^7-10^8 A/cm^2. With the coupling existing between the arc-cycle processes and the large amount of nonlinearity in the cold-emission and metal-evaporation effects, this situation demands automatic contraction of the dimensions of the active region of the discharge progressive in time, as has already been pointed out in the literature [118]. The limit to the contraction is set by the particle and energy dissipation processes. As a result of the action of the two opposing tendencies, some equilibrium dimensions are established for the cathode spot. The inherent magnetic field of the arc will play an important role in setting up the equilibrium. Acting as it does as an evelope around the high current density region of the discharge, the field will greatly reduce the charge diffusion losses, thus tending to increase the current density in the arc. With the high current densities existing in the arc, the indirect effect of the inherent magnetic field will be very large at the cathode, so that the field must be taken due account of as one of the factors determining the properties of the arc cycle. There is no doubt that neglecting the magnetic field of the arc has been a serious omission in previous work on the cathode spot, and that it has been a source of the difficulties in explaining a number of features in the behavior of the spot.

c. Excess Electrons in the Ionization Region of the Arc. Electron Envelope of the Cathode Spot. There are considerably more electrons than positive ions entering the ionization region V_q of the cathode spot per unit of time. Actually, the number of electrons entering is equal to the sum of the number of particles $j_e/q + \Delta N$ supplied by the flux of primary electrons and those produced by ionization of the mercury vapor. The number of positive ions entering the ionization region is given by the sum $j_p/q + \Delta N$, where the first term gives the number of ions supplied by the ion component of the current. But, under any conditions, the ion component of the current in the arc will be small in comparison to the electron component. Here the ratio of the number of electrons to the number of ions entering the region V_q will be approximately equal to $1 + (j_e/q)/\Delta N$. Since the number of ionizations ΔN will in any case be less than the number of electrons j_e/q arriving from the cathode, the number of electrons entering the region will be at least a factor of two greater than the number of positive ions. Removal of excess electrons from the region V_q solely by means of the longitudinal component of the electron current would require an electric field of the same order of magnitude as the field in the cathode-drop region, and this must be regarded as completely impossible. But then the principal way of removing the excess electrons from the region in question will be the dissipation of the electrons into the surrounding space by the combined action of diffusion and the field produced by the negative space charge. The magnetic field of the arc, which, at the boundaries of the high current density region, will reach a strength of the order of 10^3 Oe, will, by greatly reducing the transverse diffusion rate, make it very difficult for the electrons to get out. Under these condtions, a region of relatively high electron concentration will inevitably be formed around the cathode spot, and it will have an appreciable negative space charge. By performing the function of an electron block or barrier ahead of the ionization region V_q of the cathode spot, the envelope around the spot will greatly slow down the motion of charge in the radial direction. The result of this will be to raise the charge concentration in the ionization region, and increase the role played by volume recombination, which can be the cause of the bright luminescence of the cathode spot which shows a continuous spectrum.

The additional factors, consisting of the excess electrons in the ionization region of the arc and the inherent magnetic field of the discharge, which so far has not been considered in this theory, will have an appreciable effect on the potential distribution in the vicinity of the cathode spot in both the radial and longitudinal directions. Possible potential distributions in both directions are shown in Fig. 70 together with a diagram of the physical structure of the cathode spot with its four regions and the electron envelope in the zone of strong magnetic field H_i of the arc. The heavy horizontal line in the figure is the cathode. The cross-hatched surface portion of the cathode shows the emission region S_e, which coincides with the region of intense metal evaporation S_T. The cross-hatched vertical column above this area corresponds to the high current density discharge channel. Immediately adjacent to the cathode part of the channel is the positive space-charge region V_ρ, where the cathode drop is concentrated. Above this is the mercury vapor ionization region with its highly concentrated plasma. Around the channel, at the same height as the ionization region V_q, in the zone of the strong magnetic field H_i bounded by the vertical lines at the sides, is the electron envelope of the cathode spot.

The curves above the diagram of the cathode spot show two possible radial potential distributions at the level of the ionization region. The thin horizontal line is approximately the level of potential corresponding to zero space charge. The solid curve is a more probable potential distribution corresponding to a negative space charge in the ionization region, which will be one of the results of a greater number of electrons than positive ions entering the region under the restraining action of the arc field. The broken curve is for the less probable case in which there is a positive space charge in the ionization region. The curves at the right of the diagram show the respective potential distributions in the longitudinal direction. It is easily seen from the diagram that the potential distribution at the cathode favors concentration of positive ions at the cathode spot, and permits us to assume that the actual cathode drop is considerably in excess of the values measured in practice.

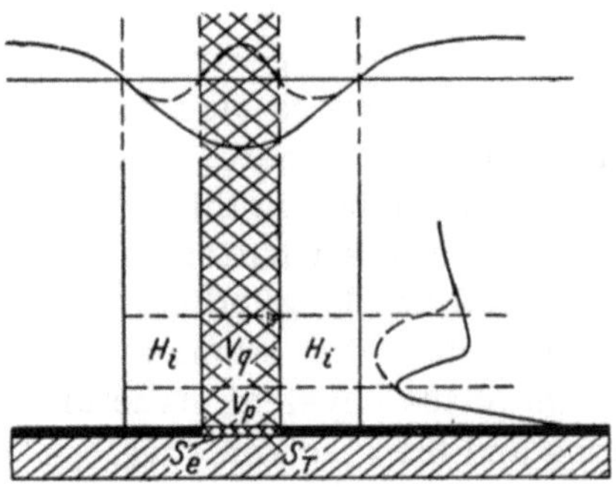

Fig. 70. Schematic representation of cathode-spot model and radial and longitudinal potential distribution. S_T) Intense cathode evaporation region; S_e) emission region; V_ρ) space-charge region; V_q) ionization region; H_i) region of intense magnetic field of the arc and electron envelope of the cathode spot.

d. Simplified Dynamic Model of the Cathode Spot. It has already been noted in Sec. 43 that the particle concentrations and the intensities of the processes in the arc cycle will be quite unequally distributed over the cross sections of the cells in the cathode spot and variable in time. This is to an even greater extent true of the cathode spot as a whole. One of the sources of inhomogeneity in the spot is the focal structure of the spot itself. But, in each cell, the particle concentration and the intensity of all the processes will decrease monotonically from the center of the cell to the periphery, as follows if the cell is interpreted to be an equilibrium state between two opposite tendencies, namely, automatic compression of the emission region and dissipation of particles and energy. An equilibrium of this sort can, however, only be thought of on the average. As a result of disturbances of the equilibrium between the arc-cycle processes, there will be continuous disturbances of the equilibrium state of each of the cells, which can be reflected as a continuous change in dimensions and disturbance of the symmetry of the four principal regions of the cell.

In view of the exceeding complexity of the cathode spot and the changes occurring in it, it would be useless to try to describe either in all its details. The theoretical problem consists in isolating the essential features of the phenomenon and giving a description of the basic laws governing the behavior of the cathode spot. This purpose is served by the simplified dynamic model of the spot given below. In setting up the model we shall distinguish two opposing principles in the cathode spot and its individual cells. The active principle, in the sense of the role it plays in re-forming the spot, consists of the related, practically inertialess electrical processes, namely: electron emission by the cathode, ionization of the metal vapor in the ionization space, and motion of charge in the electromagnetic field. It is precisely the changes in these processes and the associated regions that cause re-formation of the cathode spot. On the other hand, the passive principle in re-forming the spot is provided by the sluggish processes of heating and evaporating the cathode. Since they are the final stage of re-formation, the changes in these processes limit the re-formation rate of the spot. The most sensitive indication of changes in the active principle is provided by the emission region S_e moving over the cathode and changing its shape. It must be assumed that the location of the emission region on the cathode at any instant of time is determined by the location of the maximum in the charge concentration in the regions V_q and V_ρ, which in turn depends on the distribution of electron concentration in the electrode envelope around the cathode spot. In view of the continuous uncontrolled changes in the distribution of charge concentration in all the regions, the emission region must be continually undergoing rapid changes. The direction of re-formation depends on having the most favorable location on the cathode during the spot re-formation cycle, as is shown in the final analysis by the change in position and shape of the intense metal evaporation region.

It is impossible for us to get accurate knowledge of the instantaneous position and state of the emission region because of the rapid, uncontrolled changes in charge distribution in the regions lying above it as well as for the reason that we simply do not have at our disposal methods for observing it directly under the arc-discharge conditions considered here. Accordingly we can not attempt to describe all the changes which the emission region undergoes. We shall only be interested below in the changes in shape and location of the evaporation region, which in the final count are the determining factors in re-forming the cathode spot, and can be checked by direct observation. But giving a description of these changes does not require an accurate knowledge of the location of the emission surface on the cathode at every instant of time. For this purpose we need only have at our disposal information on the probability distribution $W(xy)$ of finding the emission region at different parts of the cathode during each spot re-formation cycle. As a matter of fact, this probability field uniquely determines the distribution of the heat energy liberated at the cathode, which alone determines the direction of cathode spot re-formation. In order to simplify the problem, we can limit ourselves to investigating the probability that the center of an idealized, perfectly circular emission region will be found at a given part of the cathode. Also, to simplify the problem, we can assume that the center of the evaporation region of each of the autonomous cell groups, considered as a whole, simply shifts at the end of the re-formation cycle to the point corresponding to the maximum probability of finding the emission center. If two or more such maxima occur, it may mean that the group will start dividing into smaller autonomous regions or spots.

Thus, part of the problem of describing the re-formation of the cathode spot and the associated changes in shape and location on the cathode reduces to investigating some field of the probability distribution W(xy) for finding the emission center at some part of the cathode within the evaporation region. Here it will be assumed that the probability field is given by some value, averaged over the cycle, for the charge distribution in the ionization region V_q, which in turn depends on the distribution of concentration in the electron envelope of the spot. It is obvious that the maximum concentration will always move with respect to the center of the evaporation region in a direction such as to give the largest electron concentration in the envelope.

In this form, the model only gives a description of one aspect of the spot re-formation process, which consists in simple motion of the parts of the spot over the cathode, and has a characteristic cycle length of the order of 10^{-7} sec. For us to be able to use the model at the same time to describe the above mentioned re-formation process where the current is redistributed among the autonomous spots, we must assume in addition that the number of cells in each autonomous spot can increase or decrease with time, which implies a change in dimensions or in the current carried by the spots. Here the total number of cells in the spot and, accordingly, the total current will remain constant. The duration of the cycle, as found previously, will be of the order of magnitude of 10^{-4} sec.

45. The Dominant Role Played by the Magnetic Field in Regular Cathode-Spot Re-Formation. The Maximum Field Principle

The above simple model of the cathode spot enables us to predict in what direction spot re-formation will develop under various experimental conditions. Here, the problem of giving a description of the behavior of the cathode spot reduces to finding the stability of the cells on different parts of the cathode and to analyzing the distribution of electron concentration around the boundaries of the spot. The solution of the problem is facilitated by the fact that below we shall be interested exclusively in regular changes in shape and position of the spot on the cathode. The reason for the changes can only be that the conditions under which the cells exist are always changing and that the distribution of electron concentration around the various groups of cells is always asymmetric. As a result of this, it is no longer necessary to give detailed consideration to sources of various kinds of temporary random changes in the conditions under which the cells exist or in the charge distribution in the spot.

Turning now to the first part of the problem, finding the stability of the cells on different parts of the cathode, we must take into consideration all the observable factors that can influence the formation and stability of the cells. From this point of view, the case discussed below of the low-pressure arc with a homogeneous liquid cathode is the simplest, since here the problem reduces to a consideration of the effect on the cells of the particle concentration and the unequal distribution of the electric and magnetic fields in the vicinity of the cathode spot. We always know the general features of the distribution of neutral and charged particles in the vicinity of the spot: the concentration is a maximum in the spot itself, and drops off rapidly with distance outside the boundary. The irregularity in the electric field at the spot will be similar in nature. As a matter of fact, the electric field at the cathode is produced by the space charge of the positive ions moving toward the cathode, and the field intensity decreases with decrease in space-charge density. But the density is in turn proportional to the charge concentration in the plasma. Thus, there will always be a certain parallelism between the change in particle conentration and the change in electric field in the vicinity of the cathode spot. The result of the unequal distribution of concentration and electric field on the stability of the cells is reflected in the fact that new cells can only be formed within active spots, and are maintained by the spots from then on. But these properties of the cells are already taken care of in the model under discussion by eliminating the possiblity of spontaneous formation of new cells outside the already existing spots, and by giving them the least possibility of existing separately ($\zeta \approx 1$).

There is no doubt that the arc magnetic field will also play a considerable role in the above stability of the cells when they are close to other cells, as is convincingly shown by the data of Chapter 2 on the stabilizing effect which the field exerts on the cathode spot. If the field is asymmetrically distributed around the spot, new cells are most likely to be formed, and be most stable, at the point where the field strength is a maximum. This means in particular that the larger groups of cells with the stronger field will grow with the passage of time at the expense of the smaller groups of cells, with the result that the smaller groups will sooner or later decay completely.

In passing to the second part of the problem, the sources of the consistently asymmetric electron distribution around individual groups of cells, or of the cathode spot as a whole, we can limit the discussion to the way in which the distribution is affected by the magnetic field. Actually, the consistent asymmetry that we are interested in around the spot can only be produced by the inhomogeneous magnetic field which naturally occurs from superposition of the fields from the separate current elements in the arc, or from applying an external field. In finding the effect of a magnetic field on the electron distribution in the envelope around the cathode spot and its various parts, we can make use of the following approximate scheme for analyzing the electron trajectories, based on the limitation imposed by the field on the free motion of the electrons in the transverse direction. If there are no large perturbations from the medium or the electric field, the electron trajectories for a limited portion of their length may be assumed to coincide with the lines of magnetic force with an accuracy to the mean Larmor radius R_L. As long as the field is not too distorted along the lines of force, the coincidence between the trajectories and the lines of force yields an important conclusion: over a limited portion of the field, the electron distribution along the lines of force corresponds to the distribution of magnetic field strength. For brevity we shall call this the correspondence rule. Before we apply the rule to the charged particle distribution at the cathode spot, we must decide how far the correspondence extends under possible actual conditions. To answer this question, we must find to what extent the electron trajectories depart from the lines of magnetic force, as well as the extent to which they are perturbed by the gas and the electric field.

In finding the value of R_L, we can use relation (26) given above. For a field strength of the order of 1000 Oe, such as the inherent field of the arc can have in the vicinity of the spot, the electron trajectories do not depart from the lines of force in a small region of the field by more than about 0.03 mm. If the dimensions of the region that we are interested in were of the order of 1 mm^2 or more, this degree of accuracy in describing the trajectories would be sufficient for our purposes. The question is, however, complicated by the fact that the correspondence rule is only strictly applicable to a limited region of the field, whose dimensions depend on the nature and degree of the distortion, as well as on the density of the medium and the electric field strength. These limitations are imposed since all these factors serve to make the electron trajectories depart further from the lines of magnetic force by an amount that increases with the distance covered by the electrons. This discrepancy is caused first of all by the field distortion itself, due to the curvature of the lines of force, since the deviation from the lines of force will be along the normal to the plane in which the part of the curve in question lies. No large amount of distortion of this type occurs under the conditions discussed below, except in the inherent magnetic field of the arc in the vicinity of the cathode spot. However, in this case the lines of force are principally plane curves lying in a plane normal to the cross section of the discharge. Accordingly, when the electrons depart from the lines of force as a result of the curvature of the lines, it will always be in the same direction as the axis of the discharge, and thus cannot have any appreciable effect on the particle distribution around the cathode spot.

From this point of view, a more serious source of trouble for the correspondence rule is that the electron trajectories are distorted by the electric field and by collisions with mercury atoms. On the average, each collision deflects the electron by an amount R_L. The higher the particle concentration, the greater the frequency of deflection by collision and the more the electron distribution will differ from the field distribution. The correspondence rule can only hold for a density in the medium such that the electron mean free path λ_e is greater than R_L. The maximum density in the medium at which the magnetic field can still affect the electron distribution is given by the condition $\lambda_e/R_L = 1$, which again brings us to Eq. (27), arrived at formerly for the limiting pressure at which a magnetic field can still be observed to exert a stabilizing effect on the arc. The reason the two conditions are the same is quite clear. The mechanism of the way in which the arc is stabilized by a magnetic field is based on the same laws of electron motion which made it possible to formulate the correspondence rule for the electron distribution in a magnetic field. As for the field, it follows first of all from (27) that the effect exerted on the particle distribution only extends to the part of the discharge lying outside the cathode spot with its exceedingly high particle concentration. In a pure low-pressure mercury arc, the electron mean free path everywhere else in the discharge, including the immediate vicinity of the spot, is so large[1] that there is no question of whether or not the correspondence rule applies. If an extraneous gas is present, the critical pressure at which the rule ceases to hold may easily be found from the same relation (27).

In passing now to a discussion of the effect of the electric field E of the arc, it should be noted that motion of electrons in mutually perpendicular magnetic and electric fields is a typical feature of the problem. It is well known that in crossed fields translational motion of electrons along the vector **H** has an added cycloidal motion which deflects the electrons in the direction of the normal to the vectors **H** and **E**. The deflection rate is proportional to the electric field strength and is numerically equal to 10^8 E/H, where E is expressed in volts per centimeter. Since the velocity component along the lines of magnetic force may be set approximately equal to $\sqrt{kT_e/m}$, the amount Δ by which the electrons are deflected in passing a distance of 1 cm along the lines of force is given by the relation

$$\Delta = 10^8 \frac{E}{H\sqrt{\frac{kT_e}{m}}}. \quad (36)$$

The electric field strength E is not large except in the cathode-drop region, where it can reach values of the order of 10^7 V/cm. Obviously, the correspondence rule does not apply in the immediate vicinity of the cathode drop. But the charge distribution in this region is wholly dependent on the distribution of electrons and positive ions in the adjacent plasma, in which E may be about 1 V/cm. This will also be the order of magnitude of the field strength in the vicinity of the cathode spot. At this field strength, Δ is not more than $3 \cdot 10^{-3}$ cm. It must be concluded from this that the electric field cannot in general be included among the factors limiting the applicability of the correspondence sule to the conditions in the arc.

The above analysis of possible exceptions to the correspondence rule in an actual mercury arc leads to the conclusion that the electron distribution around the cathode spot (in the electron enve-

[1]At the mercury vapor pressure of 0.001 mm Hg usual in this form of arc, λ_e is about 25 cm.

lope) depends in a definite way on the magnetic field distribution in the region. It is characteristic of the relation that a maximum in the field strength corresponds to the maximum electron concentration, while a minimum in the field strength corresponds to the minimum electron concentration. Since the electrons are the principal agent in ionizing the mercury vapor in the arc discharge, and the resulting positive mercury ions have to balance out the negative space charge, a similar distribution will also be set up automatically for the ion concentration. Thus, the charge distribution around the cathode spot in a low-pressure arc will always have the same general features as the magnetic field distribution. By exerting a direct effect on the charge distribution in the outer zone of the cathode spot, the magnetic field can exert an indirect control on the charge distribution in the spot itself.

Thus, the direction followed by the two basic forms of cathode spot re-formation will depend primarily on the distribution of magnetic field intensity in the vicinity of the cathode spot. In the type of re-formation in which the current is redistributed among autonomous spots, the field acts on the re-formation by increasing the stability of the cells to an extent that depends on the field strength at the point in question. In the type of re-formation in which the different parts of the cathode spot are simply moved around, the field breaks up the symmetry in the charge distribution around the various parts of the spot, thus bringing asymmetry into the charge distribution inside the spot itself. A very important thing is that in the low-pressure homogeneous liquid cathode arc the magnetic field is the only possible source of steady asymmetry in the vicinity of the spot, and the asymmetry is the only thing that can cause regular changes in the shape and location of the spot on the cathode. All this leads to the conclusion that when the cathode spot is re-formed at low pressures on a homogeneous liquid cathode, the whole process is controlled by the magnetic field distribution in the vicinity of the spot. There is one general feature of the effect of the field on the spot which strikes the eye. In both the first and second form of the process, re-formation in a field develops in a direction corresponding to the maximum possible increase in magnetic field strength at the cathode spot for any given distribution.

From the above conclusions, the general path may easily be laid out for solving the problem of how the cathode spot is re-formed on a homogeneous liquid cathode. It must first be assumed that at any instant of time t the direction of re-formation of the cathode spot is entirely determined in its controllable sphere by the distribution and intensity of the "acting" magnetic field H_a in the vicinity of the spot. By the "acting" field at any point in the cathode with coordinates (xy), we mean in the present case the total magnetic field from all imaginable sources, including the inherent field of the arc H_i and the external field H. However, in finding the field strength H_a at the point (xy) we must neglect the field from the elementary current passing through the cathode at the point in question and its immediate vicinity. All the complexity and detail of the problem consists in the fact that changing the shape of the spot and its position on the cathode in turn changes the distribution of the active field in the vicinity of the cathode spot, the field in general being given by the vector sum of the external field H and the inherent field of the arc H_i. For all the variety of possible changes in distribution of the inherent field H_i it has one characteristic property which it retains under all conditions: H_i is always strongest at the edge of the emission region which is active at any instant of time. There are two consequences of this. First, the cathode spot undergoes continuous regular changes in adapting itself to the field distribution, without ever reaching the optimum conditions for its existence in the maximum field region. Second, these changes must be regarded as the result of successive small shifts of the cathode spot into the region around it where the field strength is a maximum, in the course of which it may be assumed that the field distribution remains unchanged. By a shift in this case we mean not only the motion of the spot over the cathode as a whole in some one direction, but also change in shape, keeping the total surface constant. This concept includes, in particular, changing the dimensions of one autonomous cathode-spot region at the expense of another.

By making use of these observations, we can formulate a general principle to describe the behavior of the cathode spot on a homogeneous liquid cathode at low pressures: during any sufficiently small time interval Δt, the cathode spot will undergo small changes, keeping its total surface S constant, such that the magnetic field energy in the spot

$$F = \iint_S H_a^2(xy)\, dx dy \tag{37}$$

will receive the largest possible increase for the given distribution in intensity H_a of the "acting" field in the vicinity of the cathode spot. It must be emphasized that this rule only describes the behavior of the intense cathode evaporation region S_T, but not of the emission region S_e, where the changes are beyond our control. But the region S_T is precisely what we usually observe when we look at the bright luminescence of the arc cathode, and is what is associated with the term "cathode spot." It must also be made clear that by the small interval of time Δt we mean in this case an interval of nearly the same length as the duration t_r of a single spot re-formation cycle. The small change occurring will correspond to a single arc re-formation cycle. If we know the field distribution in the vicinity of the spot at some instant of time t, we can use the above rule to predict the changes in form and location of the spot on the cathode up to the instant of time $t + \Delta t$. Use of the energy instead of the strength of the magnetic field in this formulation is justified by the fact that in the final count the behavior of the spot may be predicted if we simply know the distribution of the absolute values of the field intensity. Under these conditions it is more logical to use the scalar values of field energy in formulating the problem.

It is easily seen that finding the changes which the cathode spot will undergo on a liquid cathode reduces to solving the variational problem of finding the surface S_1 approximately equal to S and satisfying the maximum functional condition for the field energy:

$$F_1 = \iint_{S_1} H_a^2(xy)\, dx dy, \tag{38}$$

with a number of additional limiting conditions. In view of the generality of this method of finding the behavior of the spot and its similarity with the familiar extremum principles of mathematical physics, it may for brevity be called the maximum field principle. This principle may be formulated mathematically be setting the first variation of the field energy functional equal to zero:

$$\delta F_1 = 0 \tag{39}$$

with

$$\delta^2 F_1 < 0 \tag{40}$$

and the limiting condition

$$S_1 = S = \frac{i}{j}, \tag{41}$$

which expresses the fact that the surface of the cathode spot is constant as well as that it is related to the discharge current i and the current density in the spot j which may be assumed constant to simplify the problem.

In considering the entire cathode-spot re-formation process to be a succession of small changes in shape and position, and finding these by means of the principle given, we can describe the behavior of the spot during intervals of time long enough to make exhaustive observations and comparisons with the predictions of theory. It must be noted that an accurate solution of the variational problem formulated above as applied to the actual conditions in the arc presents insuperable difficulties. The difficulties come first of all from the fact that we never know accurately how the emission current is distributed on the cathode, and this is what determines the intensity distribution of the arc magnetic field in the vicinity of the cathode spot. Accordingly, we are obliged to resort to simplifications of various sorts, based on the fact that in this case we need only know the averaged field distribution for each cathode-spot re-formation cycle considered. As a simplification of this sort it may be assumed that the averaged current distribution in each autonomous spot or group of closely bound cells is always symmetric about the center of the evaporation region of the spot, as long as it is unaffected by other spots or by an external magnetic field. Here the effect of any external fields on a given spot may be thought of as the result of destruction of the symmetry of the current distribution in the spot. By making this type of approximate solution of the problem of the behavior of the cathode spot possible, the maximum field principle serves as a convenient basis for systematizing and generalizing the observations in the next chapter.

Chapter 4

BASIC FORMS OF CATHODE-SPOT RE-FORMATION ON A HOMOGENEOUS LIQUID CATHODE[1]

46. Problems and Importance of Investigating the Motion of the Spot. Need to Include the Effect of the Inherent Magnetic Field of the Arc

In the preceding chapter, as a result of a detailed analysis of the structure of the cathode spot and the way in which the various parts of the spot behave, we came to the conclusions that, as long as it exists, the spot is being continuously re-formed as a result of the instability of the arc cycle. Then this idea was used to construct a dynamic model of the cathode spot, at the same time pointing out the general principle enabling one to determine the direction of spot re-formation under any experimental conditions in a low-pressure arc with a homogeneous liquid cathode. The basic problem of the present chapter is to obtain experimental verification of these conclusions by studying the behavior of the spot under its most typical conditions, as found on a homogeneous liquid cathode.

The solution to the problem may be greatly simplified by reducing all the changes undergone by the spot in the course of time to three basic forms of motion, as follows:

a. Forced regular motion of the spot in an external magnetic field.
b. Spot division.
c. The inherent random motion of the spot over the cathode, observed in the absence of any external influence.

The motion which the spot actually executes under any experimental conditions may be thought of as some combination of

[1]Preliminary results of the investigations were published in part in [184-188].

the above three forms. It could easily be seen from reading Chapter 1 that the only form that can be considered to have been more or less studied at the present time is the first. However, even in this point, the investigations are very one-sided and have not been brought to a conclusion. In setting up the investigations, as well as in analyzing the results, a number of factors were left out of consideration which were of the greatest importance to the problem, such as the instability of the cathode spot, the effect of a magnetic field on the instability, and, finally, the essential role played in the mechanism of the motion by the inherent magnetic field of the arc. When it comes to making an experimental study of the motion, the one-sidedness shows up in the fact that as a rule the authors only distinguished between two diametrically opposed possibilites for the direction of the motion, namely, the "normal" motion with a direction given by the laws of electrical engineering, and the "retrograde" motion in the opposite direction. Then, any study of the direction of motion simply reduced to finding out which of the two types of motion is observed under any given set of experimental conditions. But this is a long way from exhausting the problem of the direction followed by the regular spot motion. In the general case of an inhomogeneous magnetic field, the cathode spot describes a complicated trajectory on the cathode surface, during which the velocity vector must be continuously changing its orientation with respect to the field intensity vector, which is very important when it comes to understanding the nature of the motion. Nevertheless, this fact has been completely ignored in previous investigations, and no serious attempts have yet been made to investigate the trajectory followed by the spot in its motion in an inhomogeneous magnetic field. The fact that due attention has not been given to this question is related to a defect existing in the approach made to a solution of the problem of the spot motion. So far, in solving the problem, essentially no thought has been given to the inherent field of the arc in the cathode-spot region, which was shown to be incorrect by the author [184]. It is true that the assumption that the field takes part in the mechanism of the motion was made by Robson and Engel [91, 94]. However as was already pointed out in Chapter 1, these authors consider the effect of the inherent field exclusively in relation to the bending of the arc produced by an external field. According to their ideas, the cathode spot is moved in the opposite direction by the arc field which is strengthened on the side of the bend, which explains the origin of the retrograde motion. This picture is not in accord with either the actual field distribution around the spot or the observational data. It is shown by experiment (see Sec. 47c and Appendix 2) that at low pressures the cathode spot always moves in the direction in which the arc forms an acute angle with the cathode, i.e., exactly in the direction in which the arc is bent, which contradicts the basic assumption of Robson and Engel. In relating the effect of the inherent field of the arc exclusively with the curvature of the discharge path, these authors were actually quite far from taking account of the role played by the arc field in the mechanism of the cathode-spot motion. A constant asymmetry of the conditions around the spot, capable of producing directed motion in an external field, is inevitably produced by superimposing the external field on the inherent magnetic field of the arc independently of whether the discharge path is bent or not in the process. As far as the bending is concerned in this case, it can only produce second-order effects.

Whatever the initial assumptions of the theory of the motion of the cathode spot in a magnetic field may be, no correct solution of the problem, or mechanism of the motion, can be found without considering the inherent field of the arc. This follows from the fact that in a high current density arc, the inherent field intensity of the discharge at the edges of the spot will reach values no less in order of magnitude than the strength of the external field which is causing the sudden change in behavior of the spot. In order to get a quantitative value for the field strength of the arc at the edges of the spot, the problem must be cast in concrete terms, taking into consideration the greatly elongated form of the spot when it is moving in an external field. According to Froome's data [12], the cathode spot on mercury under these conditions becomes a continuous line or is broken up into segments, the width in any case being less than 10^{-3} cm. The total length l of the active segments in the line is related to the arc current I by the simple equation

$$l = \varkappa I. \tag{42}$$

For a mercury cathode, the constant $\varkappa$ takes on a numerical value of about 10^{-3} cm $\cdot$ A^{-1}. The linear current density in the individual segments, j_l, is a constant quantity, equal to $1/\varkappa$ A/cm. In accordance with the above form taken by the cathode spot, it may be assumed that in the part of the discharge nearest the cathode the current is distributed over some region whose surface intersects the cathode along the line of emission. It is known from field theory that on passing through a surface with a current flowing around it, the tangential component of the field intensity makes the discontinuous jump $\Delta H_t = 0.4\pi j_l$. The field intensity vector from the surface current is tangential to the surface and changes sign on passing through the surface, which is what produces the jump. Hence it follows that the field intensity from the surface current is equal to $0.2\pi j_l$. In defining the intensity of the inherent magnetic field of the arc at the cathode spot in this way, we must reduce the above value by a factor of approximately two, since the analogy with a surface current applies only to the upper half-space, bounded at the bottom by the cathode. After taking account of this latter fact, the field intensity of the arc at the edges of the spot may be evaluated as 300 Oe. At the outer ends of the line it may be appreciably higher.

This method of calculation can only give too low values for the inherent field strength of the arc in the vicinity of the spot, since it is derived from the dimensions of the luminescence, rather than directly from the dimensions of the emission region of the cathode, which determines the magnitude of the field. Nevertheless, even with this value of field intensity, the inherent field of the arc would have to be taken into consideration in giving a description of the behavior of the spot in an external field. Actually, as will be shown later on, if an external field of this intensity is present, it is practically enough to produce complete suppression of the random motion of the spot over the cathode. It must be concluded from this that a field of this magnitude produces a large change in the vicinity of the spot no matter what its origin. The inherent field of the arc deserves special attention in view of the fact that as a result of combining it with an external or foreign field, a large and permanent asymmetry is produced in the distribution of field intensity and electron concentration around the edges of the spot, the nature of which is shown schematically in Fig. 71. As a result, the field intensity and concentration are increased in one direction, but reduced in the other. A quantitative expression for the asymmetry is provided by the field intensity gradient of the

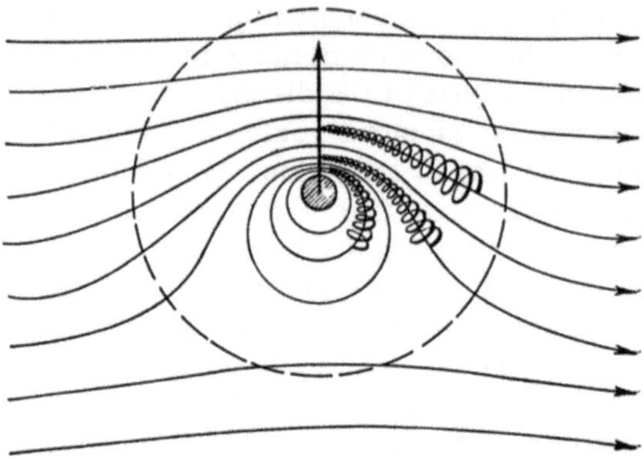

Fig. 71. Asymmetric distribution of the total magnetic field around the cathode spot. The lines of magnetic force are shown by light lines. The heavy lines are a schematic representation of the electron trajectories.

total field $\nabla|H_a|$ at the cathode spot. As a result of the small dimensions of the spot, the magnitude of $\nabla|H_a|$ in this region reaches enormous values of the order of 10^6 Oe/cm or more. This field asymmetry will play a dominant role among all the controlling factors exerting an effect on the direction in which the cathode spot is re-formed.

If these conclusions fit the facts, investigating spot re-formation takes on special interest as a means of finding the strength of the inherent field of the arc at the edges of the spot. The important thing about making such a measurement is that the field strength at the edges of the spot depends directly on the current density at the cathode of the arc, and can thus serve as the most reliable measure of this important constant of the discharge. Thus, it is certain that a detailed study of the re-formation process under a variety of conditions will make it possible to develop new and improved methods of finding the current density at the cathode. It is, however, first necessary to get direct proof of the fact that spot re-formation actually occurs as a result of the asymmetry in the magnetic field. One of the most direct ways of solving this problem is to investigate the trajectory of the cathode spot as it moves in an external magnetic field. According to the maximum field principle, the spot will always move in the direction of greatest increase in total field intensity. It is easily seen from

Fig. 71 that in the special case of a homogeneous field this direction is diametrically opposite to the one in which the current should move according to Ampere's rule. Thus, the so-called retrograde motion of the cathode spot in a homogeneous magnetic field finds a natural explanation if we include the inherent field of the arc. Nevertheless, it is not impossible that this agreement between theory and experiment in the special case of a homogeneous field is the result of chance coincidence. In order to be sure that the maximum field principle is unconditionally applicable to the problem of the motion of the cathode spot in a magnetic field, we need to show that the direction of motion of the spot always corresponds to the direction of greatest increase in field intensity, including the more general case in which the external field is inhomogeneous. The results of a study of the trajectory followed by the spot in an inhomogeneous field are given in the next section.

A. REGULAR MOTION OF THE CATHODE SPOT ON A LIQUID MERCURY CATHODE

47. Direction of the Motion

a. Theoretical Trajectory of the Spot in an Inhomogeneous Magnetic Field. The first step in finding the trajectory of the spot as it moves in a given external field H must be to find the inherent field of the arc, H_i, in the vicinity of the spot. In finding this field, we shall resort to a number of simplification. First of all, as a start, it will be assumed below that the distribution of the field H_i in the immediate vicinity of the spot is independent of the external field. This means, in particular, that we are neglecting any effect on H_i of possible curvature of the path of the discharge produced by the external field. The distribution of the arc field at the spot is determined mainly by the region of the discharge nearest the cathode, the length of which is measured in millimeters. As for the more distant parts of the arc, their effects are all second-order. Accordingly, any curvature of the discharge path outside the cathode region of the arc can produce nothing more than a relatively insignificant redistribution of the field in the critical region of the cathode spot in comparison to what happens when an external field is combined with the inherent field of the arc.

Thus, to simplify the problem, we shall assume that in the

immediate vicinity of the cathode the current is predominantly concentrated in the cathode spot and that the current is normal to the cathode. Accordingly, the field distribution in the vicinity of the spot and at distances greater than the spot dimensions will approximate the distribution of the field from an imaginary line of current passing normal to the cathode through the center of the cathode spot. The strength H_i of the real arc field will be less in absolute value than the strength of the field from the equivalent line of current. Outside the cathode spot, to lower the calculated strength of the real field somewhat, we can set it equal to half the field intensity due to the line of current, i.e.,

$$H_i = \frac{0.1I}{r}, \tag{43}$$

where I is the discharge current in practical units and r is the distance from the center of the spot to the point in question on the cathode.

The cathode spot is not in reality anything monolithic, but usually consists of several active portions or autonomous spots, separated from one another by inactive parts of the cathode, as shown schematically in Fig. 72. These active regions are in rapid motion with respect to one another, alternatly decaying and forming again by division. At the same time, if an external magnetic field is present, they execute transitional motion as a group, with their "center of gravity" describing a more or less regular trajectory. It is not necessary to go into the details of the structure of the spot to find the trajectory. The trajectory of the spot taken as a whole may be found by averaging the processes in time and space. The averaging is possible when applied to the present problem because the regular motion of the cathode spot as a whole at any point in the trajectory depends in the final count on the magnetic field distribution in the vicinity of the spot, where the inherent field of the arc is given with sufficient accuracy by Eq. (43), independently of the details of structure of the emission surface. All that we need to predict the trajectory of

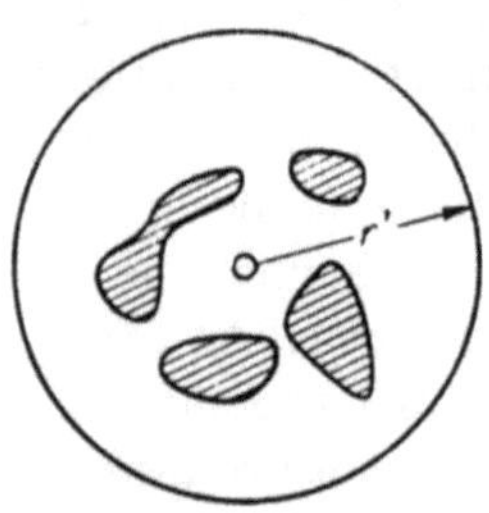

Fig. 72. Schematic representation of a cathode spot consisting of several autonomous emission regions.

the spot is knowledge of the distribution of the total field on a circle described around the spot at some radius r' for which the direction of greatest increase in field intensity coincides with the direction of motion of the spot at each point of the trajectory. The value of r' may be called the effective radius of interaction of the cathode spot with the magnetic field. In the sense of the present problem, the radius r' in absolute value will be somewhat greater than the linear dimensions l of the active parts of the cathode, which makes it legitimate to apply Eq. (43) to the circle in question. Along with this r' cannot be more than one order of magnitude greater than l since, according to the assumption, the direction of motion of the spot depends on the field distribution in the immediate vicinity of the spot. Thus, if the above ideas on the mechanism of the spot motion are correct, the value of r' will lie within the limits

$$l < r' < 10l. \tag{44}$$

That there is a definite effective radius r' for each current is shown by the results given below of investigating the trajectory of the spot in an inhomogeneous field. The exact value of r' for any given current may be found subsequently by comparing the theoretical and actual trajectories of the spot. If this gives reasonable values, this fact in itself may be regarded as a confirmation of the theory. Up to that point we shall find the inherent magnetic field of the arc around the edges of the spot by means of the equation

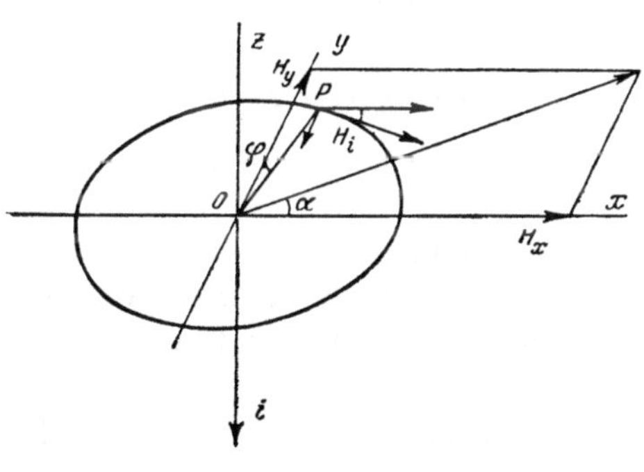

Fig. 73. Construction for combining the external field H and the inherent magnetic field of the arc, H_i, on a circle described around the cathode spot at radius r'.

$$H_i = \frac{0.1I}{r'}, \qquad (43a)$$

leaving the value of r' undetermined.

These simplifications make it possible to determine the trajectory of the spot without solving the difficult variational problem which use of the maximum field principle leads to in the general case. Introducing the concept of the effective radius of interaction between the spot and the magnetic field reduces the problem to investigating the distribution of the total or acting field H_a on the circle of radius r' described around the spot at the point of the trajectory in question. In passing directly to the solution of the problem, we locate a rectangular system of coordinate axes in such a way that the xy plane coincides with the plane of the cathode and the z axis is directed upward toward the anode. In the cathode region let there be a constant inhomogeneous magnetic field H with the components H_x, H_y, and H_z fixed on the surface as a function of the coordinates xy, the derivatives of the components being everywhere continuous and finite. Consider the distribution of the total field H_a on the circle of radius r' drawn around the spot, which is located at the arbitrary point O with the coordinates x_0y_0. In Fig. 73, the inherent field H_i at some point P on the circle is shown as a vector tangent to the circle. Since the field of the arc is assumed to be symmetric with respect to the spot, the field cannot by itself cause any motion of the spot in any one particular direction φ. The necessary asymmetry of the field occurs as a result of combining the fields H and H_i. Expressing the Cartesian coordinates of the circle by the relations

$$x = x_0 + r' \sin\varphi;$$
$$y = y_0 + r' \cos\varphi,$$

the components of the external field on the circle may be represented as a function of the azimuth φ, after which it is a simple matter to find the fields at any point on the circle:

$$H_a = \{[H_x(\varphi) + H_i \cos\varphi]^2 + [H_y(\varphi) - H_i \sin\varphi]^2 + H_z^2(\varphi)\}^{1/2}. \qquad (45)$$

The angle φ is read clockwise from the positive direction of the y axis. The components of the field intensity are assumed positive if the corresponding vectors are turned in the positive direction of the axes. The direction φ_m in which the total intensity takes on its extreme value may be found from the condition

$$\frac{\partial H_a}{\partial \varphi} = 0, \tag{46}$$

which leads to a trigonometric equation.

In investigating the trajectory followed by the cathode spot in its regular motion, the only case of interest is that of motion in a strong magnetic field where the effect of random deviations of the spot from the trajectory is small. It must also be kept in mind that under actual experimental conditions the inhomogeneity of the external field is relatively small. Under these conditions, after differentiating (45) with respect to φ, we can neglect a number of terms containing H_i and derivatives of H, as a result of which Eq. (46) takes the simple form

$$H(\varphi)\frac{\partial H(\varphi)}{\partial \varphi} - H_x(\varphi) H_i \sin\varphi_m - H_y(\varphi) H_i \cos\varphi_m = 0.$$

In this equation, H, H_x, and H_y are the intensities of the external field and its components on a circle described at the small radius r' around the point x_0y_0. Without making any serious error, we may replace them by the values of these quantities at the point x_0y_0. Since, further, the field H is given as a function of the Cartesian coordinates, we can make use of the transformation

$$\frac{\partial H}{\partial \varphi} = \frac{\partial H}{\partial x} r' \cos\varphi - \frac{\partial H}{\partial y} r' \sin\varphi,$$

after which the equation, when solved for φ_m, gives the symmetric formula

$$\tan\varphi_m = \frac{\dfrac{\partial H}{\partial x} - \dfrac{H_y}{H}\dfrac{H_i}{r'}}{\dfrac{\partial H}{\partial y} + \dfrac{H_x}{H}\dfrac{H_i}{r'}}. \tag{47}$$

For each value of the fraction on the right-hand side of the equation there are two directions of φ_m, one of which is the direction of the maximum while the other is the direction of the minimum in the field. They are diametrically opposite to one another. Since r' is by assumption the effective radius of interaction of the cathode spot with the magnetic field, the direction of the maximum will also be the direction of the motion of the spot at the point in question on the cathode, $x_o y_o$. It should be noted that the direction of the maximum can change within very wide limits, depending on the radius of the circle. Accordinlgy, the question of whether or not the interaction assumed between the spot and the magnetic field really exists reduces to whether or not the entire trajectory of the spot in an arbitrary inhomogeous field can be described by means of the same constant value of r' for the whole trajectory, and whether or not this value satisfies condition (44).

If the coordinate axes are located in such a way that the denominator in the right-hand side of Eq. (47) remains positive under all conditions, the maximum in the field occurs at acute angles, and we can get from (47) to a differential equation for the trajectory of the spot by writing the derivative $\partial x/\partial y$ in place of $\tan \varphi_m$. In a homogeneous field we have the relation

$$\tan \varphi_m = -\frac{H_y}{H_x} = -\tan \alpha. \tag{48}$$

Accordingly, in the special case of a homogeneous field, the spot will move at a right angle to the direction of the tangential component of the applied field, since the angles φ and α are read from mutually perpendicular axes. It is easily seen that the direction of the motion is opposite to the ponderomotive force acting on the current; i.e., the motion is of the retrograde type.

In an inhomogeneous field, the direction of the spot motion is determined by the more complicated rule (47); i.e., it depends on the ratio between the values giving the inhomogeneity of the applied field, $\partial H/\partial x$ and $\partial H/\partial y$, to the values of $(H_y/H) \cdot (H_i/r')$ and $(H_x/H) \cdot (H_i/r')$, which give a quantitative expression for the asymmetry of the total field along the axes x and y. The asymmetry is introduced by the inherent magnetic field of the arc. For a homogeneous field, the only source of asymmetry is the asymmetry of the combined fields as given by (48). In the general case of an inhomogeneous field, two different types of behavior of the cathode spot are, generally speaking, possible. If the inhomogeneity of the external field were greater than the asymmetry of the combined fields, the spot would simply move in the direction of increase of the external field. This hypothetical case can occur with an exceedingly large inhomogeneity in the applied field, or in an arc with low current density. In the actual low-pressure mercury cathode arc under discussion, the current density in the spot is so large that the asymmetry of the combined fields will predominate over any other sources of field inhomogeneity. Accordingly, a mercury cathode will show the so-called retrograde type of spot motion, with other complications due to the inhomogeneity of the field. This shows up in the fact that at the same time as it is executing retrograde motion the spot is deflected in the direction of increasing intensity of the applied field. Under the usual conditions, the result of the deflection is always that the spot seems to seek out the maximum field region and stays there for an unlimited time, which provides the basis for electromagnetic anchoring of the spot.

It is easily seen from (47) that there is another important feature of the regular motion of the spot which is supported by direct observation. The equation does not explicitly contain the field component H_z normal to the cathode, and this means that a longitudinal field alone cannot exert any effect on the motion of the spot. The reason for this is that large asymmetry, with the resulting regular motion of the spot, can only be produced by the component of the external field tangential to the cathode as a result of combination with the inherent field of the arc. Thus, a natural explanation is found for the fact that a longitudinal field has no effect on the spot motion.

The good agreement between these qualitative conclusions of the theory and the various features of the regular motion of the cathode spot as observed experimentally gives a clear illustration of how well the theory fits the fact. Nevertheless, the principal problem in establishing the practical basis of the thory is to make a quantitative check of Eq. (47) for some concrete field distribution. The equation of the spot trajectory takes on a simple form for the case where the field depends only on the x and z coordinates and has its tangential component directed along the x axis:

$$\frac{dx}{dy}=\frac{dH}{dx}\Big/\frac{H_l}{r'}\cdot\frac{H_x}{H}\,. \tag{49}$$

A field of this sort may be produced with an ordinary horseshoe magnet with pole faces in the form of parallel plates separated by a gap oriented along the y axis. A characteristic property of the trajectory (49) is that the tangent of the angle between the tangent line at any point of the curve and the direction of the retrograde motion, which coincides with the y axis, is equal to the ratio of the inhomogeneity of the field of the magnet, dH/dx, to the inhomogeneity $(H_i/r') \cdot (H_x/H)$ introduced by the inherent field of the arc. This property of the trajectory reflects the relation existing between the retrograde motion of the spot and the way it is displaced in the direction of increasing intensity of the applied field. Both these motions are due to the same general cause, namely, the asymmetry of the field in the vicinity of the spot. The asymmetry resulting from combining the external field with the inherent field of the arc provides the reason for the retrograde motion, while the inhomogeneity of the external field is in itself sufficient to produce the deflection of the spot along the x axis. It is easily seen that the angle between the tangent and the trajectory will be strongly dependent on the value of the effective radius of interaction of the cathode spot with the magnetic field, r', the square of which is implicitly contained in the right-hand side of Eq. (49). Having such a relationship enables us to count on being able to determine the value of r' for different currents with the necessary accuracy by comparing the theoretical trajectory with the one actually found.

b. Experimental Study of the Trajectory of the Cathode Spot in an Inhomogeneous Magnetic Field. The experimental work on the trajectory of the spot in an inhomogeneous field was carried out with the discharge device shown in section in Fig. 74 consisting of a metallic cathode part, or basin, and a glass bulb holding the hollow cylindrical anode. The two parts were held together with a flat ground joint cooled, along with the cathode, by running water maintained at a temperature of 20-22°C. The magnetic field in the cathode region was produced by an electromagnet welded into the double bottom of the cathode, which was made of nonmagnetic chrome nickel steel. The magnet itself consisted of two rectangular iron bars fastened together at the bottom with a yoke holding the magnetizing winding. The magnetic circuit was closed inside the tube at the part of the cathode under study, the flux passing between iron pole faces made of two parallel plates separated by a gap 1 cm wide and 6 cm long. The discharge took place in mercury vapor at a pressure of about 0.0015 mm Hg. Enough mercury was put into the tube so that the pole faces were completely immersed in the liquid to a depth of about 0.6 cm. In order to prevent the

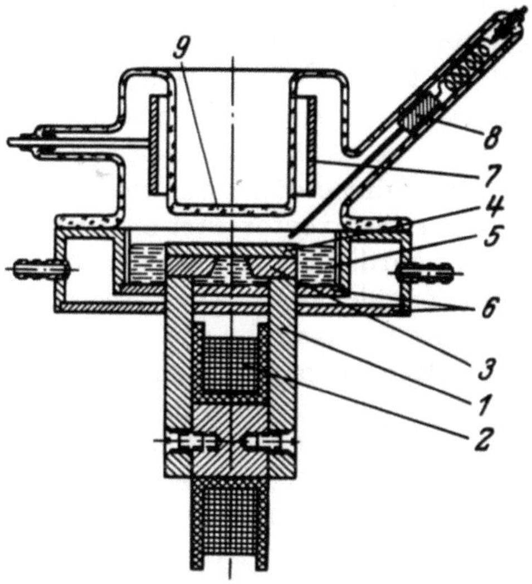

Fig. 74. Construction of tube for investigating the trajectory of the cathode spot in an inhomogeneous magnetic field. 1) Electromagnet; 2) exciting winding; 3) pole faces; 4) amalgamated copper plate; 5) mercury; 6) cathode basin of nonmagnetic steel; 7) anode; 8) arc exciting device; 9) glass window for photographing the trajectory of the cathode spot.

cathode spot from moving onto the shell of the cathode, the shell was protected by a glass cylinder.

The trace of the cathode spot was photographed through the plane upper wall of the glass bulb, which was free of optical defects. The cathode spot could be excited at any instant of time at a definite point O with an adjustable contact needle or semiconducting igniter put into operation at the same time as the shutter on the objective was opened. The position of the starting point of the trajectory with respect to the gap between the pole faces and the polarity of the electromagnet was chosen such that the spot was formed to one side of the axis of symmetry of the field Y, almost at the edge of the gap, and then was guided by the field all the way along the gap, thus describing a curved trajectory. The mutual location of the pole faces, the point where the spot is first excited, O, and the trajectory of the motion away from this point are shown schematically in Fig. 75.

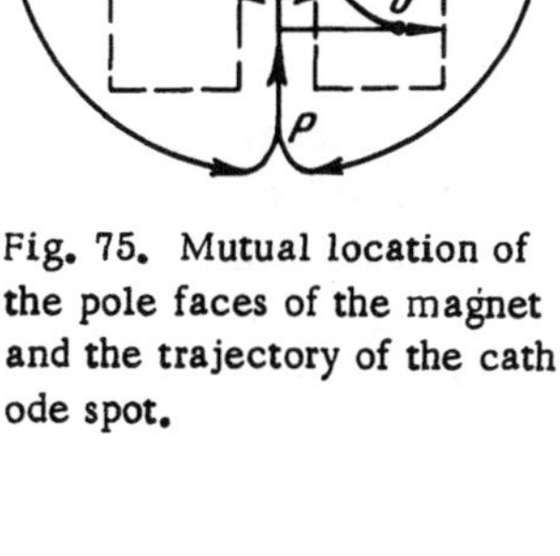

Fig. 75. Mutual location of the pole faces of the magnet and the trajectory of the cathode spot.

At the part of the cathode surface above the gap between the poles of the magnet, the inhomogeneity in the magnetic field corresponds completely to the case for which the trajectory equation (49) was derived. Actually, the direction x of the tangential component of the field intensity in this region is the same as the direction of the inhomogeneity. Obviously, the x axis is directed across the gap. Since the gap is relatively long, over the greater part of it the field distribution is independent of the y coordinate, with the exception of the ends, which are not taken into consideration. Thus, the problem reduces to finding the field in the xz plane for any value of y in the middle part of the gap, i.e., solving Laplace's equation for a magnetic potential that is a function of two coordinates. The solution of this problem for the boundary conditions applicable to the case gives the field distribution as a sum of functions of the form $b_n e^{-\lambda_n z} \cos \lambda_n x$. The only thing needed to solve the trajectory equation (49) is information on how the field is distributed in the plane of the cathode, i.e., as a function of the single coordinate x. The distribution of the field intensity H that we are interested in and its tangential component H_x on the cathode surface within the gap may be approximated with a sufficient degree of accuracy by the algebraic equations

$$H = H_0(1 - C_1x^2); \quad H_x = H_0(1 - C_2x^2), \tag{50}$$

where H_0 is the field strength at the axis of symmetry. The constants C_1 and C_2 depend on the level of the cathode. The values for the level Z = 0.6 cm used were found experimentally by measuring H and H_x for different values of x.

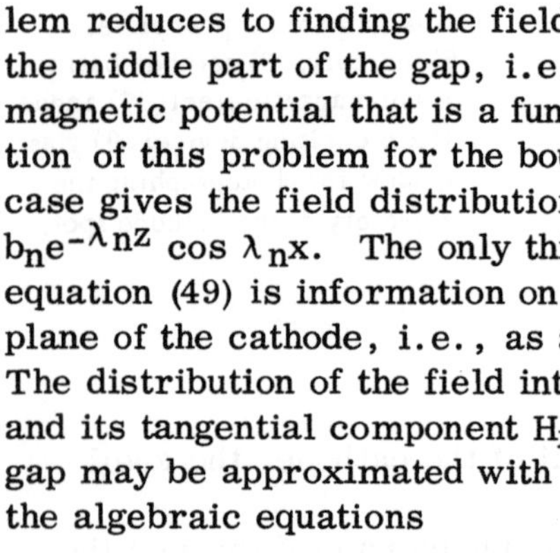

Fig. 76. Results of measuring the magnetic field intensity distribution on the cathode surface. O) Distribution of the field component H_x; Δ) distribution of field intensity H.

These measurements were performed before the tube was filled with mercury and were made by means of a miniature winding fastened to an insulated rod which was rotated around its axis by a motor at a strictly constant speed regulated by an automatic device. The axis of the winding made a right angle with the axis of rotation, and the induced emf, proportional to the field intensity, was measured with a high-resistance cathode millivoltmeter. By fixing the axis of rotation parallel to y and moving it in the x direction with a micrometer screw, the distribution of the total field intensity H could be found as a function of x. When the axis of rotation was set parallel to z, the readings of the apparatus were proportional to the tangential component H_x of the field, which allowed a similar determination of the distribution of this component to be made in the x direction. The radius of the measuring winding, which contained 50 turns of very fine wire, was only 0.1 cm. With a coil of these dimensions, the difference between the field averaged over the area of a turn and the exact value of the field strength at the center of the coil was less than 1%. A more important source of error was that in determining the xz coordinates of the center of the coil, which was moved with a special rack and pinion, as well as the error in reading the emf on the millivoltmeter (inaccuracy of reading, nonlinearity of the apparatus). The maximum error in measuring the relative field intensities in the gap between the poles was not greater than 8%. The apparatus was calibrated ahead of time by measuring homogeneous fields known to a high degree of accuracy, and, in addition, a measurement was made of the inhomogeneity of the field in the test gap in the direction of the axis of symmetry Y. In the center part of the gap, within ±2 cm from the center, the inhomogeneity was not greater than 3%.

The results of measuring the field at the level of the cathode are shown in Fig. 76. The squares of the distance x from the axis of symmetry of the field are plotted along the horizontal axis, while the values of $(1 - H/H_0)$ and $(1 - H_x/H_0)$, found from the measurements of H and H_x, are plotted along the vertical axis. For the range x = 0.7 cm, the experimental points lie on two inclined straight lines, the one with the smaller slope being for the total field intensity H. The slope of the lines gives the values of the constants

$$C_1 = 0.28 \text{ cm}^{-2};$$
$$C_2 = 0.76 \text{ cm}^{-2}.$$

For a fixed field distribution, the equation for the cathode-spot trajectory to be expected from the differential relation (49) takes the form

$$\ln\left[\frac{x}{x_0}\left(\frac{1}{C_1}-x^2\right)^{\frac{1}{2}\left(\frac{C_2}{C_1}-1\right)}\right]=-Ay, \tag{51}$$

where

$$A=\frac{20\,(r')^2}{I}C_1H_0, \tag{52}$$

and x_0 is a constant of integration.

The equation holds for $x < 0.7$ cm, but under these conditions no important change in form of the trajectory is produced if it is replaced by the simpler equation

$$\ln\frac{x_0}{x}=Ay. \tag{53}$$

Thus the problem reduces to comparing the expected exponential trajectory of the cathode spot, (53), with the trajectory actually found in the part of the cathode located along the axis of symmetry and limited by the condition

$$|x|<0.7 \text{ cm}.$$

In order to make possible a quantitative treatment of the pictures taken of the spot trajectory, it must be known exactly how it is located with respect to the axis of symmetry of the field Y. Marking the axis of symmetry on the pictures did not present any particular difficulties since it is along the axis that the field intensity at the cathode surface reaches its maximum values. According to (49), the spot will gradually approach the axis of symmetry of the field, and once it has reached the axis will hold to it steadily throughout the rest of the motion, as long as it does not encounter any mechanical obstacles. This property of the spot made it not only possible to obtain an image on the pictures of the curved trajectory and the inhomogeneous field, but at the same time to obtain an image of the axis of symmetry of the field drawn in when the spot went through later on. The usual path of the spot is shown in Fig. 75. From the place at which it started, O, within the gap between the pole faces, the spot moved in the positive direction of the Y axis, at the same time constantly approaching it. Once it reached the point Q at the edge of the cathode, the spot went either to the right or to the left around half the circumference of the cathode until it intersected the axis of symmetry of the field at the point P, after which it continued exactly along the axis. If in taking the picture, the exposure was chosen so that at least the above stages of motion of the spot were caught, the picture would contain all the necessary data, including the fundamental curved part of the trajectory and the base line for reading distances from the trajectory to the axis of symmetry of the field.

In the first experiments, the cathode spot made its run on a pure mercury cathode, where the form of the trajectory could be distorted as a result of the surface waves excited by the spot itself. This is not so much true of the fundamental curved part of the trajectory, OQ, where the spot is moving ahead of a comparatively slowly propagating wave, as of the base line PQ which it describes when it goes around again. In order to eliminate these distortions, in the rest of the experiments a 0.58-cm-thick copper plate was screwed onto the pole faces; the plate surfaces had been ground to an exact plane and then amalgamated (see Fig. 74). The level of the mercury surrounding the plate was set several tenths of a millimeter below the level of the plate, so that the mercury spread over it in a very thin uniform film. With the cathode constructed in this way, the required mercury level could be maintained at the test gap without appreciably departing from planarity.

Figure 77 shows two pictures of the cathode-spot trajectory taken at a field intensity H_0 in the center of the gap of about 1300 Oe and for two different values of discharge current. The form of the trajectory is in complete agreement with the theoretical curve (53). As had been assumed, in the retrograde type of motion the cathode spot is simultaneously deflected in the direction of increasing intensity of the applied field, getting steadily closer to the axis of symmetry. Once it gets there, the spot follows the axis exactly, the whole time being in the region where the field intensity is a maximum. More far-reaching conclusions as to the similarity between the theoretical and actual spot trajectories can be drawn by making a quantitative treatment of the pictures. For analysis, the Y axis was divided into a number of equal segments, and, at the end of each segment, the distance between the curved part of the trajectory OQ and the base line was measured. The results of such an analysis of the pictures for a current of 5 A are shown in Fig. 78. The horizontal scale in the graph gives the distances in centimeters along the Y axis from the zero point y_0, which was selected so that

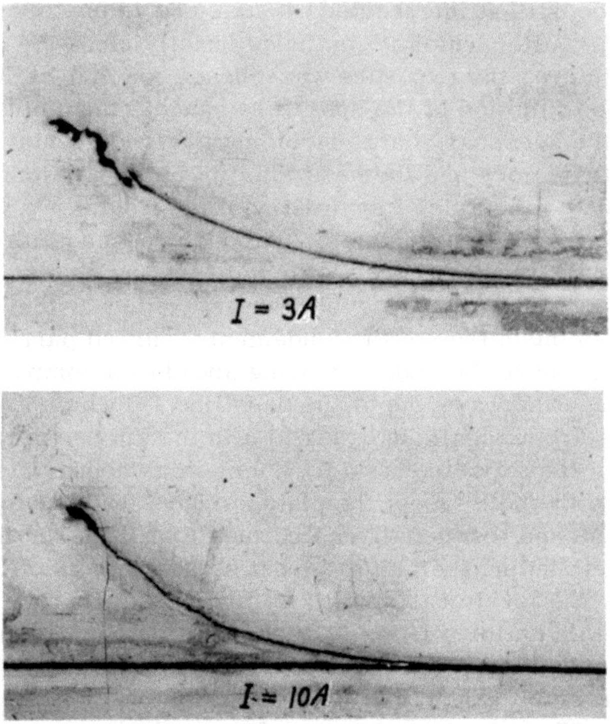

Fig. 77. Pictures of the cathode-spot trajectory at H = 1300 Oe and at different currents.

the condition $x < 0.7$ cm was satisfied along the trajectory. The vertical scale gives the values of $\ln (x_0/x)$. The experimental data for different values of field intensity H_0 are shown by different types of points. With a certain amount of spread, depending partly on the inaccuracy in reading the distances, the points are grouped about an inclined straight line. Thus, the trajectory is nearly exponential in form, in accordance with the deductions of the theory. Apparently it changes somewhat with change in absolute value of the intensity at the axis of symmetry of the field. However, the observations made in this relatively narrow range of field intensities are not sufficient to draw reliable conclusions as to the form of the function.

That the trajectory of the spot is approximately exponential in form was also shown by an analysis of the pictures taken at currents of 3, 10, and 15 A. The data from the analysis of the pictures plotted in the form $\ln (x_0/x) = f(y)$ in all cases gave a quite close fit to an

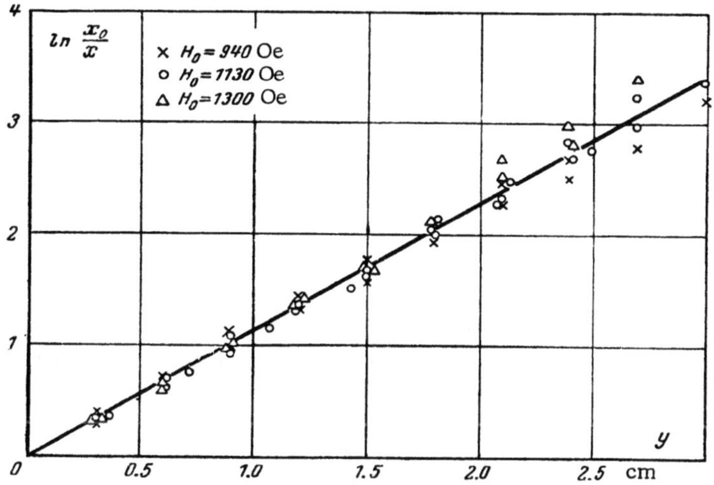

Fig. 78. Results of analyzing the pictures of the cathode-spot trajectory at a current of I = 5 A.

inclined straight line passing through the origin of coordinates. The only effect of a change in the current was a change in slope of the line. The slope of the line is numerically equal to the constant A in Eq. (53), from which we may find, using Eq. (52), absolute values for the effective radius r' of interaction of the cathode spot with the magnetic field for the currents used. The results of the measurements made in this way are shown in Table XII. The first column of Table XII gives the discharge currents at which the observations were made. The second column gives the values of the constant A in Eq. (53) as found from an analysis of the pictures, while the third column gives the corresponding values of the radius r' in centimeters found from Eq. (52) using the value H_0 = 1100 Oe on the axis of symmetry; the latter was found by averaging the values of the field strength used. The fourth column gives the values of intensity of the inherent field of the arc at the circle of radius r' described about the cathode spot as calculated from Eq. (43a). Finally, the fifth column gives the dimensions of the cathode spot from Eq. (42) derived from Froome's experimental data.

In analyzing the data given, it must be noted first of all that the values found for the radius r' are of a reasonable order of magnitude. Actually, r' is always somewhat greater in absolute value than the spot dimensions l as found by optical methods, but by less than one order of magnitude. This is completely in accord

TABLE XII

i, A	A, cm^{-1}	r', cm	H_i, Oe	l, cm	A_δ, cm^{-1}	Corrected r', cm
3	1.1	$2.32 \cdot 10^{-2}$	12.9	$0.3 \cdot 10^{-2}$	0.1	$0.7 \cdot 10^{-2}$
5	1.15	$3.06 \cdot 10^{-2}$	16.3	$0.5 \cdot 10^{-2}$	0.15	$1.1 \cdot 10^{-2}$
10	1.45	$4.86 \cdot 10^{-2}$	20.6	$1.0 \cdot 10^{-2}$	0.45	$2.7 \cdot 10^{-2}$
15	1.65	$6.35 \cdot 10^{-2}$	23.6	$1.5 \cdot 10^{-2}$	0.65	$4.0 \cdot 10^{-2}$

with the assumption made earlier that the asymmetry of the magnetic field in the immediate vicinity of the cathode spot plays a dominant role in the mechanism of the regular motion, as expressed, in particular, in the condition (44). It may be seen further that for these values of r' it is entirely legitimate to use Eq. (43a) to find the intensity of the inherent field of the arc on the circle of radius r'. The figures of the fourth column of Table XII show further that the intensity of the inherent field at the circle in question is already so small as to provide complete justification for the simplifications made in deriving the general formula for the cathode-spot trajectory (47).

An important detail of Table XII is the regular way in which the radius r' increases with increase in current. This relation is illustrated by the graph (Fig. 79) plotted from the data of the table. In the range of currents investigated, the effective radius of the spot varies linearly with the current, while reducing the current to zero[1] corresponds to reducing r' to the definite finite value $r_0' = 1.23 \cdot 10^{-2}$ cm.

The way in which the radius of the spot varies with current may be represented by the empirical formula

$$r' = r_0' + \beta I \tag{54}$$

with the constant $\beta = 0.36 \cdot 10^{-2}$ cm$\cdot$A^{-1}.

This relation shows that there are two different effects, one depending on the current, the other constant. The nature of these effects can be determined from the results given below of a more detailed analysis of the factors which determine the direction of motion of the cathode spot in a magnetic field.

[1]Actually the current can only be reduced to the threshold value $I_0 = 0.07$ A.

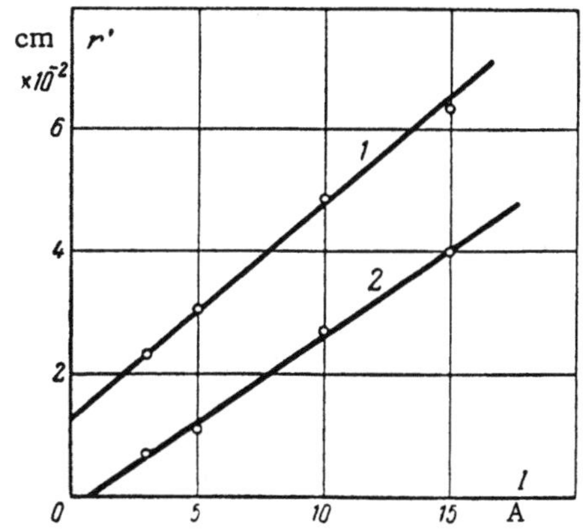

Fig. 79. Effective radius r' of interaction of the cathode spot with the magnetic field as a function of current. 1) Uncorrected values of r'; 2) values of r' after correcting for the effect on the trajectory of the slope of the magnetic lines of force.

c. The Trajectory of the Cathode Spot as Affected by the Slope of the Magnetic Field Vector Relative to the Cathode. The investigation of the trajectory of the spot in an inhomogeneous magnetic field which we have just described far from exhausts the question of the direction of the regular spot motion, in spite of the encouraging results obtained. Moreover, the preliminary interpretation of the results of the investigation that we have made from considerations of logical sequence requires serious corrections and refinement, since the question of the direction of motion of the spot in an inhomogeneous magnetic field is actually considerably more complicated than was assumed in deriving (47). So far, in order to simplify the problem, we have assumed the inherent field of the arc to be identical with the field of an equivalent line of current passing through the cathode at the center of the spot and normal to the surface of the metal. Solving the problem with this simplification leads to the conclusion that there is a unique relationship (47) between the direction of motion of the spot at any point on the cathode and the distribution of the absolute values of the intensity of the external field at the surface of the metal. But the actual field of the arc may differ appreciably from the idealized field of a line of current normal to the cathode as given by Eq. (43). This difference is reflected first of all in that, in addition to the component normal to the cathode, the actual current in the arc almost always has a more or less large tangential component. As a matter of fact, in a low-pressure discharge with relatively small electric field gradients in the plasma, the electron current will flow principally along the lines of force of the external magnetic field. Thus, the applied magnetic field will to some degree fix the direction of the current immediately above the cathode spot. Any departure of the current from a direction normal to the cathode will disturb the symmetry and the distribution of the inherent magnetic field of the arc in the vicinity of the cathode spot; this can serve as an additional reason, not taken into account in (47), for the spot's departure from the retrograde motion. It is easily shown that the departure will always occur in the direction of the smallest angle between the current and the cathode surface, since the arc field has its maximum intensity in this direction. If we denote the angle by γ, then the difference ΔH_i between the field intensity of the inherent field of the arc at the edges of the spot in the direction of γ and in the opposite direction may be expressed approximately as $2H_i \cos\gamma$, where H_i is the intensity of the arc field when the current is in the normal direction. If we assume that the direction of the current at every point of the cathode coincides with the direction of the intensity H of the external field, it is a simple matter to take account of the effect of the angle γ on the trajectory of the spot. It is only necessary to replace $\cos\gamma$ by the ratio H_x/H and to add to the numerator of (49) the term $(H_i/H) \cdot (H_x/r')$, which takes care of the asymmetry of the inherent field of the arc in the vicinity of the spot produced by changing the slope of the current. This type of theoretical treatment of the question of how the direction of motion of the spot depends on the angle that the field intensity vector makes with the cathode is, however, hindered by a number of secondary effects which complicate the picture. Thus, at small values of γ, to which the results of investigating the spot trajectory in an inhomogeneous field particularly apply, a deviation of the current from the direction of the magnetic lines of force will be observed which it is impossible to take account of. In the field tangential to the cathode, the current will be split into two equal parts running in opposite directions from the cathode spot, which will tend to restore the symmetry of the field.

In view of the above complications in a theoretical determination of the way in which the direction of spot motion varies with the angle γ, there was some point in turning to experiment for help. A simple experiment, shown schematically in Fig. 80, will easily convince one that we have arrived at correct conclusions concerning

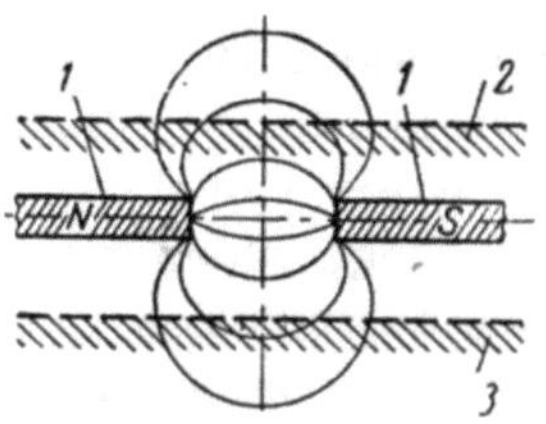

Fig. 80. Arrangement of an experiment to show how the trajectory of the cathode spot in an inhomogeneous field is affected by the inclination of the field vector at an angle to the cathode. 1) Pole faces; 2) cathode placed above the pole faces; 3) cathode placed below the pole faces.

the effect that the inclination of the magnetic lines of force at an angle to the cathode has on the spot trajectory. The figure gives a sectional view of the pole faces of a magnet consisting of two parallel horizontal plates producing a symmetrical field uniform in the direction normal to the plane of the figure. The direction of the lines of force in the gap between the poles is shown by fine lines. The two horizontal dotted lines show two different positions of the level of the mercury cathode, symmetrical with respect to the field. As a result of the symmetry, the distribution of absolute values of field intensity at the cathode surface will be exactly the same in both cases. However, the direction of the magnetic lines of force with respect to the cathode surface will be quite different in the two cases. At the cathode level corresponding to the upper horizontal line the acute angle γ will always be turned toward the center of the gap, which is the axis of symmetry of the field. In the lower position of the cathode, the acute angle is turned away from the center of the gap toward the poles. If we compare the form of the cathode-spot trajectory for the two cathode levels, we can easily separate the effect on the direction of motion of the spot due to inhomogeneity of the field from that due to the inclination of the lines of force. Quite simply, if the effect of field inhomogeneity is predominant, the spot will in both cases move toward the center of the gap, which is the axis of symmetry of the field, where the field strength of the magnet is a maximum. In the opposite case, the spot in the upper position will move toward the center, while in the lower position it

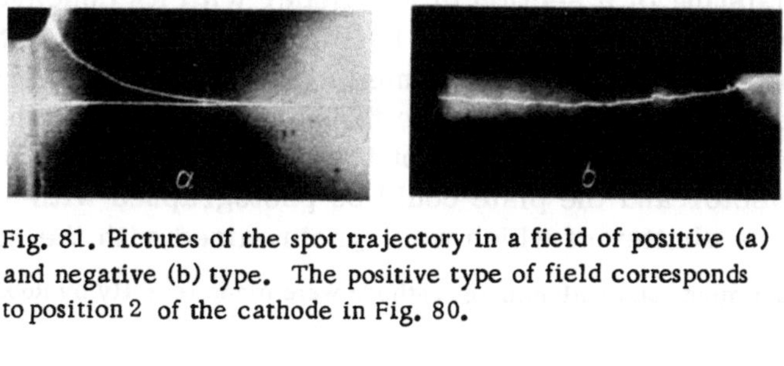

Fig. 81. Pictures of the spot trajectory in a field of positive (a) and negative (b) type. The positive type of field corresponds to position 2 of the cathode in Fig. 80.

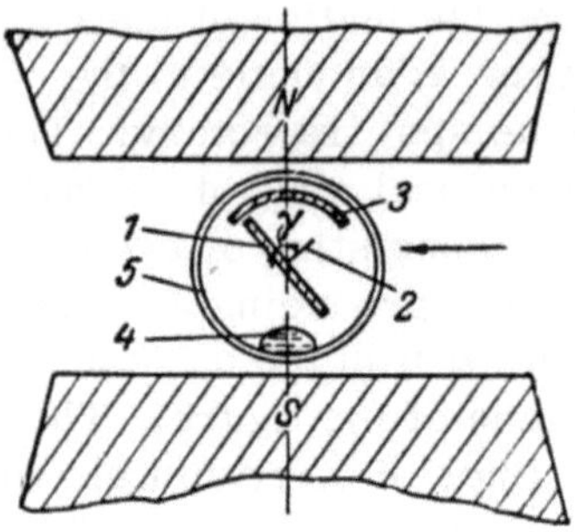

Fig. 82. Arrangement of an experiment to show how the trajectory of the cathode spot in an inhomogeneous field is affected by the inclination of the magnetic lines of force at an angle to the cathode. 1) Cathode; 2) igniter; 3) anode; 4) mercury; 5) tube. The arrow shows the direction in which the photographs were taken.

will move away from the center. The pictures of the spot trajectory for the two cases are shown in Fig. 81. Here the gap between the poles was about 1 cm wide and 6 cm long, which gave a quite uniform field distribution along the length. In the upper position, the cathode level was 0.4 cm above the upper edge of the poles, while in the lower position it was 0.4 cm below the lower edge. It may be seen from the pictures that in the upper position the cathode spot is deflected toward the center of the gap, the center being shown by the line where the spot went across again. With the cathode in the lower position, the deflection is in the opposite direction, thus showing that under these conditions the direction of motion of the spot is affected more by the inclination of the field vector than by the inhomogeneity of the field.

In order to make a more exact quantitative separation of the effects of the two factors on the spot trajectory, some additional experiments were set up to determine how the direction of the spot motion is affected by the angle which the lines of magnetic force make with the cathode surface in a homogeneous field.[1] In these experiments, the cathode spot was excited with a semiconducting igniter on an amalgamated copper plate placed in a homogeneous magnetic field, as shown in Fig. 82. The plate was fastened to a ground joint which could be rotated around an axis normal to the magnetic lines of force, thus changing the orientation with respect to the magnetization vector H. Above the plate there was a fixed anode consisting of a section of a cylinder with its concave side toward the plate. The discharge between the cathode and the anode took place in a straight line oriented along the field. There was a small pool of mercury in the tube to prevent drying out the plate. The trajectory of the cathode spot for different angles γ between the field vector and the plate could be photographed with a film camera, the shutter of which was synchronized with the igniter.

[1]Similar experiments on an aluminum cathode were made recently by Robson.

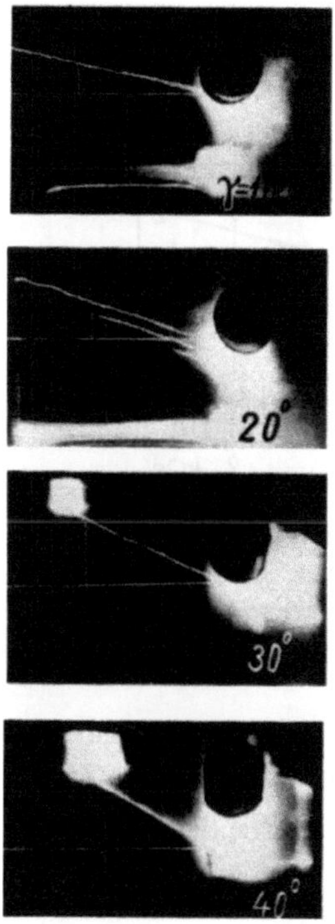

Fig. 83. Pictures showing the trajectory of the cathode spot in a uniformly inclined magnetic field, for different values of the angle γ at which the vector H is inclined to the cathode.

Several of the pictures obtained in this way are shown in Fig. 83. The trajectory is an inclined straight line making an angle φ with the direction of the retrograde motion, which increases with increase in the angle γ. The results of measuring the angle φ for different values of γ and a number of currents and field intensities are given in Fig. 84. To separate the experimental points for different values of γ, the results are shown in two different graphs, one giving the curves for $\gamma = 10$, 30, 60°, the other the curves for the intermediate angles 20 and 40°. In both graphs, the measurements for 1 A are shown by dots, while the results for 5 A are shown by crosses. The independent variable plotted on the horizontal axis is the magnetic field intensity, while the vertical axis gives the values of tan φ. Within the limits of accuracy of the measurements, the angle of the trajectory is independent of the current and increases somewhat with increase in field intensity. There is a more pronounced relation between the angle of the trajectory and the angle γ which the magnetization vector makes with the cathode surface. This relation is shown separately for a field intensity of 1 kOe in Fig. 85, where the abscissas are tan γ and the ordinates are tan φ. At first, the angle φ is approximately proportional to γ, but then it increases more slowly, and the curve approaches an asymptotic value at about tan $\varphi = 0.7$. At higher field intensities, the maximum value approaches 1. It is interesting to note in this connection that the theoretical value of tan φ for strong fields is equal to 1, and is independent of both current and γ, since changing either of these parameters produces the same changes in the components of velocity in the two mutually perpendicular directions.

From the experimental relation that we have found between tan φ and tan γ, we can find what quantitative effect the angle between the lines of force and the plane of the cathode has on the spot

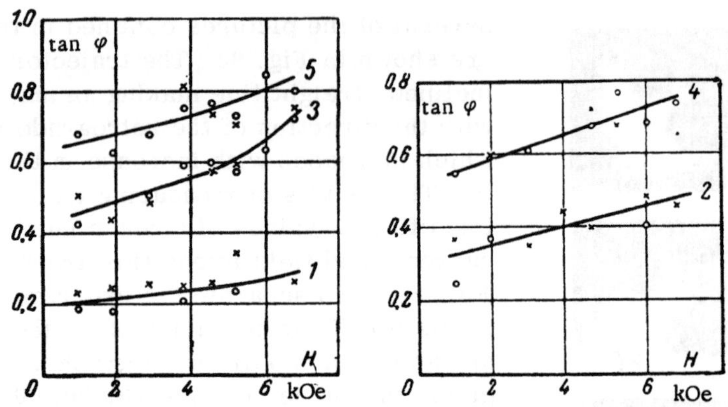

Fig. 84. Results of measuring the angle of inclination φ of the trajectory for different values of the angle γ between the vector H and the cathode as a function of field intensity for currents of 1 and 5 A. 1) $\gamma = 10°$; 2) 20°; 3) 30°; 4) 40°; 5) 60°.

trajectory in an inhomogeneous field and thus establish the true role played by field inhomogeneity in curving the trajectory. The relation in Fig. 85 is quite well described by an exponential function of the type

$$\tan \varphi = G(1 - e^{-\varphi \tan \gamma}). \tag{55}$$

If we limit ourselves to the solution of the problem for small values of γ, which is sufficient for our case, where the spot is moving close to the axis of symmetry of the field, we can replace (55) by the simple expression

$$\tan \varphi = \tan \gamma, \tag{56}$$

which is suitable for the whole range of currents investigated at H = 1 kOe. The quantity $\tan \varphi$ is the value of the derivative dx/dy at each point on the trajectory of the spot. The quantity $\tan \gamma$ is equal to the ratio of the field components H_z/H_x and, by using (50), may be expressed as a function of x. To a first approximation, $\tan \gamma$ may be replaced by the expression $x\sqrt{[2(C_2 - C_1)]}$. Using the values of the constants C_1 and C_2 found for this field distribution, we finally obtain the equation for the trajectory:

$$\frac{dx}{dy} = -x. \tag{57}$$

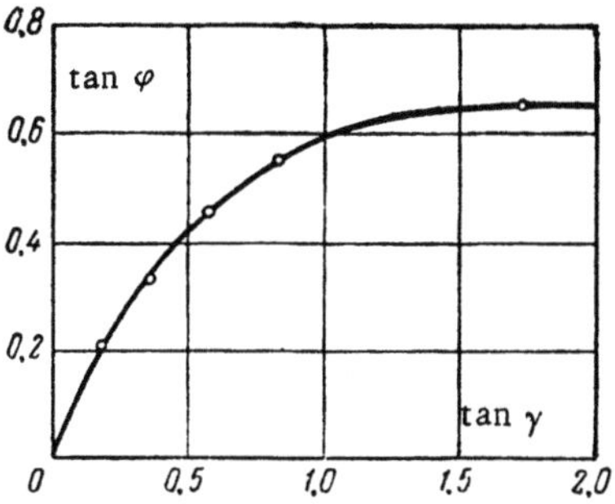

Fig. 85. Angle between the trajectory of the cathode spot and the direction of the retrograde motion as a function of the angle between the magnetization vector and the cathode for H = 1 kOe.

In contrast to Eq. (53), in deriving this last equation, the only relation considered was that between the direction of motion of the spot and the angle between the lines of force and the cathode; no consideration whatever was given to the effect of nonuniform distribution of the absolute values of the field intensity. The values found experimentally for the constant in Eq. (53) include both the effect on the trajectory of field inhomogeneity and the effect of angle of inclination. This means that, in essence, the constant A may be broken up into two terms, A_δ and A_γ, the first of which takes account of the field inhomogeneity factor, while the second takes account of the angle of inclination of the lines of force. Then the general equation, including both factors, will be of the form

$$\frac{dx}{dy} = -(A_\delta + A_\gamma)\, x. \tag{58}$$

For a fixed distribution, we have $A_\gamma = 1$, in accordance with (57). Accordingly, the values of A given in Table XII are actually the sum

$$A = 1 + A_\delta, \tag{59}$$

where A_δ is related to the effective spot radius r' by the expression

$$A_\delta = \frac{20\,(r')^2}{J} C_1 H_0\,; \tag{52a}$$

we formerly mistakenly used this value for the constant A, which includes the effect of the angle between the lines of force and the cathode. The values of A_δ are given in Table XII in the next to the last column. If the current is increased from 3 to 15 A, these values change from 0.1 to 0.65 cm^{-1}. This means that for this field distribution the effect of field inhomogeneity on the trajectory as such is small in comparison to the effect of the angle between the lines of force and the cathode. However, the effect of the slope

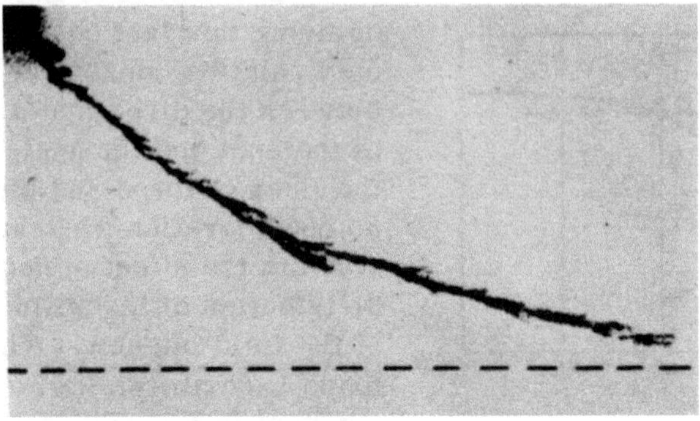

Fig. 86. Picture of the curved portion of the trajectory of the spot in an inhomogeneous field at a current of 15 A, showing additional deflection of the spot as a result of the high stability of the branches sent out in the direction of increase in field. The dotted line is the axis of symmetry of the field, where the field strength is a maximum.

of the lines is independent of the current, while the effect of inhomogeneity increases with the current. Thus, as the current is increased, the ratio between the two effects changes in favor of field inhomogeneity. The increase in the effect of field inhomogeneity on the spot trajectory at large currents is the result of increasing the effective spot radius r'. It should be clear from the above that the values of r' given in the third column of Table XII are too large. They were obtained from Eq. (52), which takes no account of the factor involving the inclination of the field vector to the cathode surface. The correct values of r' may be found from Eq. (52a), using the data on the values of the constant A_δ. The corrected values are given in the last column of Table XII. They are already very close to the true dimensions of the cathode spot, and are proportional to the current as given by the relation

$$r' = 0.25 \cdot 10^{-2} I, \quad (60)$$

where I is expressed in amperes and r' is expressed in centimeters.

The corrected effective radius of interaction of the cathode spot with the magnetic field is only a factor of 2.5 greater than the spot dimensions given by Eq. (42), which was derived by Froome from optical measurements. This fact must be considered to prove directly that the motion of the cathode spot depends on the field distribution around the boundaries of the spot.

The fact that the values of the radius r' which we have found by investigating the trajectory are so close to the true dimensions of the cathode spot, and that they are proportional to the current, leaves no doubt as to the correctness of the original assumption that the asymmetry of the total field at the spot plays a dominant role in the regular motion of the spot. Nevertheless, without limiting ourselves to these proofs, the role played by field asymmetry in the motion of the spot may easily be demonstrated by a direct experiment. For this purpose, let us turn to the picture in Fig. 86, which shows an enlargement of the trace of the spot in an inhomogeneous magnetic field at a current of 15 A. The approximate location of the axis of symmetry of the field, where the maximum intensities occur, is shown by the dotted line. In the region where the field is highly nonuniform, the greater part of the traces of the decaying unstable branches of the spot are turned toward the weaker field. The spot moves ahead by virtue of the more stable branches, which are turned predominantly in the direction of increase in field strength. Accordingly they experience an additional deflection in the direction of increase of the field. The picture gives a clear illustration of the stabilizing action which a magnetic field exerts on autonomous spots, at the same time revealing one of the reasons why the radius of interaction of the spot with the magnetic field increases at large currents. Apparently, division of the spot plays a certain role in this effect, with the result that the zone in which the spot interacts with the field is appreciably broadened.

In summarizing the investigations that have been made on the trajectory of the cathode spot in a magnetic field, we can note the following. Under all conditions considered, including the so-called retrograde motion of the spot in a uniform tangential field, as well as motion under the more complicated conditions where uniform and nonuniform fields are inclined at an angle to the cathode, the direction of motion of the spot at any point in its trajectory coincides with the direction of greatest increase in total field intensity at the boundaries of the spot. This result provides a sufficient basis for the more general conclusion that the maximum field principle gives a correct description of the motion of the cathode spot on mercury at low pressures in the medium, as well

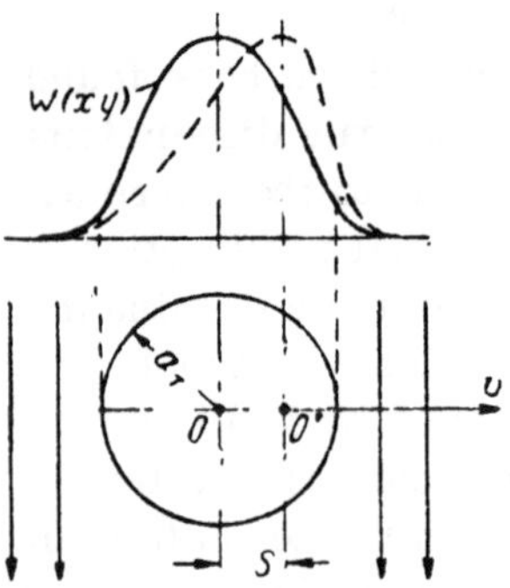

Fig. 87. Schematic representation of displacement of the cathode spot resulting from distorting the symmetry of the conditions existing at the boundary of the spot.

as under any other conditions, independently of whether or not the distribution of the total magnetic field is subject to our control. But that the maximum field principle applies to the problem of the motion of the spot means in essence that the motion is caused by asymmetry in the distribution of the total field at the boundaries of the spot. Since the principal source of such asymmetry is the inherent magnetic field of the arc, the fact that the maximum field principle holds serves at the same time to confirm the idea that the inherent field of the arc plays an essential role in the regular motion of the cathode spot.

48. Rate of Motion of the Spot and New Ways of Finding the Current Density at the Cathode of the Arc

It has been shown above that the direction of the regular motion of the spot at low pressures in the medium is always in accordance with the idea that the motion results from disturbing the symmetry in the magnetic field distribution around the spot. But the direction of motion of the spot is only the first half of the problem. The second important characteristic of the motion is its velocity. Accordingly, to be able to say that we have proved that field asymmetry plays the dominant role in the motion of the spot, we must also convince ourselves that the theory gives a correct description of both the absolute values of the rate of motion and its variation with the field intensity and the gas pressure in the medium.

a. Theoretical Discussion of the Question. Limits of Applicability of the Theory. If one component of the velocity and the trajectory of the spot are accurately known for a given field distribution, the second component and the total velocity may easily be found at any point in the trajectory. Accordingly, in discussing the rate of motion of the spot we can limit ourselves to the simplest case of motion in a uniform tangential field, where the velocity is completely described by the single component normal to the lines of force in the magnetic field. This case corresponds to the retrograde type of motion in its pure form.

In solving the problem of the rate of motion of the spot in a magnetic field, we must return to the dynamic model of the spot discussed in its general features in the preceding chapter. As already pointed out in discussing this model, the motion of the spot over the cathode as a whole must be related primarily to the cell-displacement type of re-formation, which has a considerably shorter cycle than re-formation of the type where the current is redistributed among the various parts of the spot. With this in mind, the slower process of changing the make-up of the cathode-spot cells may be completely neglected in the present problem.

Let us now discuss in detail how the spot is displaced from its original position, with the center of metal evaporation at the point O. For simplicity, we shall imagine the cathode evaporation region S_T to be the perfect circle shown in Fig. 87, with the center at O and with some radius a_T. It must be assumed in accordance with the dynamic cathode-spot model that the center of the emission region moves rapidly about inside the evaporation region as a result of uncontrollable disturbances in equilibrium among the various processes in the arc cycle. Here there will be some probability field W(n) for the location of the emission center about the point O, depending on the time average of the charge distribution n(x, y) above the part of the cathode in question. Excluding from consideration any possible irregular motion of the spot due to random distortion of the symmetry of the charge distribution in the region S_T, we are justified in assuming here that if there is no perturbation of the spot by an external magnetic field, the function W(x, y) is symmetric about the center O, and decreases monotonically from the center to the circumference like a normal distribution curve. The perturbation from the external field H will be reflected as a steady distortion of the symmetry. We will recall that the idea of the model is that the magnetic field does not act directly on the distribution of concentration in the spot, but by introducing a constant asymmetry into the electron distribution around the edge of the spot in the region where the inherent field of the arc is strong. The result of the asymmetric field distribution around the spot is that the maximum in the probability distribution W(x, y) will shift, along with the maximum charge concentration, from the center O in the direction of greatest increase in field intensity through some distance s, which may be regarded as a quantitative measure of the perturbing effect of the field. The result of the change in the probability distribution W(x, y) will be that at the end of the re-formation cycle the center of the metal evaporation region is shifted in the same direction to the new point O', removed from the initial position by approximately the distance s. Then the rate of motion of the spot v may be found as the ratio of the displacement s of the center of evaporation of the metal after a single re-formation cycle to the duration of the cycle, t_r, i.e.,

$$v = \frac{s}{t_r}. \tag{61}$$

In accordance with the point of view adopted here that the dominant role in the cathode-spot motion is played by asymmetry in the distribution of the surrounding magnetic field, it must be assumed that the value of s depends exclusively on the difference between the conditions existing at opposite edges of the spot. A quantitative measure of the difference is the difference in absolute values of the total field intensity, ΔH_a, at the opposite sides of the spot in the direction of most rapidly increasing field. Thus, the magnitude of the spot displacement s per cycle will be some increasing function of the field intensity difference ΔH_a. Since the probability distribution W(x, y) always decreases rapidly outside the region S_T, an additional limitation must be put on the quantity s, namely that, even for the largest possible values of ΔH_a, the displacement of the spot cannot exceed the radius of the metal evaporation region, a_T, i.e.,

$$s < a_T. \tag{62}$$

With this condition, the desired relation between s and ΔH_a is conveniently represented in the form

$$s = a_T \Phi(\Delta H_a). \tag{63}$$

The function $\Phi(\Delta H_a)$ will have a positive derivative for all values of ΔH_a,

$$\frac{d\Phi(\Delta H_a)}{d(\Delta H_a)} > 0, \tag{64}$$

and will take on values from 0 to 1 as ΔH_a changes from 0 to ∞. Further, s will vary linearly with ΔH_a for small perturbations, in the first or linear approximation to the solution of the problem. All these conditions are satisfied by the function

$$\Phi(\Delta H_a) = 1 - e^{-\varepsilon(\Delta H_a)}, \tag{65}$$

where the coefficient ε has dimensions of reciprocal field intensity, and expresses the sensitivity of the cathode spot to the perturbing effect of the magnetic field.

Considering all that has been said about the relation between s and ΔH_a, we may express the velocity of the retrograde motion of the cathode spot by the relation

$$v = \frac{a_T}{t_r}(1 - e^{-\varepsilon \Delta H_a}) \sum k n_k, \tag{66}$$

where k is the number of charges on the mercury ion and n_k is the relative concentration of ions having this charge.

The expression under the summation sign on the right-hand side of the equation takes account of the acceleration of the spot re-formation process which occurs if doubly and triply charged mercury ions are present, as a result of their supplying more energy to the cathode.

According to the above formula, in a relatively weak external field, the rate of motion of the spot will increase approximately in proportion to the field intensity H, since under these conditions ΔH_a may be replaced by twice the value of H, the expression in parentheses having approximately the value $2\varepsilon H$. However, with further increase in field intensity, more and more departure will be observed from the simple linear relationship in the direction of lower velocities right up to saturation of the v(H) curve at large field intensities, as long as doubly and triply charged mercury ions do not appear on the scene. One of the reasons for this departure is the limitation placed by Eq. (62) on the maximum value of the displacement s of the center of evaporation of the metal per reformation cycle, which is in turn the result of the fact that the probability of getting out of the emission center drops off rapidly outside the evaporation region. In addition, the difference between the total field intensities on opposite sides of the spot, ΔH_a, will increase with increase in the external field intensity H only as long as H does not become greater than the inherent field of the arc, H_i, at the edges of the spot. After this, ΔH_a becomes equal to $2H_i$ and, accordingly, does not change with further increase in intensity of the external fields; this produces a more or less sharp break in the v(H) curve in the direction of saturation. This behavior of the total field at the cathode spot is clearly visible in Fig. 88, which is a schematic representation of superimposing the inherent arc field H_i on different values of the external field H. The two vertical lines in the figure are the boundaries of the cathode spot. The dotted lines show the distribution of the absolute values of intensity of the inherent field of the arc and of the external field, while the solid curve is the intensity distribution of the resultant field H_a. On the right of the center line of the figure, H_a is the algebraic sum of H and H_i, while on the left H_a is the difference of the two fields. The external field intensity increases from top to bottom in the figure, as shown by the increase in the level of the horizontal dotted line. As long as H remains less than H_i, increasing H produces a proportionate increase in the difference ΔH_a between opposite sides of the spot. As soon as the external field intensity becomes equal to the intensity of the inherent field of the arc at the edges of the spot, ΔH_a stops increasing.

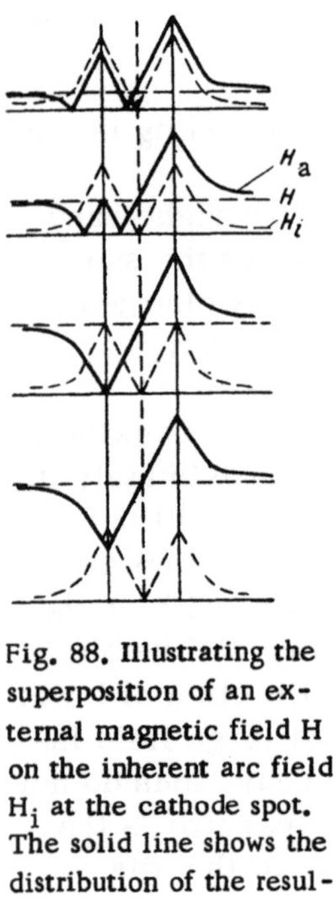

Fig. 88. Illustrating the superposition of an external magnetic field H on the inherent arc field H_i at the cathode spot. The solid line shows the distribution of the resultant field H_a.

The break in the v(H) curve for $H \geq H_i$ in the direction of saturation, if it could be observed experimentally, would be a valuable way of measuring the field intensity of the arc, H_s, at the edges of the spot, and it could serve as a basis for a new and improved method of measuring the current density j_c at the cathode. Actually, H_s is a quantity which is directly dependent on j_c. In order to find this dependence, at least approximately, we can imagine a simplified region of high current density consisting of a regular circular cylinder with a uniform distribution of the current density j_c inside it. If we designate the current concentrated in the spot as I, we will have, for the above type of current channel, the relation

$$j_c = 100 \cdot \frac{H_s^2}{\pi I}, \tag{67}$$

which can be used to find the value of j_c from the inherent field intensity of the arc, H_s.

Returning now to the question of the velocity at which the cathode spot executes its regular motion in a magnetic field, we need to calculate the maximum velocity to be expected from Eq. (66) for the maximum difference in conditions existing on opposite sides of the spot. The maximum possible velocity of the spot is determined by the dimensions of the evaporation region a_T and the duration t_r of the re-formation cycle, which depends principally on the sluggish process of heating the metal at another part of the cathode. The maximum velocity may be calculated from (66) using the simple relation

$$v_{max} = \frac{a_T}{t_r} \sum kn_k, \tag{66a}$$

with $a_T = 10^{-3}$ cm and $t_r = 10^{-7}$ sec, in accordance with the data of the preceding chapter. If there are only singly charged ions in the discharge, the maximum velocity will be about 10^4 cm/sec. If doubly charged ions are present, the velocity will increase to twice this value for 100% doubly charged ions in the ionization region of the discharge nearest the cathode, due to the metal being heated more rapidly by ion bombardment; i.e., the re-formation time t_r becomes shorter.

As we shall see below, all these conclusions are in complete agreement with the experimental data. However, before we get to the experiments, we must consider some additional predictions of the theory regarding the way in which the rate of motion of the spot varies with the gas pressure in the discharge gap, and in addition point out the limits of applicability of the theory. As far as the first of these problems is concerned, we shall limit ourselves to pointing out in a purely qualitative way that the velocity at which the spot moves in a magnetic field will always decrease if an extraneous gas is present: the denser the gas, the greater the decrease. This follows from the fact that the asymmetry in the

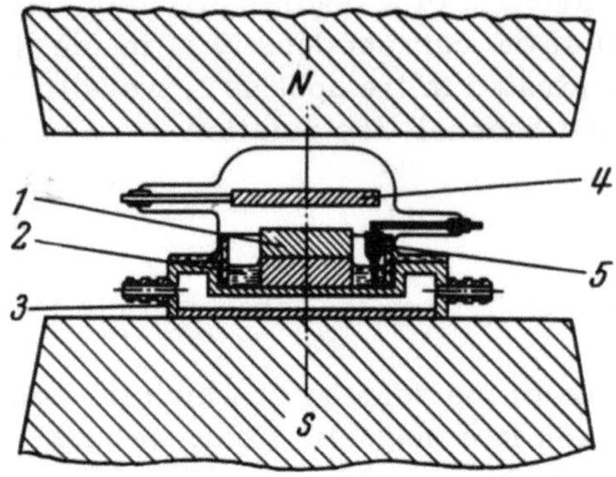

Fig. 89. Construction and arrangement between magnetic poles of tube for measuring the rate of motion of the cathode spot in a magnetic field. 1) Metal cylinder around which the spot rotates; 2) mercury; 3 cathode reservoir ; 4) anode; 5) igniter.

charge distribution of the spot produced by the field, which is the cause of the motion, decreases if a gas is present and continues to decrease as the gas density is increased. The reaons for this effect have already been discussed in the last chapter, so that it is unnecessary to repeat them here. The effect of the medium also determines the limits of applicability of the theory. It is obvious that the asymmetry of the magnetic field at the cathode spot can only continue to be the dominant cause of the spot motion as long as the density in the medium is such that the mean free electron path λ_e remains greater than the mean Larmor radius R_L of the rotational motion of the electrons in the magnetic field. Thus, the upper limit of applicability of the theory may be found from Eq. (27) (see Chapter 2).

b. Results of Measuring the Rate of Motion of the Spot as a Function of Magnetic Field Intensity and Current. The rate at which the cathode spot moves in a magnetic field was measured on two widely different experimental setups. In one case, the spot rotated around a cylinder fastened in the center of a tube having the mercury cathode in a metal reservoir, as shown in Fig. 89. A plane anode of considerably larger diameter was located above the cylinder in the upper glass part of the tube, through which the observations were made. The entire tube was placed between the poles of a large electromagnet which produced a uniform field in the discharge gap, parallel to the axis of the cylinder. The cylinder consisted of two different sections fitting tightly together. The lower part of the cylinder wet by the mercury was made of pure copper, while the upper part was made of nonmagnetic chrome nickel steel. The two parts of the cylinder were turned, fastened together, on a lathe so that the outside surfaces were generated by the same line. The lower copper cylinder was 2-3 mm higher than the depth to which it was immersed in the mercury. Thus the dividing line between the copper and the steel cylinders was only a little above the level of the mercury marking the top of the region where the mercury wet the surface of the cylinder. This held the line where the mercury wet the copper in place in a perfect circle, and this was where the cathode spot moved. The anode and the metal reservoir in the tube were all made of nonmagnetic steel, which eliminated any possibility of distortion of the magnetic field in the gap by the metal parts. During the experiments, the temperature of the cathode was maintained at the desired level by continuously circulating water heated to the required temperature through the cavity in the reservoir. The frequency of rotation of the cathode spot around the cylinder was determined with a photomultiplier, the signals from which were fed to an oscillograph and compared with an alternating emf supplied by an audio-fre-

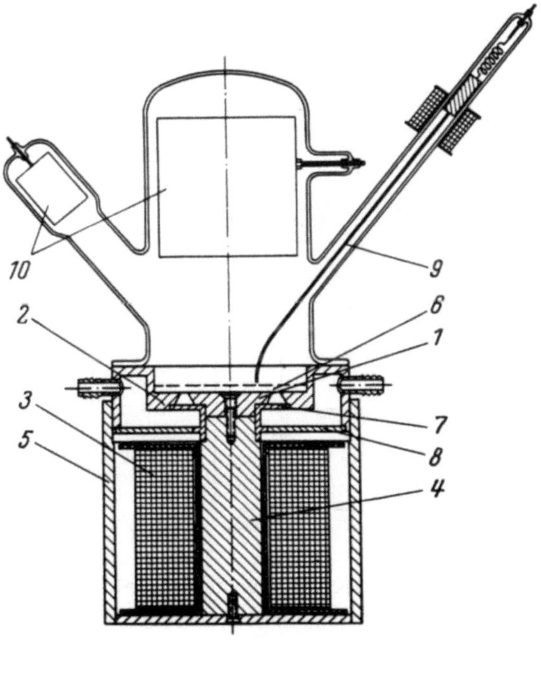

Fig. 90. Construction of tube used to measure the rate of motion of the cathode spot in a radial magnetic field on a uniform mercury surface. 1, 2) Concentric pole faces; 3) winding; 4) core of magnet; 5) iron case; 6) mercury; 7, 8) cathode reservoir of nonmagnetic steel; 9) igniting needle; 10) anode.

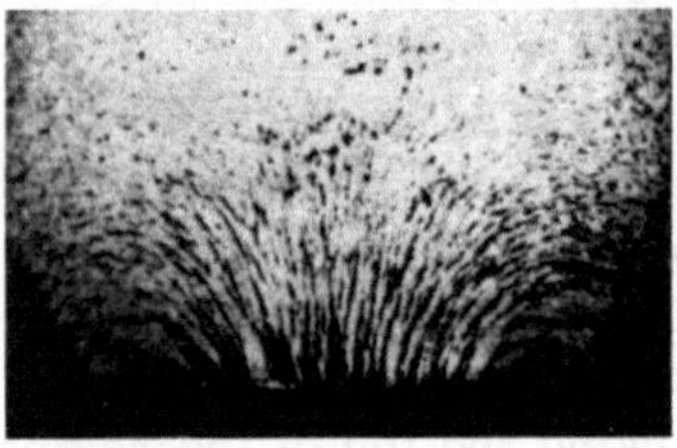

Fig. 91. Illustrating the direction of the lines of magnetic force in the gap between the pole faces of the tube shown in Fig. 90.

quency signal generator. It may be seen from this description that the experimental conditions in the present case were very nearly the same as those obtained previously in [89, 93]. The difference between the two cases is principally that we had rigid cathode temperature control, as well as a high degree of uniformity in the magnetic field instead of maximum intensity, and finally that we used copper as the auxiliary metal instead of molybdenum.

In the second series of experiments, the velocity of the spot was measured as it moved over a uniform mercury cathode in a circular zone containing a radial magnetic field, produced by an electromagnet with concentric pole faces separated by a ring-shaped gap.

The experimental tube with a cathode of the second type is shown in section in Fig. 90. The pole faces forming the bottom of the cathode are several millimeters below the level of the mer-

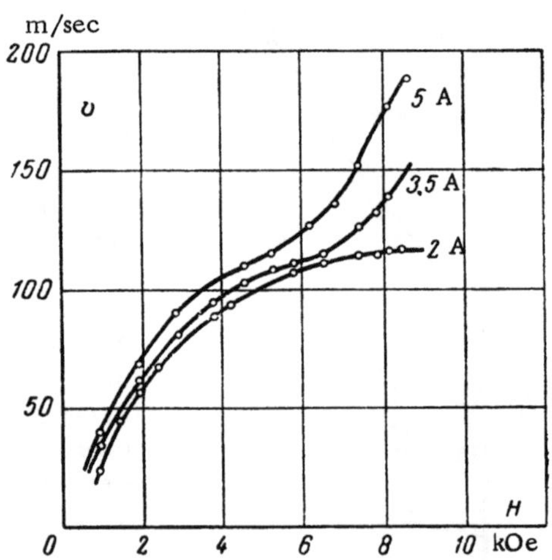

Fig. 92. The rate of motion of the cathode spot as a function of field intensity for three currents, determined with the tube shown in Fig. 89.

cury, which is shown by the dotted line. In order to avoid closing the magnetic circuit with metal, the pole faces were joined together with a disc of nonmagnetic chrome nickel steel. The magnetic field was excited by an external winding, mounted on the central iron bar welded to the bottom of the cathode. The winding was covered on the outside by an iron shell, forming the outside link in the magnetic circuit. Inside the tube, the magnetic flux leaves the iron at the ring-shaped gap, thus forming an annular zone on the surface of the mercury where the lines of force are in a radial direction as shown in the picture (Fig. 91) of the pattern of iron filings. In this zone of the cathode, the acute angle between the lines of force and the mercury surface points toward the middle of the gap, with the result that the spot is anchored in the middle while executing rapid motion around the circumference. With the experiment arranged in this way, the relation between the velocity and the field intensity will be distorted at relatively weak fields as a result of the appreciable departure of the cathode spot from its regular trajectory due to the effect of random components in the motion, which must be allowed for in interpreting the results of the measurements. In considering the results of the measurements, we shall first of all discuss the curves giving the rate of motion of the spot as a function of magnetic field intensity, as shown in Fig. 92. These curves are for the case where the cathode spot is rotating along the mercury meniscus surrounding the copper cylinder. The curves in the figure differ for different values of discharge current, as shown on the graph. As the field intensity is increased, the velocity at first increases almost linearly, but then bends more and more toward lower velocities, showing a tendency to saturation, in accordance with the theoretical relation (66). Practically complete saturation is reached at a field intensity of about 8-9 kOe in the lower curve for a current of 2 A. It is noteworthy that here the velocity is only a little greater than our theoretical figure of 10^4 cm/sec. Unfortunately, the field intensity shown was also the maximum for the magnet used in the experiments, so that the rest of the curve can only be guessed at. It is to be expected from the form of the upper curves that a sudden rise would occur at somewhat stronger fields. The point at which the velocity starts to increase again shifts regularly toward lower field intensities as the current is increased, in accordance with the data obtained in [93] which we have cited. How-

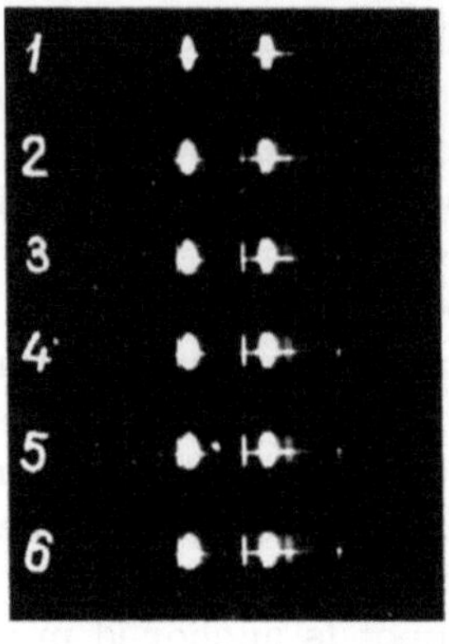

Fig.93. Spectrogram of luminescence from the cathode region of the arc with the cathode spot moving rapidly in a magnetic field under conditions corresponding to Fig. 89; I = 3.5 A.
1) H=1.9 kOe 2) H=4.5 kOe
3) H=6.5 kOe 4) H=7.4 kOe
5) H=8.1 kOe 6) H=8.7 kOe

ever, in our case, the second rise in the curve is considerably smoother, which may result from the fact that we had rigid control of the cathode temperature.

In finding the reasons for the second increase in spot velocity in strong fields, the following two facts are of particular importance. First it follows from the observations of the authors of [93], who had a stronger magnet, that the result of the second rise at high field intensities is to double the velocity of the spot motion, after which the curve saturates again. Second, as the field intensity is increased, the spectrum of the discharge begins to show intense spark lines, as may be seen from the spectrograms given in Fig. 93. The spark spectrum is particularly intense in the region of the spot itself, which is shown in the spectrograms by the place where the lines are thickest and the band of continuous spectrum occurs. It becomes obvious in view of these facts that the second increase in cathode-spot velocity in strong fields arises from the formation of a large number of doubly charged mercury ions in the cathode region of the arc. This supports the earlier conclusion that, if multiply charged ions are present, the duration of the cathode-spot re-formation cycle becomes smaller, as a result of the increase in the energy flux to the cathode, which also gives a simple explanation for the second increase in spot velocity in strong fields.

Returning to the curves of Fig. 92, we can now make a more detailed commentary on the relations which they describe. It is clear from what has been said above that each of the curves gives the increase in spot velocity with increase in magnetic field intensity occurring as a result of superposition of two independent effects. One of these is the increase in displacement s of the cathode evaporation region per re-formation cycle produced by the in-

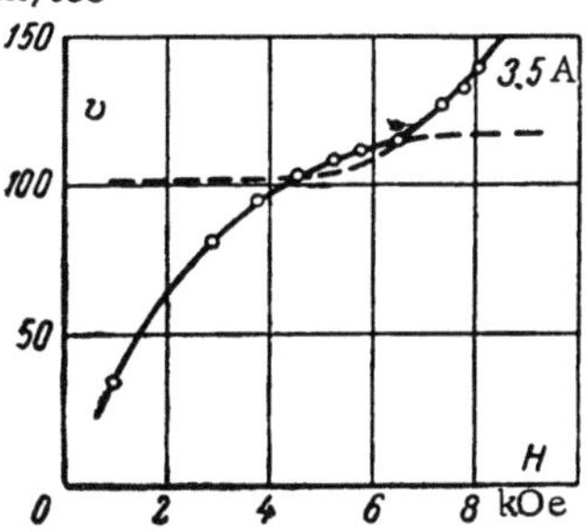

Fig. 94. Interpretation of the v(H) curve from Fig. 92 for I = 3.5 A.

crease in difference between the conditions on opposite sides of the spot. This effect produces the initial approximately linear increase in velocity which gradually slows down as the field intensity is increased, and ends with saturation of the curve, as shown by the dotted line in Fig. 94. The other effect is that the spot re-formation time becomes shorter in strong fields as a result of the formation of multiply charged ions at the cathode under these conditions. This effect leads to the additional increase in velocity observed at a definite field intensity and consisting in a second rise of the curve, coming, in the range of currents studied, in the region where the curve saturates. The initial portion of the curve of the velocity vs. field strength, corresponding to the effect of multiply charged ions on the spot re-formation time, is shown as a dotted line in the left-hand part of Fig. 94. As a result of the small number of multiply charged ions at weak fields, the re-formation time changes very little in this part of the curve, and has practically no effect on the form of the resulting velocity curve. If the discharge current is increased, the first of the functions that we have singled out undergoes no important changes, as shown by the very small changes in the initial portions of the curves before the second velocity increase starts. It must be concluded from this that the asymmetry in the distribution of the total field at the spot, which produces the motion, increases very little with increase in current under these conditions. In contrast to this, the role played by the second effect, namely, the reduction in spot re-formation time resulting from formation of multiply charged ions, increases rapidly with increase in current, as follows from the observed shift toward lower field intensities occurring in the second rise of the curves. It is obvious that as the current increases, there is a large increase in the relative number of multiply charged ions at the cathode spot, as shown further by the fact that the intensity of the spark spectrum increases at large currents.

It becomes obvious, in view of the effect that multiply charged

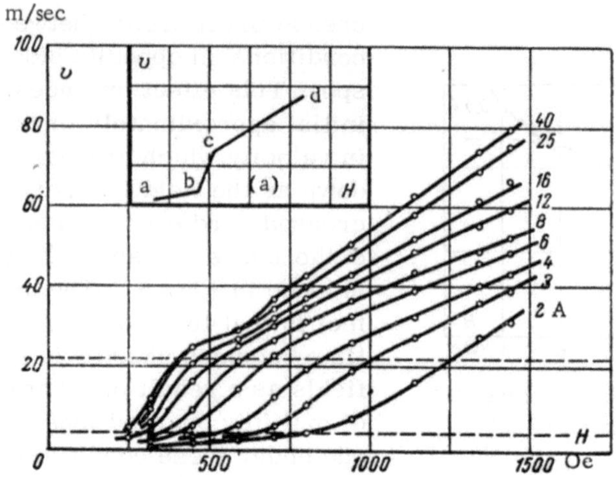

Fig. 95. Spot velocity on the uniform mercury cathode shown in Fig. 90 as a function of field strength for a number of currents.

ions are observed to have on the form of the relation between spot velocity and field intensity, that if the second effect were excluded, the curves in Fig. 92 would undergo saturation beginning at approximately 5 kOe. Here an important question arises: which of the two causes that we have been discussing theoretically does the saturation come from? It must be kept in mind that one of the reasons may be that the displacement s of the center of evaporation of the spot per re-formation cycle is limited by the small dimensions of the region and the rapid decrease in probability of the emission center leaving the region. The other cause of saturation may be the attainment of the greatest possible difference between the conditions on opposite sides of the spot when the external field and the inherent field of the arc are of equal intensity at the edges of the spot. If the saturation of the curves is due to the second cause, this means that the intensity of the inherent field of the arc, H_s, reaches about 5 kOe at the edges of the spot. In the opposite case it would be even higher. If these conclusions are in accordance with the facts, then from Eq. (67) the current density in the cathode spot reaches values of the order of 10^8 A/cm^2. We are immediately struck by the fact that this is exactly the order of magnitude of the current density nominally required in the cold-emission theory of the cold arc.

It is of course obvious that before we draw any far-reaching conclusions as to the mechanism of the arc, a careful check must be made to see whether or not there is any justification for measuring the current density in the cathode spot from the characteristics of the spot motion.

The relations noted above apply to conditions where the spot is moving around the cylinder in the tube shown in Fig. 89. Everything is somewhat complicated by random departures from the circular trajectory and by division of the spot when it is moving in a radial field on a uniform mercury cathode (see Fig. 90). The curves of velocity vs. field strength for these conditions are given for a number of currents in Fig. 95. The striking aspect of this graph is that the curves are not linear at velocities below 25 m/sec, which gives the curves the characteristic double bend shown schematically in the upper part of the figure. It is easily shown that the distortion in the form of the curves comes from the effect on velocity of the departure of the spot at random from the part of the cathode where the field strength is a maximum. This phenomenon may be clearly seen from the pictures (Fig. 96) taken at different field intensities. At low fields, the departures from a regular trajectory are very large, and thus for a certain part of the time the spot is in a region of the cathode where the tangential component of the field intensity is appreciably reduced, since the maximum occurs approximately above the middle of the ring gap between the pole faces (see Fig. 90). Since the forward motion of the spot on its curved trajectory is produced by the tangential component of the field intensity, any departure of the spot from the circle will reduce the rate of rotation. As the field intensity is increased, the random departures from the circular trajectory be-

Fig. 96. Pictures of the circular trace of the cathode spot in a radial magnetic field illustrating the reduction in random departures from the circumference as the field intensity is increased. 1) v = 2.5 m/sec; 2) v = 14 m/sec; 3) v = 36 m/sec.

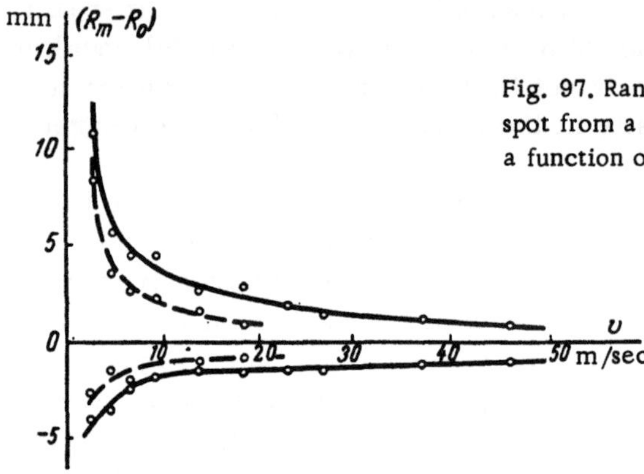

Fig. 97. Random departures of the spot from a circular trajectory as a function of spot velocity.

come smaller, as shown in Fig. 97. Here a quantitative measure of the extent to which the spot departs from a circular trajectory is provided by the difference in the quantities $(R_m - R_0)$, where R_m is the maximum or minimum radius of the circle drawn from the center of the cathode and bounding the region of departure of the spot from a perfectly circular trajectory of radius R_0. Departures toward the outside are considered positive, and in this case R_m is the radius of the circumscribed circle; departures toward the inside are considered negative, and in this case R_m is the radius of the inscribed circle. The solid curves in Fig. 97 give the results of an investigation of the departures at a current of 20 A; the dotted curves are for a current of 6 A. It follows from the graph that there is an obvious preponderance of departures toward the outside, due to the special properties of the field distribution over the ring gap between the pole faces. This causes an additional reduction in the rate of rotation of the spot around the circle, this being used to calculate the linear velocity. The random departures from the circular trajectory decrease regularly as the field intensity and the rate of forward motion are increased. They practically cease to exist at a velocity of about 25 m/sec, and beginning at this instant the width of the annular zone into which the cathode spot is compressed is determined by the process of spot division. The result of this reduction in random excursions of the spot is a large increase in velocity in the range 5–25 m/sec, which corresponds to the part bc of the curve in Fig. 95.

Another feature of the curves in Fig. 95 as compared with those in Fig. 92 is that the velocity of the spot is fairly sensitive

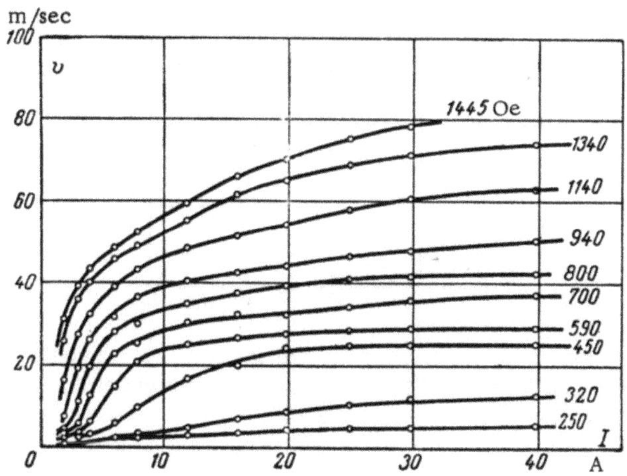

Fig. 98. Rate of motion of spot in a radial field as a function of current for different values of field intensity.

to the current. In order to show this in more detail, Fig. 98 gives curves of the velocity as a function of current for a number of field strengths. The initial portion of the curves in the range from 7 to 10 A shows a large increase in velocity with increase in current, after which the slope of the curves decreases and almost complete saturation sets in. These features of the relation between the spot velocity and the discharge current must come from peculiarities in the structure of the spot in an arc with a uniform mercury cathode. As a result of the large bond strength between the cells of the spot under these conditions, they form tightly bound groups which get larger as the current increases. The intensity of the arc magnetic field at the edge of the group will also increase, with the result that the spot interacts more strongly with the external magnetic field. There will, however, be some value of current at which the spot starts to divide into a number of autonomous spots. Each autonomous group of cells reacts with the external field almost independently of the other groups. Since the average current carried by each autonomous group does not increase any further with increase in discharge current, the rate of motion of the spot as a whole remains almost unchanged. Thus, it is possible to tell the structure of the cathode spot from the form of the characteristics giving the velocity at which the spot moves in a magnetic field.

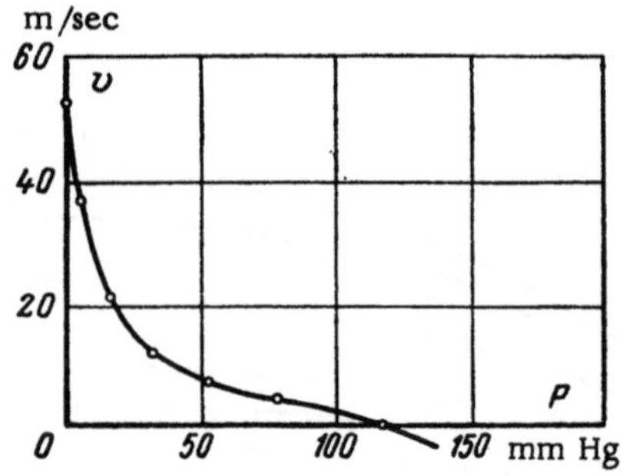

Fig. 99. Rate of motion of the cathode spot in a magnetic field as a function of helium pressure at a current of 2 A and a field intensity of about 2 kOe.

c. Rate of Motion of the Spot in a Magnetic Field as a Function of Density of the Gas Medium, and the Motion Inversion Phenomenon. Thus far, cathode-spot motion has been discussed under the natural conditions existing in a vacuum mercury arc which develops in the metallic vapor from the cathode at an equilibrium pressure corresponding to a temperature of about 20°C. In order to complete the picture, as well as to check the conclusions arrived at above as to the limits of applicability of the theory, we must further clear up the question of how the rate of motion of the spot is affected by an extraneous gas.

As a result of experiments made with various inert gases, it has been established, in accordance with the data of other authors [84, 89], that the velocity of the spot in a magnetic field always decreases if an extraneous gas is present, the more so the higher the gas pressure. The decrease in velocity of the spot motion when different amounts of helium are introduced into a tube of the type shown in Fig. 89 at constant currents and field intensities is illustrated by the curve of Fig. 99. The forward motion of the spot stops at a helium pressure of about 120 mm Hg. If the pressure is increased further, the spot begins to move in the opposite direction in accordance with Ampere's rule. Obviously the pressure p_i becomes a sort of boundary marking the limits of applicability of the theory.

In view of the importance of the question of the conditions under which inversion of the spot motion occurs in a gaseous

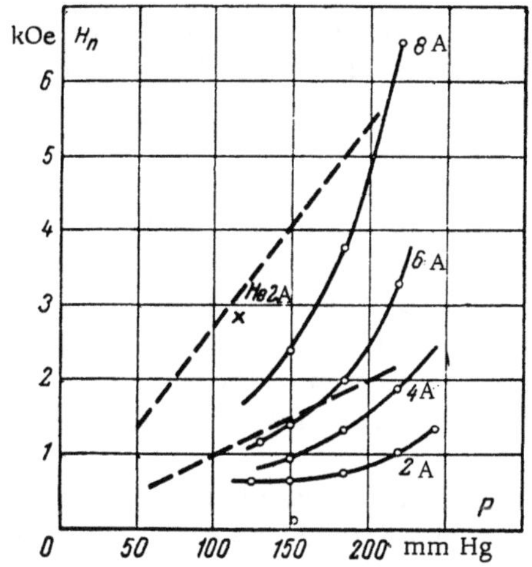

Fig. 100. Conditions for inversion of the cathode spot motion in the presence of xenon and helium.

medium, it is interesting to compare the values of p_i found experimentally with the theoretical predictions of the limiting conditions under which asymmetry in the magnetic field still affects the spot re-formation process, as derived earlier in Eq. (27). The results of a comparison for two gases are given in Fig. 100. The more detailed experimental data on the inversion of the spot motion are those for xenon. They are plotted as a family of solid curves showing, for a series of currents, the functional relationship between the field intensity and the pressure for the values of the parameters at which a change is observed in the direction of the spot motion around the cylindrical cathode (see Fig. 89). To the right and below each of the curves is a region where the spot executes a normal Ampere motion, while to the left and above lies the region of retrograde motion. The conditions under which inversion occurs depend on the discharge current. Increasing the current greatly broadens the region of normal motion at the expense of narrowing the retrograde motion region. The only datum on the graph for helium is given by the single point borrowed from Fig. 99 and shown by a cross at p = 120 mm Hg. In comparison with the curves for xenon, this reading was shifted a long way in the direction of the retrograde motion, which is somewhat unexpected.

In contrast to the solid curves, the sloping dotted lines give the theoretical predictions (27) of the limiting conditions under which the retrograde cathode spot motion may still be observed. The steeper of the lines is for xenon, and the other is for helium.

Although the actual curves depart appreciably from those expected, both the form of the experimental curves and the absolute values of gas pressure and field strength at which inversion of the spot motion is observed are quite close to those given by theory. Thus, the limits of applicability predicted for the theory developed to explain the motion of the spot in a magnetic field, as given by Eq. (27), are in accord with the facts, in their most important features.

In making a brief summary of the results of investigating the regular motion of the cathode spot in a magnetic field, it may be stated that all the material on the direction of motion, the velocity, and the conditions under which inversion occurs are completely in accord with the point of view developed in this book: that the regular motion of the spot is one of the forms of re-formation, occurring principally as a result of constant asymmetry existing in the total magnetic field in the cathode region of the arc.

B. DIVISION OF THE CATHODE SPOT AND INTERACTION BETWEEN AUTONOMOUS SPOTS

49. Fundamental Concepts

In the preceding section on the type of spot re-formation occurring during regular motion in a magnetic field, we were interested principally in the motion of the spot over the cathode as a whole, which did not require a detailed discussion of the make-up of the spot. In passing now to a study of spot division, we must first of all give the problem concrete form by determining how the spot is constituted.

a. The Form of the Cathode Spot and What Factors Determine It. The starting point for a theoretical determination of the configuration of the spot on a mercury cathode is provided by the maximum field principle, which has once again been verified by the phenomena observed in the regular motion of the spot. In applying the principle to the special problem of configuration we need only consider in what direction the process of forming and

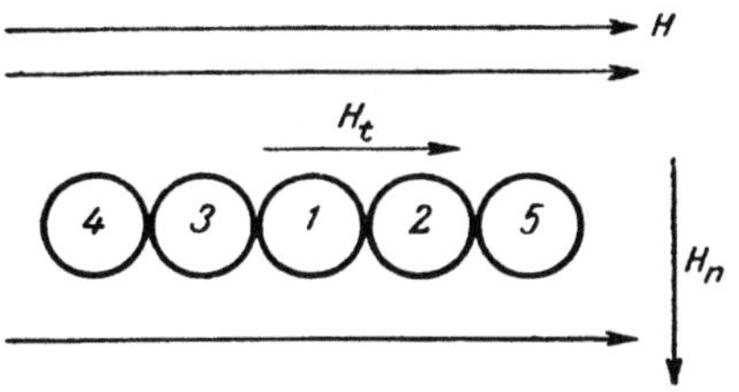

Fig. 101. Illustrating the process of forming and adding to the cathode spot.

adding to the cathode spot will occur as the discharge current is gradually increased.

As long as the current is close to the threshold value I_0 for an arc discharge on a mercury cathode, it will all be concentrated in a single cathode-spot cell, which we shall represent here as a circle with a uniform cross-sectional distribution of current density. If the current is increased to approximately twice the threshold value, a second elementary cell will be formed at the edge of the first cell, the form being similar to that of the first cell, as shown in Fig. 101. As the current is increased further, more and more new cells will be added to the group. Here in accordance with the maximum field principle, the cells will be arranged in such a way that the spot takes the form of an extended chain composed of cells in contact with one another. Actually, the inherent field intensity of the arc is always a maximum at the ends of the chain, so that each new pair of cells will be fastened to the ends, thus serving to make the chain longer. The field intensity at the edge of each of the cells may be found from the relation $H_i = 0.1\, i_1/a$, where i_1 is the mean current carried per cell and a is the radius of the cell. After n additional pairs of cells have been added to the original cell, the field intensity at the edges of the cells in the center of the chain may be written as the convergent series

$$H_t = \frac{0.1 i_1}{a}\left(1 + 2\sum_{n=1}^{k} \frac{1}{(2n)^2 + 1}\right), \tag{68}$$

while the field intensity at the ends of the chain under the same conditions is given by the divergent series

$$H_n = \frac{0.1 i_1}{a}\left(1 + 2\sum_{n=1}^{k} \frac{4n}{(4n)^2 - 1}\right). \tag{69}$$

Each term of the second series is greater than the corresponding term of the first series, and as the total number of terms is increased the difference $H_n - H_t$ rises rapidly. Thus, applying the maximum field principle to the process of forming and adding to the cathode spot leads to the conclusion that the spot has the form of a thin chain or line, the thickness of which is approximately equal to the diameter of a single cell. It is probable that the shape of the chain undergoes continual changes as a result of various types of rapidly varying forces acting on the cathode spot, as well as random factors. In the process it may bend, wind up, or break in different places. With all this, however, the spot maintains its generally extended configuration. These conclusions are completely in accord with Froome's observations [12] and thus there can be no doubt that the cathode spot on mercury has a long drawn-out shape.

In investigating spot division, much importance attaches to the observations made when the spot is executing regular motion in a magnetic field tangential to the cathode. It is important to establish in this connection whether or not such a field will affect the orientation of the chain, and what effect is to be expected. In answering this question, we can start with the fact that the orientation which the spot maintains with respect to the field vector is the one which is most stable. Obviously, the most stable orientation of the chain will be that in which there is the least difference between the conditions under which all the cells in the chain are existing in the external magnetic field. This condition corresponds to the orientation of the chain in the direction of the field vector, as was shown by Froome in the paper of his which we have already quoted several times.

b. Distribution of the Inherent Magnetic Field of the Arc at the Cathode Spot, and the Role Played by the Components of the Field. Before applying the maximum field principle to the problem of cathode-spot division, we must find the distribution of the inherent magnetic field of the arc in the vicinity of the spot. In approaching this problem, we shall neglect the distortions in form of the spot, possible in practice, where the chain is found to be curved. In Fig. 102, the cathode spot is shown schematically as a linear segment of length l symmetrically located at the origin of coordinates and extending along the X axis. The positive direction of the axis coincides with the intensity vector of the external field,

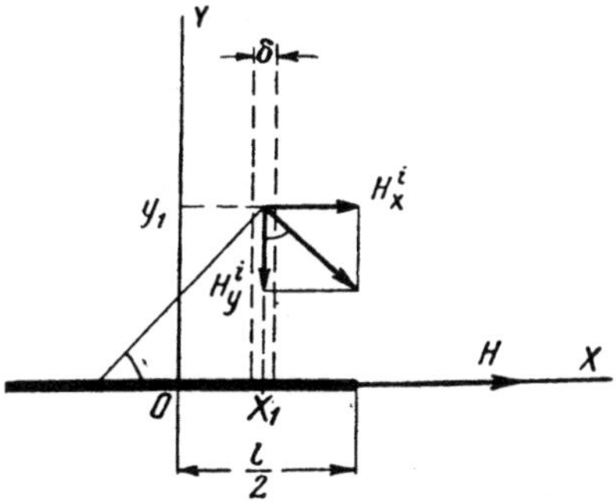

Fig. 102. Construction used in finding the inherent magnetic field of the arc at the cathode spot.

while the motion of the spot as a whole is along the Y axis. In order to simplify the problem of finding the inherent field of the arc in the vicinity of the spot, it is permissible to assume that the transverse dimensions are infinitely small in comparison to the length l. Outside the cathode spot the field of an actual arc is described with sufficient accuracy by this idealized model. Using this model, and introducing the concept of linear current density $j_l = I/l$, the components of the arc field at any point on the cathode with coordinates x_1y_1 may be found by summing the fields from the elementary linear currents $j_l\,dx$ passing through the cathode in the segment l:

$$H_x^i = 0{,}1 j_l \int_{-\frac{l}{2}}^{\frac{l}{2}} \frac{y_1 dx}{(x - x_1)^2 + y_1^2}; \tag{70}$$

$$H_y^i = 0{,}1 j_l \int_{-\frac{l}{2}}^{\frac{l}{2}} \frac{(x - x_1)\, dx}{(x - x_1)^2 + y_1^2}. \tag{71}$$

In determining the field "acting" at any point, we must, in summing the fields, neglect the currents passing through the cathode in the small region δ extending around the point. As long as the point in question is outside the line of emission this requirement is satisfied automatically, and the components of the "acting" field are the same as the values given above. If, however, the point of observation is inside the cathode spot $\left(-\frac{l}{2} < x_1 < \frac{l}{2};\ y_1 = 0\right)$, the integration must not be extended over the region of x from $(x_1 - \delta/2)$ to $(x_1 + \delta/2)$.

An analysis of the expressions given above for the components H_x^i and H_y^i of the inherent field of the arc shows that the two components play a fundamentally different role in the behavior of the cathode spot. To see why this is so, we must consider in detail what parts of the planar current in the chain produce the magnetic field in the narrow region lying between the parallel lines $(x_1 - \delta/2)$ and $(x_1 + \delta/2)$, shown as dashed lines in Fig. 102. If we integrate separately over the narrow segment δ and then over all the rest of the chain, and designate the result of integrating over the first segment by the symbol Δ, each of the field components may be represented as the sum of two terms:

$$H_x^i = 0.1 j_l \Delta_x + 0.1 j_l \left[\arctan \frac{(l/2 + x_1)}{y_1} + \arctan \frac{(l/2 - x_1)}{y_1} - \Delta_x\right]; \tag{72}$$

$$H_y^i = 0.1 j_l \Delta_y + 0.1 j_l \left[\frac{1}{2} \ln \frac{(l/2 - x_1)^2 + y_1^2}{(l/2 + x_1)^2 + y_1^2} - \Delta_y\right]. \tag{73}$$

Here Δ_x and Δ_y are given by

$$\Delta_x = 2 \arctan \frac{\delta}{2y_1}; \tag{74}$$

$$\Delta_y = \int_{x_1 - \frac{\delta}{2}}^{x_1 + \frac{\delta}{2}} \frac{(x - x_1)\, dx}{(x - x_1)^2 + y_1^2}. \tag{75}$$

We first find what current elements produce the component H_x^i. If the point y_1 approaches the chain, then, as $y_1 \to 0$, the second term in Eq. (72) decreases to zero, while the first term approaches $0.1\pi j_l$. This result means that at the edge of the cathode spot the tangential component of the inherent arc field is produced exclusively by the currents passing through the cathode right at that part of the chain. In other words, this component, by its very nature, cannot add anything to the interaction of different cells in the chain with one another. Since it is precisely the combination of this component with the external field that produces the large asymmetry in the resultant field intensity distribution at the edges of the spot, it is this component which makes it possible

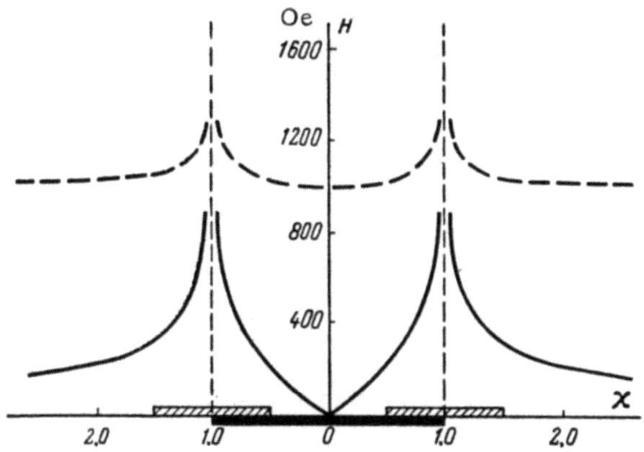

Fig. 103. Distribution of "acting" magnetic field along a chain of cathode-spot cells.

for the cathode spot to interact with an external or applied field, which causes the forward motion of the spot along the Y axis. The tangential component H_x^i at the edge of the spot is practically independent of the length of the chain of cells and, accordingly, of the current. This explains why the rate of the retrograde motion is only slightly dependent on the discharge current under the conditions corresponding to Fig. 92.

An entirely different role is played by the component of the arc field H_y^i normal to the chain. The quantity Δ_y in Eq. (73) is identically equal to zero, as is easily seen from (75). Hence it follows directly that the normal component H_y^i at any point x_1 is produced exclusively by the currents passing through the cathode in the region δ around the point. It is also obvious that this component cannot be responsible for the asymmetry in the resultant field occurring at the edge of the cathode spot when an external field is applied. Thus the component of the inherent arc field, H_y^i, normal to the chain is precisely the "acting" field which produces the interaction between the individual cells making up the cathode spot. Since the problem of the present part (B) consists in investigating the interaction between the various parts of the cathode spot, we shall be interested exclusively in the normal component of the arc field.

c. Irregularity in the Inherent Field of the Arc and the Role Which It Plays in Regular Spot Division. Effect of an External Field on Division. In view of having established above that the role of the component H_y^i is that of the field which produces the interaction between the different parts of the spot, if we are going to predict the behavior of the various parts of the spot, we must consider how the normal component of the field is distributed along the chain. Setting $y_1 = 0$ in Eq. (73) and introducing the new variable $\chi = x/(l/2)$, the field "acting" at the chain may be written in the form

$$H = 0.1 j_l \ln \frac{|1-\chi|}{|1+\chi|}, \tag{76}$$

which shows that the field intensity is a function solely of the fractional distance of the point in question from the center of the chain O, expressed in units of chain length. This fact gives Eq. (76) a universal character, so that it can be applied without change to a chain of any length. The distribution of the absolute values of the acting field intensity along the chain and outside it, as given by Eq. (76), is shown by the solid curve in Fig. 103. The distances from the center of the chain, χ, are plotted as abscissa, in relative units. The chain itself is represented by the heavy horizontal line symmetrically located about the origin. In the center of the chain, the field intensity is nearly zero. It increases rapidly from the center to the edge, reaching its maximum values at the ends of the chain. The fact that the function (76) is discontinuous at $\chi = 1$ is the result of simplifications that we have made consisting in neglecting the transverse dimensions of the chain in comparison with its length, which, however, are of no particular importance in view of the way in which we have defined the acting field. If we apply the maximum field principle to the distribution just found, we can expect that, since the cells move in the direction of increasing field, the chain will break into two parts, as shown explicitly by the cross-hatched strips. Since the field intensity at the ends of the chain is always higher than in the break which is formed, mutual repulsion of the two parts continues even after the spot divides.

In evaluating the experimental data given below on cathode-spot division under conditions where it is executing regular motion in an auxiliary magnetic field, much importance attaches to the question of what kind of an effect an external field can exert on the division process. In order to give a clear picture of how such a field acts, the heavy dashed line in Fig. 103 shows the result of combining the component H_y^i with an external field of 1000 Oe. In its most general features, the effect of the field is to increase the field strength along the whole length of the chain and to smooth out the difference between the conditions existing in the center and at the ends. The result of smoothing out the difference in this way is that spot division occurs with less energy when an external magnetic field is present. It can also be foreseen that an external field will reduce the probability of symmetric spot division, i.e., division into two equal parts. Apparently, a very strong field can completely suppress the division process. However, for this to occur, the external field must have an intensity many times greater than the intensity of the inherent field of the arc at the cathode spot.

50. Interaction Between the Autonomous Parts of the Cathode Spot Formed by Division

The autonomous groups of spots formed when the cathode spot divides should, after formation, move further apart as a result of the same factors which caused the spot to divide into parts in the first place. Accordingly, one of the most direct ways of seeing what division of the cathode spot consists of is to observe the interaction between the autonomous spots. It is then natural to turn first of all to this source of accurate information on spot division.

The same methods of continuous photographic scanning of the enlarged image of the spot as were used to investigate the structure of the spot may be used to make an experimental study of spot division. They have already been described briefly in Secs. 33 and 48. Of all the methods, the most complete information on the division process as well as on the interaction between autonomous spots is provided by photographing the spot directly on a stationary film, while the spot is executing rapid regular motion in an auxiliary magnetic field. This is true for the following two reasons. First, as has already been pointed out in the preceding section, the chain of cells is always oriented along the lines of force in the magnetic field, while the motion of the spot is normal to the lines of force. When a chain divides, the two parts separate in a direction normal to the direction of motion, so that the pictures can show all the acts of division, including those in which the separated parts of the spot cannot get further from one another during their lifetime than a distance comparable to their dimensions. This limiting condition on recording the division is actually the most sensitive of any criteria for division of the cathode spot that can be thought of in practice. Hence it follows that this method of investigation has the maximum possible resolving power. Second, using an auxiliary magnetic field permits a quantitative study of the interaction between the autonomous spots, as will be shown below. A drawback of this method of observation is that the auxiliary magnetic field required is capable of producing a certain amount of distortion in the division process. In order to evaluate the possible distortions, data obtained in this way must be compared with the results of observations made without using an auxiliary field, as is also done below.

Several typical pictures of the cathode spot showing how it behaves when it is moving regularly in a magnetic field of about 0.5 kOe are given in order of increase in current in Fig. 104. Each branching of the line in the figures corresponds to a single act of division, and the individual branches are the traces of the images of moving autonomous spots. Pictures of this type contain quite complete information on both spot division and the interaction between the parts of the spot. As a matter of fact, it is possible, by examining the form of the individual branches, to get some idea of the nature of the interaction between the autonomous spots and of how they are formed, while the length of the branches in the direction of scanning make it possible to find the duration τ of the spots. Further, by measuring the distances between starting points of neighboring branches, we can find the distribution of the time intervals t between successive acts of division. In the present section we shall be interested almost exclusively in the form of the branches.

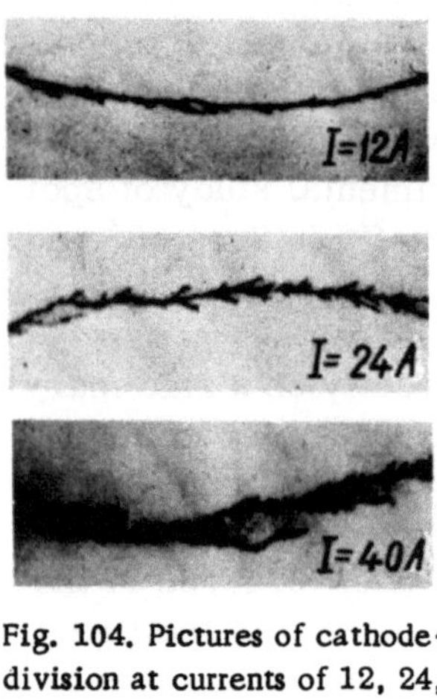

Fig. 104. Pictures of cathode-spot division at currents of 12, 24, and 40 A.

If we exclude from consideration the various sorts of irregularities in form of the branches due to random causes, there is one general

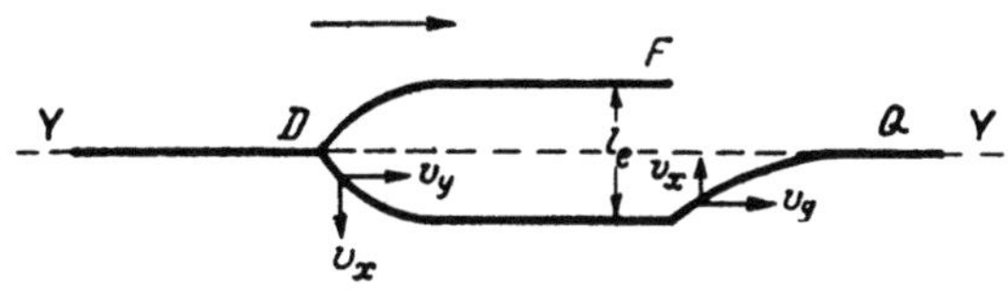

Fig. 105. Schematic representation of symmetric division of a cathode spot moving in a nonuniform magnetic field with the maximum intensity on the axis of symmetry.

governing principle that strikes the eye. The branches diverge steadily in opposite directions as they get farther from the starting point. As a rule, the angle between the branches is a maximum at the starting point and gradually decreases toward the end. This is shown particularly well in the case of symmetric branching, where the spot divides into two approximately equal parts. The trajectories of the autonomous spots in symmetric division are shown schematically in Fig. 105. Immediately after dividing at the point D, the two parts of the spot quickly begin to diverge from one another in a direction normal to the forward motion shown by the arrow, along the Y axis. After a while, their divergence becomes less rapid. If the magnetic field intensity is a maximum at the axis, and the vector H is oriented with respect to the cathode in such a way that the acute angle points inside toward the axis, sooner or later some equilibrium distance l_e is reached between the spots, after which they move parallel to one another. If one of them decays at the point F, the other gradually returns to the axis of symmetry of the field Y, reaching it at some point Q.

Our discussion of the trajectories of the autonomous spots in symmetric division proves beyond any doubt that a repulsive type of interaction exists between them. Whatever lies at the basis of the interaction does not cease to operate even when the two spots are moving parallel to one another on the segment DF, as can be seen from the behavior of the more stable of the spots, which is left alone on the segment FQ of the trajectory after the other spot has gone out. Obviously, in the first part of the trajectory the mutual repulsion between the spots is counteracted by the action of the external magnetic field, which forces them to move toward the axis of symmetry Y. In this case, l_e provides a quantitative measure of the interaction between autonomous spots, which may be used to aid in determining the nature of the interaction. From the fundamental considerations given above, it may be assumed with a considerable degree of certainty that the repulsion observed between the different parts of the spot after they divide is produced by the mutual magnetic fields of the currents concentrated in each of the two parts. But this assumption can be verified quantitatively if we determine theoretically the variation of l_e with the discharge current I and the field intensity H_0 on the axis of symmetry, using the general relation (47). All we have to do is set $\tan \varphi_m = 0$, which gives the equation

$$\frac{\partial H}{\partial x} - \frac{H_y}{H} \cdot \frac{H_i}{r'} = 0. \qquad (77)$$

In accordance with the conditions of the problem being solved here, we must in the present case understand the inherent field H_i to be the magnetic field of the current concentrated within the autonomous spot under discussion, while H must be defined as the sum of the field external with respect to the arc and the field of the second current. Expressing the values of H_y and H_i by the relations

$$H_y = \frac{0.1I}{2l_e}, \qquad H_i = \frac{0.1I}{2r'},$$

and using (50) with the values of the constants C_1 and C_2 found for this external field distribution, we easily obtain the approximate formula

$$l_e = \frac{0.1I}{r'H_0}. \qquad (78)$$

It must be remembered that, in Eq. (47), r' represents some effective radius of interaction between each of the autonomous spots in the magnetic field. Under the assumption that the direction of motion of the spot depends exclusively on the distribution of the absolute values of field intensity on the cathode surface, the values of r' for different currents were found from the trajectory of the spot, and may be expressed by Eq. (54). Actually, the direction of motion of the spot was found to depend not only on the field intensity distribution, but also on the angle at which the field vector was inclined to the cathode. Therefore, the values of r' found in this way must be regarded as too high. Nevertheless, since, for

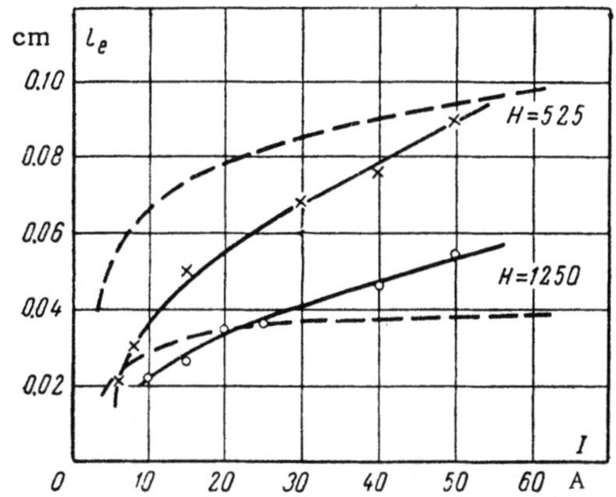

Fig. 106. Theoretical (dashed) curves and experimental curves showing the equilibrium distance between autonomous spots as a function of current for two values of field intensity.

a given field distribution, Eq. (47) actually gives a correct description of the motion of the spot for the values of r' found, it is permissible in the present case to use them without change.

It is to be expected from Eq. (78) that the equilibrium distance between the parallel branches on the photographic scanning pictures of the spot will increase with increase in current and decrease in field intensity, the dependence on the current showing saturation at high currents. These conclusions are completely confirmed by the pictures. More than that, the theoretical values of l_e are very close to those found experimentally, as shown by the curves of Fig. 106, which give the equilibrium distance between the branches as a function of discharge current for two values of field intensity. The experimental curves are shown by solid lines, the theoretical curves by dashed lines. The similarity in form and absolute values of l_e between the two sets of curves shows that, as was assumed, the interaction between the autonomous spots is caused by the magnetic field of the current located in the spots and is correctly described by the maximum field principle. But this type of interaction will be particularly large when the two interacting parts of the cathode spot are still in contact with one another, forming a single chain of cells. Hence it follows directly that the cause of division is the nonuniformity of the inherent magnetic field of the arc at the cathode spot, as we have concluded from fundamental considerations.

There is an interesting possibility of quantitative measurement of the current density at the cathode which suggests itself in connection with the role that has been shown to be played in spot division by the inherent magnetic field of the arc. It is easy to see how to do this if we recall that each one of the separate autonomous spots executes a complicated motion under the simultaneous action of two forces. One force is the interaction of each of the spots with the external field, which forces the spot to move in the direction of the Y axis. The other is the interaction with the field from the current of the second spot. The effect of this field is to move the spot in question in the direction of the X axis. In a uniform field tangential to the cathode, the derivative dx/dy at every point in the trajectory of one of the autonomous spots is equal to the ratio of the intensity of the field from the current in the second spot to that of the external field, i.e.,

$$\tan\varphi = \frac{dx}{dy} = \frac{0.1I/2}{2xH}. \tag{79}$$

This relation is the equation of the trajectory of the autonomous spots for symmetric division in a uniform field. It is easily shown that the trajectory is a parabola of the form

$$x^2 = \frac{0.1I/2}{H} y + C.$$

Here the maximum angle of divergence between the spots will come at the instant of division, when the distance between centers, 2x, is equal to the dimensions of each spot, as is in fact observed. By measuring the initial angle of divergence of the branches, $2\varphi_0$, it is a simple matter to find the field intensity of each of the spots at the point where the center of the second spot is located at the instant of division:

$$H_i = \frac{0.1I/2}{2a},$$

and along with this the current density j in the spot. As a matter of fact, after eliminating the spot radius a and the field H_i from the last relation by means of the equations

$$j=\frac{I/2}{\pi a^2}$$

and

$$\tan \varphi_0=\frac{H_i}{H}$$

we obtain the formula which correctly gives the desired quantity j in terms of the angle φ_0:

$$j=\frac{4\cdot 10^2 H^2 \tan^2\varphi_0}{\pi I/2}. \tag{80}$$

Quite clear pictures of spot division can be taken on metallic cathodes consisting of thin films applied by evaporating the metals in vacuum. At an external field intensity H = 1300 Oe and discharge current I = 4 A, the angle φ_0 in a number of metals is close to 45°, which leads to a value of j of about $1\cdot 10^8$ A/cm^2, using the method based on finding the variation of the rate of motion of the spot with field intensity.

By applying the maximum field principle to the problem of cathode-spot division, we can without difficulty explain a number of other laws that have been found to govern the behavior of the autonomous spots. Thus, it is easily shown that, in the unsymmetric type of division, the smaller current always goes with the larger angle of deflection of the branch. Although we never know in practice exactly how the current is distributed among the autonomous spots, this conclusion can easily be verified in its general features from the pictures, since a spot with a smaller current is always less bright, and its lifetime is shorter. It is also easy to understand why this is true. Actually, the magnetic field intensity will be higher at the spot which has the larger current concentrated in it. Under these conditions, the re-formation process in which one cell is replaced by another will develop in such a way that the number of cells in a larger spot increases at the expense of a smaller spot. The result is that the smaller spot goes out earlier. The time required to redistribute the current will be, in order of magnitude, quite close to the mean duration τ_1 of the individual cells. Under conditions where the spot is fixed, τ_1 is about 10^{-4} sec. As a result of the lower stability of the cells on a uniform liquid cathode, it is to be expected that under these conditions the type of re-formation in which the current is re-distributed among autonomous cells will proceed considerably more rapidly.

It can thus be foreseen that the mean duration of the autonomous spots on a uniform mercury cathode will be less than 10^{-4} sec. The result of unsymmetric spot division and the above general law, according to which the current is redistributed from the smaller autonomous spots to the larger, will be some statistical distribution of the duration times τ of the spots. How correct these ideas are will be shown by the results of the next section.

51. Distribution of Duration Times of the Autonomous Spots

In order to get a correct understanding of the laws governing the division process, we must have at our disposal data on the distribution of durations τ among the autonomous spots formed. As has already been pointed out, the autonomous spots formed during division of the cathode spot appear on the pictures as separate branches which usually diverge immediately after division. This fact makes them easier to observe. The problem of finding the duration distribution of the spots reduces to measuring the length of each of the branches in the direction of branching. Before we proceed with this, we must turn our attention to some of the general laws governing division

A preliminary qualitative study of the pictures taken at different currents and scanning rates reveals the following characteristic features of division.

In the range of currents studied, from 3 to 40 A, the trace of the cathode spot on the pictures is, in the great majority of cases, divided into not more than two main branches at the same time, while each of the branches can later undergo division again. This fact shows that the term "cathode-spot division" used in this book corresponds exactly to the nature of the process observed. One of the newly formed autonomous spots always decays after some length of time τ_1. In the majority of cases, the spot that decays carries the smaller fraction of the current, so that it may be called the secondary spot. The other one, which we shall call the fundamental or main spot, carries the larger fraction of the current, and maintains the arc until division occurs again. As the current is increased, the division process is complicated by the increase in duration of the secondary spots and the division which they undergo on their own, thus increasing the number of spots existing on the cathode at one time.

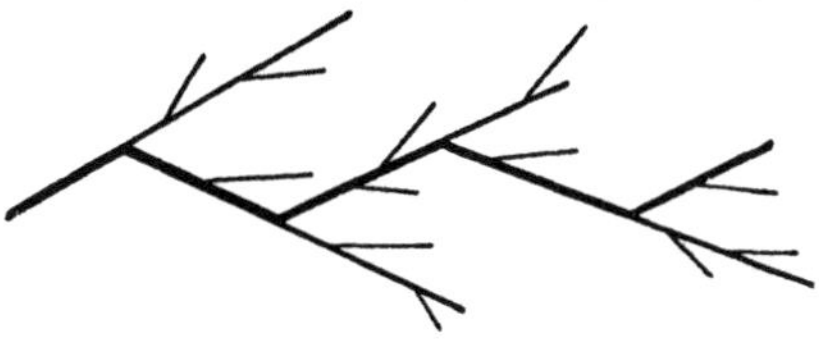

Fig. 107a. Scheme of cathode-spot division.

These characteristic features of spot division are shown schematically in Fig. 107a. Here each act of spot division corresponds to the splitting of the line into two branches. This is essentially an idealized picture of the image of the spot obtained by photographic scanning. The main spot, which is responsible for the prolonged existence of the arc, is shown by the somewhat heavier lines. The finer straight-line segments are the secondary, decaying spots. The finest lines show small autonomous spots formed by additional division of the main and secondary spots, but of short lifetime. The end of each of the lines corresponds to disappearance or "decay" of the autonomous spot giving the image. At small currents, this additional division of the main and secondary spots may be neglected, because of the short lifetime of the spots thus formed. Thus, the division scheme shown in Fig. 107a may be replaced by the simplified scheme in Fig. 107b. Since the data given below are for moderate currents, all of our subsequent discussions will be based on the simplified spot-division scheme.

a. Method and Observational Errors. The Criterion for Spot Division. The material presented below and in the next section was obtained by measuring the lengths of two types of segments on the pictures of the moving image of the spot; the method of taking the readings is shown schematically in Fig. 107b. The first type of quantity measured was the projection ξ_i of each branch on the direction of scanning, while the second type of quantity was the projection s_i of the distance between the starting points of successive branches. Knowing the scanning rate v_f, these measurements could be used to find the duration τ_i of each of the autonomous spots and the time interval t_i between successive acts of division, as given by the relations

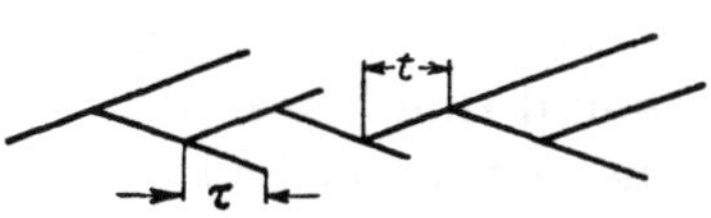

Fig. 107b. Simplified scheme of spot division.

$$\tau_i = \frac{\xi_i}{v_f}; \qquad t_i = \frac{s_i}{v_f}. \tag{81}$$

The results of the measurements were tabulated as a series of fractions τ_1/t_1, $\tau_2/t_2, \ldots, \tau_n/t_n$, where the numerator was given by the duration of the isolated spots in chronological order and the deonominator was the time interval between successive divisions. This method of making and tabulating the measurements permitted subsequent use of the method of artificial statistical selection of observations, which has led to a number of important conclusions.

Before we present the results of the measurements, we must evaluate the errors made in the measurements, and describe the criterion used to define division of a spot. The principal source of error, in comparison with which all other errors of measurement could be neglected, is the conversion from the distances to time intervals by Eqs. (81). The reason for this is the difference between the mean scanning rate of the spot image, v_f, used in the conversion, and the true rate of motion v_i of the autonomous spot observed. Replacing the true value by the mean scanning rate is nevertheless necessary since there is no accurate information on the values of v_i. The nature and magnitude of this error show up in different ways in the two scanning systems used, with the result that it must be evaluated separately for each system. In photographing a spot moving around a circular trajectory in a radial magnetic field, the mean scanning rate is given by $\mu(2\pi R/T_0)$, where μ is the magnification of the image, R is the radius of the circular trajectory, and T_0 is the period of rotation of the spot. This rate is actually the result of averaging the instantaneous velocities v_i of the image of the main spots on their way around the closed circumference, with the result that we have the relation

$$v_f = \frac{1}{T_0} \sum v_i \Delta t_i.$$

The summation extends over a single circuit of the spot around the circumference, and Δt_i is the small time interval during which the rate of motion may be considered constant. The instantaneous velocities v_i depend on the current concentrated in the main spot. It may be assumed that the velocity changes in proportion to the current. The current in the main spots can vary from some maximum value i_m to approximately half this value, $i_m/2$. It follows from this that the instantaneous velocities of the main spots lie within the limits

$$v_m > v_i > \frac{v_m}{2},$$

where v_m is the maximum possible velocity, corresponding to the largest amount of current in any one spot. With the relation existing between v_i and v_f, v_f will also lie between the above limits. Accordingly, the maximum error introduced in determining the intervals of type t_i, resulting from replacing the instantaneous values of the velocity v_i by the mean value v_f, will in any case not be greater than 100%. Errors of this order will only occur in rare cases and will not play any large role in the conclusions drawn from the statistical treatment of the material.

Matters are somewhat worse when making an accurate measurement of intervals of type τ_i. In contrast to the main spot, the secondary spots corresponding to the decaying branches in the pictures can be carrying a current considerably less than $i_m/2$. Here the velocity of the decaying branches can differ by a larger amount from the mean scanning rate, with the result that the systematic error in finding the intervals of type τ_i is on the low side and can reach 100% or more.

These errors are of a somewhat different nature in mirror scanning of the image of a spot moving at random over the surface of a uniform mercury cathode with no external magnetic field. In this case, v_f is the rate of scanning of the image of any stationary point on the object, which is constant during the time the observations are being made, and is known to a high degree of accuracy. An error in determining the time intervals can only occur as a result of the inherent motion of the autonomous spots. If the maximum rate of the motion is v_s' and the magnification of the optical system is μ, the true velocity with which the image of the spots moves on the film will lie within the limits $v_f \pm \mu v_s'$. As will be seen from what follows, v_s' is of the order of magnitude 10^2-10^3 cm/sec. For $\mu = 3$ and the frequently used scanning rate of $5 \cdot 10^3$ cm/sec, the maximum error in determining the intervals t_i and τ_i with mirror scanning is of the order of $\pm$ 60%. An additional analysis of spot motion shows that with this type of scanning, the error of measurement will, in fact, be much less than the maximum value given above.

In studying cathode spot division, we have to deal not only

with errors in measuring the time intervals, but with uncertainties of another sort which are of a more fundamental character. As will be seen from what follows, the results of the measurements depend to a considerable extent on what criterion is chosen to define division of the spot. The reason for this is that the duration of the autonomous spots formed by division varies over a wide range of values, starting at the very small values at which one of the spots decays before it can get away from the other spot by a distance greater than its own dimensions. Under these conditions lack of a sensitive enough criterion of what constitutes division can result in a large number of observations being missed, which gives a false conception of how the division works. For the investigation to be as complete as possible, we must select the most sensitive criterion of division provided by the method used. In the present work, the criterion for division used was that the separate parts of the cathode spot diverge by an amount comparable to their own dimensions. This is apparently the most sensitive of all physically meaningful criteria for division. Actually, from the point of view of the physics of the cathode spot in our model, only those spots may be considered autonomous which have both their own space-charge region and a separate center of evaporation for the cathode material. This definition is based on recognition of the fact that evaporation of the cathode is one of the fundamental processes in the arc cycle. But the bright luminescence which we photograph and call the cathode spot is in all probability nothing more nor less than a region of intense mercury evaporation. From this it follows that, as long as the distance between the two parts of the radiating region remains less than the dimensions of the radiating parts themselves, it is, by the nature of the case, impossible to call them autonomous spots. In the range of currents investigated, from 3 to 40 A, the minimum distance required to record a spot division varies from $2 \cdot 10^{-3}$ to $2 \cdot 10^{-2}$ cm, according to the actual dimensions of the chain of cells, as given by Eq. (42). With the multiplication by a factor of six used in the pictures, the smaller of the above values corresponds to a distance between branches on the pictures of only about 0.1 mm, which is actually at the limit of resolution. Thus, using this minimum amount of divergence as the criterion for autonomous spots corresponds for practical purposes to the resolving power of the method of observation used.

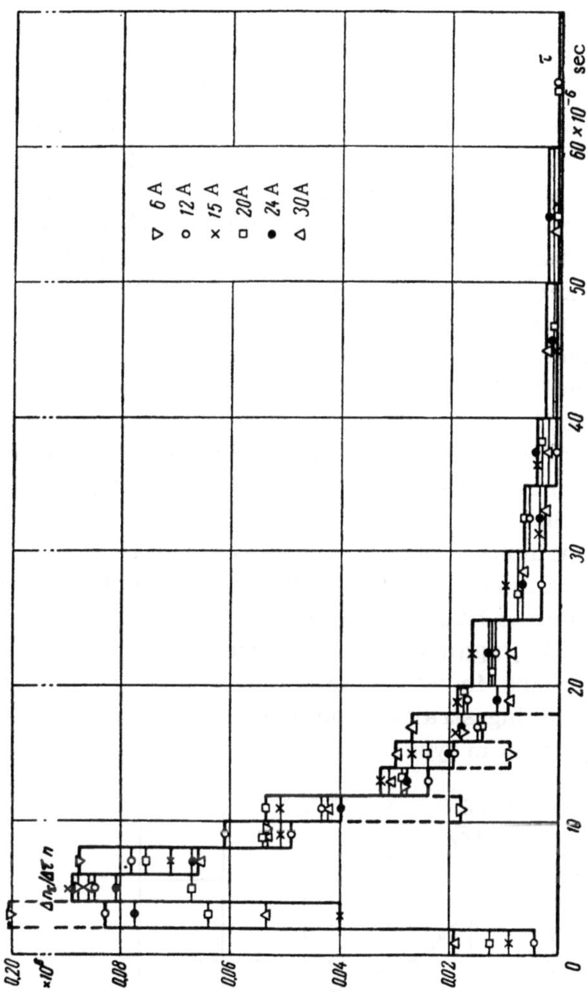

Fig. 108. Distribution of autonomous spots in duration τ for a number of currents with directed motion.

b. Results of Measurements during Directed Motion of the Spot in a Magnetic Field. For a statistical treatment of the durations τ_i of the autonomous spots, the range of times was divided into a series of intervals $\Delta\tau$ and then the number of readings Δn_τ falling within each of the intervals was found. The results of grouping the observations in this way for a field intensity of about 1 kOe and various values of current are shown in Fig. 108. The values of spot duration τ are plotted along the horizontal axis, while the relative number of readings in each spot age group, $\Delta n_\tau/n$, divided by the auxiliary interval $\Delta\tau$ is plotted along the vertical axis. The histograms for different currents are shown by points of different shape. The total number of observations n used in constructing the individual histograms varied from 230 to 480. The only exception is for a current of 6 A, for which the number was less than 100.

What strikes the eye in Fig. 108 is that the individual histograms are irregular in shape. These irregularities come from statistical fluctuations, which are perfectly natural in making a limited number of observations. The distortions become less noticeable if we combine the histograms for different currents. This clearly reveals that there is a common distribution of the duration $V(\tau)$ of the autonomous spots for all currents. The distribution may be expressed by a monotonically falling curve which is nearly exponential. In order to find the function $V(\tau)$, we have, in Fig. 109, plotted $\log(\Delta n_\tau/\Delta\tau)$ against τ for a current of 30 A. In the 5-40 μsec range, the experimental points are grouped with a certain spread about a sloping straight line which corresponds to a distribution of purely random type:

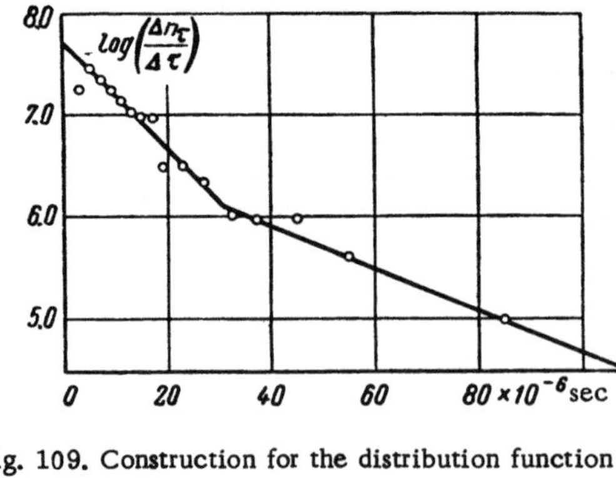

Fig. 109. Construction for the distribution function $V(\tau)$.

$$\frac{dn_\tau}{d\tau}=n_0\Omega e^{-\Omega\tau}. \tag{82}$$

In the large time range, the points are too high, which shows that there is an excess of spots having long duration. They are, in all probability, also located on a straight line, which, however, has a somewhat smaller slope. This may mean that the long-lived spots form a special group having their own special random distribution with some smaller value of the parameter Ω. In contrast with long times, at short times the deviations of the points are definitely on the low side, which shows that there is a considerable deficit of spots with a short lifetime. This fact must have something to do with the missed observations already mentioned. Actually, a deviation is observed at values $\tau \leq 3\cdot10^{-6}$ sec, which corresponds to the threshold of sensitivity of the method of observation. In view of the obvious nature of the relation between the deviations at short times and missed observations, it must be assumed that the desired distribution for each of the two groups of spots is actually given by Eq. (82); i.e., it is purely random. A distribution function of this type is uniquely determined by the parameter Ω, which is the reciprocal of the mean duration of the spots. By comparing (82) with the straight line in Fig. 109 we can find not only the value of the parameter Ω, but at the same time calculate the number of observations missed as a result of the short dura-

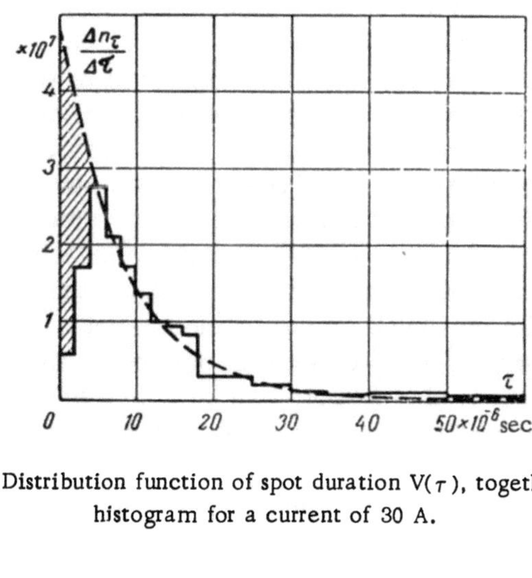

Fig. 110. Distribution function of spot duration $V(\tau)$, together with the histogram for a current of 30 A.

tion of the parts of the dividing spots. By taking the logarithm of Eq. (82) we can see that $\log(n_0\Omega)$ is numerically equal to the ordinate intercept of the line in Fig. 109, while Ω is found from the slope of the line by the relation

$$\Omega = \frac{\log(n_0\Omega) - \log\left(\frac{\Delta n_\tau}{\Delta\tau}\right)}{\tau \log e}. \quad (83)$$

The value of $\log(n_0\Omega)$ is equal to 7.675, which gives the desired quantities separately:

$$\Omega = 1.19 \cdot 10^5 \ \mathrm{sec}^{-1};$$
$$n_0 = 398.$$

From the meaning of the quantity n_0 in (82), the last figure corresponds to the total number of side branches on the pictures of a given series of observations, independently of duration. The number of branches n actually observed in this series of readings was 315. The number of readings missed, n', is the difference:

$$n' = n_0 - n,$$

i.e., 83 misses, or about 21% of the total number of branches expected.

The function found for the distribution of spot duration τ for a current of 30 A is shown as a continuous curve in Fig. 110, together with the histogram for this current. The number of branches not included in the histogram is equal to the cross-hatched part of the area, which is the difference between the areas bounded by the smooth curve and the histogram. It is useful to know this when analyzing the distribution of the intervals between successive spot divisions, as discussed below. However, since the criterion for spot division used in the work is the same as the threshold of sensitivity of the method of observation, not all these 83 missed branches should be recorded as autonomous spots, so that they cannot be truly classified as misses in observing the division of the cathode spot. Only those spots can be considered separate that have a duration lying outside the line drawn to the point $\tau = 3$ μsec.

At first glance, the distribution of the duration of the autono-

TABLE XIII

I, A	6	12	15	20	24	30
$\bar{\tau}$, sec $\cdot 10^{-6}$	6.8	11.1	12.5	12.9	14.2	13.4

mous spots seems to be the same for all the currents studied. A more careful analysis of the histograms of Fig. 108 shows that the values of the ordinates drop off somewhat more steeply with increase in τ at low currents. The amount of difference visible between the histograms for low and high currents obviously does not correspond to the large differences which strike the eye when making a visual comparison of the pictures for the two cases (Fig. 104). This is due to the fact that in visual observation one's attention is fixed principally on the long branches, the importance of which, however it may seem, is actually extremely small. Accordingly, the appearance of long-lived spots at large currents produces only a barely appreciable rise in the curve for large values of τ, and thus produces only a small increase in the mean spot duration $\bar{\tau}$, as illustrated by the figures in Table XIII. Since these figures are practically constant, there is no great loss of accuracy if we take an overall mean value $\bar{\tau}$ for all the range of currents studied. The values of all the quantities derived from the measurements are given in Table XIV. It must be explained that the values given in the first three columns are for the distribution $V(\tau)$, found for the basic group of spots, and is approximately the same for all the currents investigated. Here the value of $1/\Omega$ must be interpreted as being the mean lifetime of the spots in the distribution, including those which fail to satisfy the criterion for autonomous spots. In contrast to this, the fourth and fifth columns give values of the parameter Ω' and its reciprocal, derived for the long-lived group of spots from the data for a current of 30 A. The value

TABLE XIV

$n_0\Omega$, sec^{-1}	Ω, sec^{-1}	$1/\Omega$, sec	Ω', sec	$1/\Omega'$, sec	$\bar{\tau}$, sec	δn	$\delta n/n_0$
$4.73 \cdot 10^7$	$1.19 \cdot 10^5$	$8.4 \cdot 10^{-6}$	$0.46 \cdot 10^5$	$21.6 \cdot 10^{-6}$	$13.4 \cdot 10^{-6}$	83	0.21

of τ in the sixth column is the overall mean lifetime of the autonomous spots recorded at all currents. Finally, the last two columns give the absolute and relative number of spots not counted because of their short lifetime.

In using the data of the tables we must keep in mind that there can be a large error introduced in the transition from distances measured on the pictures to time intervals. The conclusion that there are two groups of spots with a different distribution $V(\tau)$ may, in particular, be a result of this error. Considering the nature of the error, the data given on the mean duration of the autonomous spots must be regarded as somewhat too low. With all this, it is beyond doubt that the falling characteristic of the distribution curve $V(\tau)$ which we have found does not arise from observational errors, but truly reflects the law that is operating.

c. Results of Measurements in the Absence of an External Field. The data for the preceding point are for the artificial conditions under which the arc is subjected to the effect of a magnetic field. As has already been pointed out, the observations made under these conditions are of interest in view of the fact that they provide a very complete picture of the autonomous spots and make possible a quantitative study of the interaction between them. Along with this, the auxiliary magnetic field used in the observations can introduce distortions into the spot division and the way in which the current is continually being redistributed among the autonomous spots formed. In order to be sure that the same phenomena occur in the absence of a field, and to determine the degree and nature of the distortions which the field introduces, we must consider the behavior of the spot under the natural conditions in the arc, where it is not perturbed by a magnetic field of external origin.

Figure 111 gives the results of a statistical treatment of three series of observations, made at discharge currents of 5, 10, and 15 A, in an arc where the spot was moving at random over the mercury, using the mirror scanning method. As in Fig. 108 discussed previously, the abscissa in Fig. 111 is the spot duration time, while the ordinate is the relative number of spots in the different age groups, referred to the length of the auxiliary intervals. Each of the three currents has its own histogram. The number of readings used to construct the histograms were 557, 597, and 128 respectively. As in the case where the cathode spot is undergoing

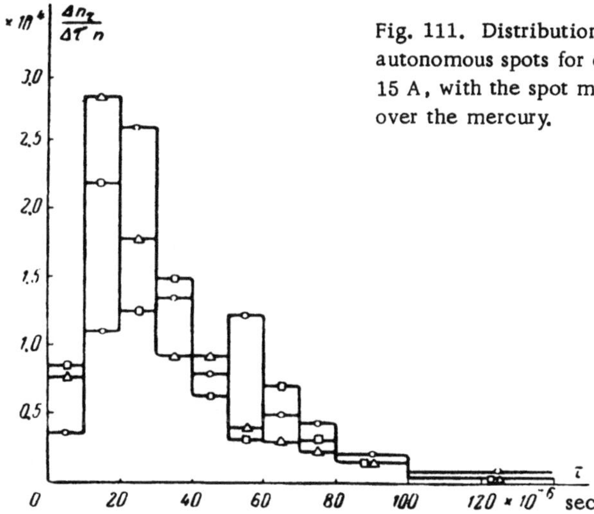

Fig. 111. Distribution of the duration of autonomous spots for currents of 5, 10, and 15 A, with the spot moving at random over the mercury.

forced motion in a magnetic field, the distribution and duration of the separate spots is of the same form for all currents. When the spot is executing random motion, the distribution function may also be expressed by a monotonically decreasing curve of exponential type. In order to find the function, a construction has again been made in Fig. 112 similar to that in Fig. 109. Here use was made of the extremely complete data from the series of observations made at a current of 10 A. As before, in the intermediate range of times, extending this time approximately from 10 to 80 μsec, the experimental points fall on a sloping line, showing that the distribution of the spots in duration is purely random. For large times, the points are again above the line, which cannot in this case be explained by errors in measurement or in converting

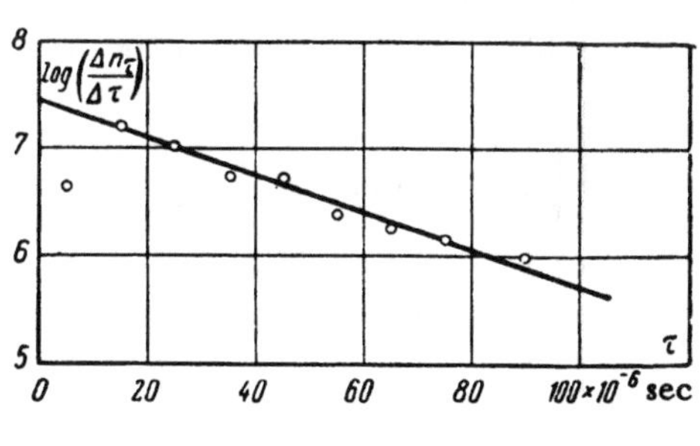

Fig. 112. Construction used in finding the distribution function $V(\tau)$ for random spot motion.

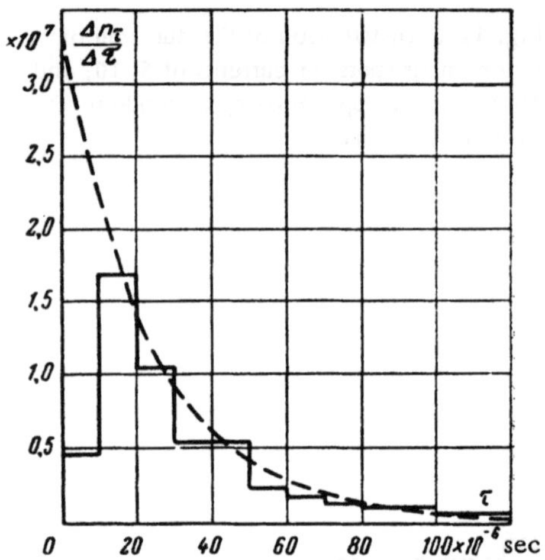

Fig. 113. Distribution function V(τ) and histogram for a current of 10 A with the spot moving at random over the mercury.

distances to time intervals. They are grouped along a sloping line which this time also corresponds to a lower value of the parameter Ω. Having such complete qualitative agreement between the results of observations made by two different methods with sources of error that are different in nature and absolute value shows that the two groups of spots with a random distribution of lifetimes, as already noted, do in fact exist. By analogy with the case of forced motion of the cathode spot in a magnetic field discussed previously, the departure from the random type of distribution at small times must be ascribed to failure to record spots with short duration. The values of the parameters Ω and Ω' for the two distribution groups, as well as the absolute and relative number of spots missed in this series of observations, may be found by the method already described using the distribution function V(τ) shown in Fig. 113. All these values, together with the mean durations $\bar{\tau}$ of the autonomous spots found for each current by direct measurement, are given in Table XV.

A comparison of the figures in Tables XV and XIV shows that all the values of the mean spot duration times derived from observations made under the natural conditions of random motion with no magnetic field are several times greater than the values of the same quantities derived from experiments in which the arc

TABLE XV

I, A	$n_0\Omega$, sec^{-1}	Ω, sec	$1/\Omega$, sec	Ω', sec^{-1}	$1/\Omega'$, sec	$\bar{\tau}$, sec	δn	$\delta n/n_0$
	1	2	3	4	5	6	7	8
5						$61.5 \cdot 10^{-6}$		
10	$3.39 \cdot 10^{-7}$	$4.41 \cdot 10^{4}$	$22.7 \cdot 10^{-6}$	$1.28 \cdot 10^{4}$	$78 \cdot 10^{-6}$	$68.2 \cdot 10^{-6}$	171	0.223
15						$96.6 \cdot 10^{-6}$		

is subjected to the action of a magnetic field. This difference may to some degree reflect an actual reduction in duration of the autonomous spots due to the suppression of the division process and reduction in the probability of symmetric spot division by the action of the magnetic field. However, it must to a considerable extent be ascribed to the effect of the systematic errors already mentioned, in the direction of lowering the time intervals, made in observations on directed motion of the spot in a magnetic field when converting the distances to time intervals. This error may be considered wholly responsible for the shift toward small times of all parts of the curve in Fig. 110 as compared to the curve in Fig. 113. As far as the mean values $\bar{\tau}$ in the last table are concerned, they will be appreciably raised by the failure to record spots having short duration. Considering these sources of error, we can conclude that the reduction in mean lifetime of the autonomous spots in an external magnetic field, if there is one, is in any case not greater than a factor of two, and may be disregarded.

In giving a short summary of the results of investigating the distribution of the duration of the autonomous spots, we must note the following. Both when the spot is executing regular motion in an external magnetic field and when the arc is operating under natural conditions with no perturbing field, the distribution of the lifetimes of the autonomous spots formed in division is given by a monotonically decreasing function, consisting of two exponential segments of different slope. This shows that there are two groups of spots with purely random lifetime distributions in each group. The mean lifetime of the less long-lived group is of the order of magnitude of 10^{-5} sec, while for the more long-lived group it approaches 10^{-4} sec. The relative number of spots in the long-lived group increases with increase in current; this is related to the small increase in mean lifetime of the spots as a whole which is observed when the current is increased. It is very probable that the long-lived group is made up of spots formed by symmetric division of a spot into two approximately equal halves. Under these conditions redistribution of current among the spots will proceed slowly, as a result of the symmetric conditions existing at the boundaries. The values found for the mean lifetime of this group of spots are the same in order of magnitude as the mean lifetime for single spots (see Sec. 36), which is in accordance with the idea of re-formation of the type where the current is redistributed among autonomous spots. According to this idea, the random lifetime distribution of the spots is a result of the random distribution of the number of cells in the autonomous spots, and of the relation existing between the rate of distribution of current among the spots and the ratio of the number of cells in the groups.

52. Distribution of the Intervals Between Successive Divisions. Frequency of Division of the Cathode Spot

One of the principal problems in investigating spot division is the determination of how often division occurs and whether it is a systematic process or whether it obeys only the laws of chance. A direct answer to these questions can be found from the results given below for a study of the distribution of the intervals between successive spot divisions, W(t). In order to follow the plan of presentation that we have adopted, we must give separate consideration to the results obtained from the two different methods of observation already described.

a. Spot Division with Directed Motion in a Strong Magnetic Field. The data from statistical treatment of the observations made with forced motion of the spot in a magnetic field of about 1000 Oe are shown as a series of histograms in Fig. 114. The time intervals t between successive divisions are plotted along the abscissa; the relative numbers of readings within each segment on the axis of abscissas is plotted along the ordinate. When constructed in this way, the area bounded by any portion of the histogram between the times $t - \Delta t/2$ and $t + \Delta t/2$ is the empirical probability that the cathode spot will divide within this length of time after the preceding division. The various histograms in Fig. 114 are for different currents, decreasing from top to bottom, as shown in the upper right-hand corner of each of the graphs. Given along with the current are the total number of readings, n, used in constructing the histograms and the arithmetic mean values of the intervals between successive divisions, $\bar{t}$, calculated for the series.

The most characteristic property of the distribution is that there is a clearly defined maximum, with the measured intervals predominantly concentrated around it. At the highest currents investigated, the maximum is located in the range 6-7 μsec. As the current is decreased from 30 to 6 A, there is a regular shift in the maximum to approximately 12-15 μsec, with the statistical spread in the intervals increasing at the same time. The only distribution that does not fit into the series is the one for 12 A, which may be due to random errors. The relation involved is obscured by random fluctuations in the grouping of the intervals, which greatly affect the shape of the various histograms. In order to reduce the effect of fluctuations in the calculations there is some point in finding the quantitative characteristics of the distribution W(t) by operating with mean values of the intervals, as well as with

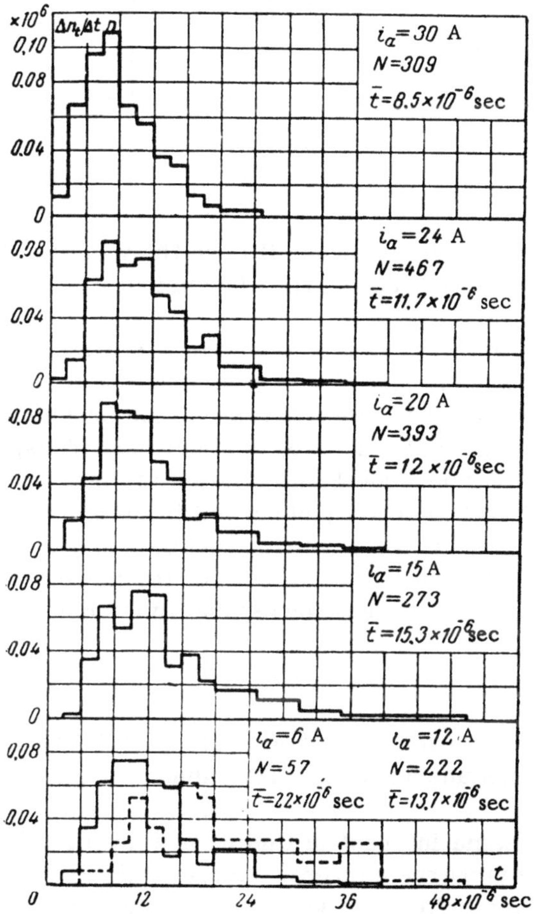

Fig. 114. Distribution of intervals t between successive divisions for a number of currents when the spot is executing ordered motion in a magnetic field of about 1 kOe.

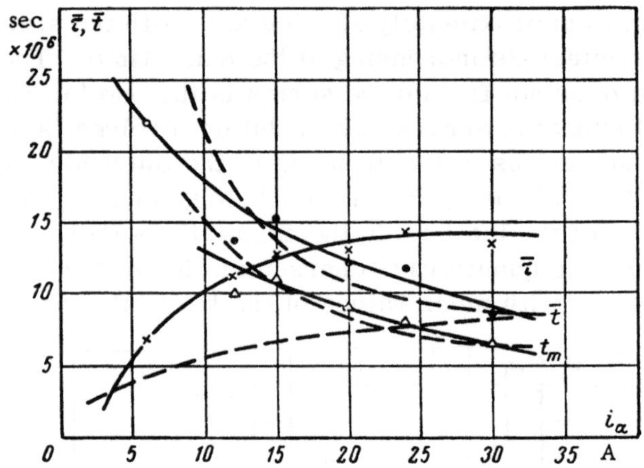

Fig. 115. The mean values $\bar{t}$, t_m, and $\bar{\tau}$ as functions of the current.

the values of spot-division probability in the vicinity of the maximum, which give some idea of the dispersion of the distribution curve.

With even a cursory glance at Fig. 114, we are struck by the fact that the great majority of intervals occur approximately in the range 4-16 μsec. At the higher currents, the probability of spot division inthe range $W(4 > t > 16)$ is about 0.8. Thus, of every five divisions, four occur 4-16 μsec after the last division of the spot. In view of the selectivity observed in the intervals, there is some point in introducing the concept of the cathode-spot division frequency f, defined as the reciprocal of the mean arithmetic value of the interval between successive divisions:

$$f = \frac{1}{\bar{t}}. \tag{84}$$

It follows from the data of Fig. 114 that the spot-division frequency reaches unexpectedly high values, of the order of 10^5 $\sec^{-1}$.

There is some interest in finding how the distribution changes with change in discharge current. To this end, we have plotted in Fig. 115 the mean arithmetic values of the intervals $\bar{t}$ and the most probable values, t_m, corresponding to the maximum on the histograms, together with the durations $\bar{\tau}$, as a function of current

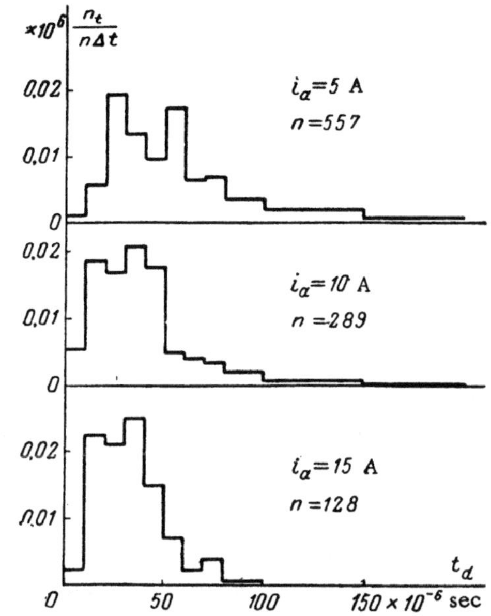

Fig. 116. Distribution of intervals between successive spot divisions under conditions of its chaotic motion at the cathode at currents of 5, 10, and 15 A.

(solid curves). The mean values of the intervals increase monotonically as the current is decreased, while the value of $\overline{\tau}$ decreases monotonically. Complete parallelism is observed in the change in the mean values of t, the ratio t/t_m being approximately equal to 1.4.The value of this ratio may be used as a measure of the asymmetry of the curve giving the distribution of the intervals. In the present case, the asymmetry is somewhat greater than that of the Maxwell velocity distribution curve. It follows from the data of Fig. 115 that, as the current is reduced from 30 to 6 A, the spot-division frequency decreases approximately from $1.2 \cdot 10^5$ to $0.5 \cdot 10^5$ sec^{-1}.

The spot-division frequencies for regular motion in the magnetic field, calculated from the observations for each of the currents studied, are shown in Table XVI along with the probabilities of division of the spot in the range 4-16 μsec after the last division.

b. Division of the Spot with No External Field. In order to eliminate any possibility of the magnetic field having an effect on spot division as one of the unknown factors, and to be sure that the results described above are correct, we must give additional consideration to the data on the interval distribution W(t) for the normal conditions in an arc which is not being acted upon by a

TABLE XVI

I, A	6	12	15	20	24	30
f, sec^{-1} W ($4<t<16$)	$0.45\cdot10^5$ 0.30	$0.73\cdot10^5$ 0.74	$0.65\cdot10^5$ 0.67	$0.83\cdot10^5$ 0.78	$0.85\cdot10^5$ 0.78	$1.18\cdot10^5$ 0.78

TABLE XVII

I, A	5	10	15
$\bar{t}$, sec f, sec^{-1} W ($5<t<20$)	$30.7\cdot10^{-6}$ $0.32\cdot10^5$ 0.39	$17.5\cdot10^{-6}$ $0.57\cdot10^5$ 0.57	$16.5\cdot10^{-6}$ $0.60\cdot10^5$ 0.69

field. The results of a study of this sort on an arc at currents of 5, 10, and 15 A, derived from mirror scanning pictures of the image of the spot, are given in our usual form in Fig. 116. Even a cursory comparison of this figure with the figure for the previous case (see Fig. 114) serves to show that the results are qualitatively the same for either method of observation. The distribution illustrated by the histograms of Fig. 116 also has a clearly defined maximum, which shifts slightly toward larger times as the current is decreased. At the same time, there is an appreciable increase in dispersion of the distribution. A quantitative comparison of the two figures shows that the intervals in the latter figure are approximately a factor of two greater than the intervals in the first figure. This fact must be attributed to the inevitable loss in random spot motion of half of the divisions which occur in, or close to, the direction of scanning. If these losses are included, both methods give the same qualitative and quantitative results, as shown by a comparison of the data in Tables XVI and XVII. The latter table gives the values obtained from mirror scanning for the distribution of the intervals between successive cathode-spot divisions in a normal arc with no external field. Here the mean values of $\bar{t}$ were derived, including the above-mentioned loss of half of the divisions, by reducing the measured values of the intervals by a factor of two. The spot-division frequencies f shown in the table were calculated as the reciprocals of the mean arithmetic values of $\bar{t}$. The same correction for missed observations was also made in calculating the probability of spot division in the range 5-20 μsec after the last division, W(5 < t < 20). A pair-by-pair comparison of the figures given in the last two tables for the same quantities at approximately equal currents shows that they agree to a few tenths of a percent. This agreement between the results of investigations of cathode-spot division made using different methods of observation and somewhat different experimental conditions is a reliable criterion of correctness, and provides the basis for stating that the tendency toward division is an internal or inherent property of the cathode spot, independent of any external forces acting during the observations. Moreover, comparing the data of the tables leads to the conclusion that the magnetic field of about 1000 Oe used in one of the methods of observation has no material effect on the cathode-spot division process in the range of currents studied.

It must be emphasized that a magnetic field has nevertheless been observed to exert an effect on spot division under somewhat different conditions, the effect being that at low currents a strong field tangential to the cathode tends to suppress the process appreciably. Thus, at currents less than 5 A, division becomes practically unobservable with the field oriented in this way, while with no field it is clearly observed even at currents of about 2 A. A possible mechanism for the way in which the field affects spot division was discussed above.

c. Artificial Statistical Selection of Observations. The Regular Nature of Cathode-Spot Division on a Homogeneous Mercury Cathode. Before we draw any conclusions as to the nature of the laws found to govern spot division on a homogeneous mercury cathode, we must consider how, and to what extent, the distribution of intervals between successive divisions is distorted by possible misses in the observations. A study of the distortions produced by missed observations will show first of all whether the selectivity observed in the intervals t comes from actual physical processes or whether it may be attributed to misses in the observations, as was the case with the maximum in the distribution of the autonomous spot duration τ. An investigation of this sort is also necessary since in making a broader interpretation of missed observations, the question of observational errors goes over into the fundamental question of what effect the criterion chosen for cathode-spot division has on the results of the observations.

For the purpose of investigating what effect possible failure to record spot divisions has on the form of a distribution such as W(t), use was made of the method of artificial statistical selection from the observations. In making the selection, groups of readings τ_i were successively rejected when they did not satisfy a newly chosen, coarser criterion for division. Here, the result of missed divisions was inevitably to increase some of the intervals t, and it was necessary to regroup the distribution W(t). The results of successive selections of this type for a current of 30 A, as applied to regular motion of the cathode spot in a magnetic field, are shown

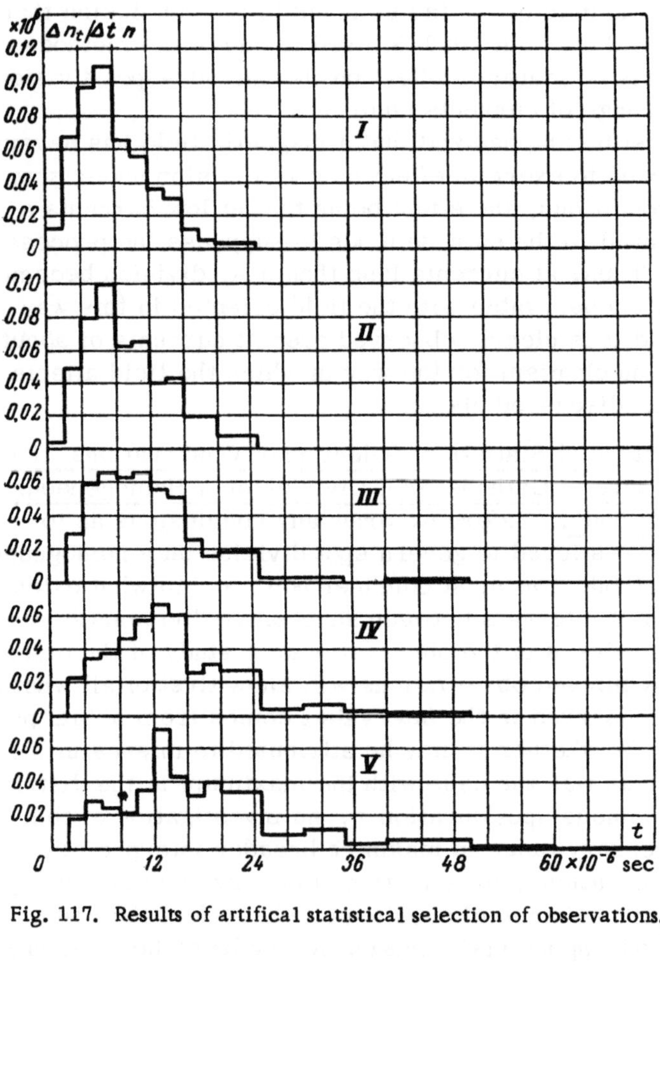

Fig. 117. Results of artifical statistical selection of observations.

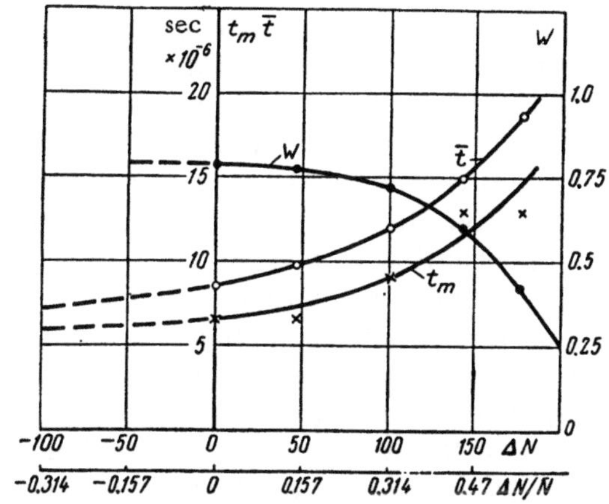

Fig. 118. Effect produced by missed observations on the values of t_m, $\bar{t}$, and W(4 < t < 16).

by the series of histograms in Fig. 117. The histograms are arranged in such a way that the number of intentional failures to record divisions increases from top to bottom along with the Roman numerals. The histogram I is the original distribution W(t). It is constructed directly from the observational data, including all the divisions satisfying the criterion selected for the work. Histogram II was obtained by throwing out 46 divisions with duration of the decaying branches $\tau < 4 \cdot 10^{-6}$ sec, which was about 14% of the total number of readings. Histogram III was constructed by ignoring 101 divisions (31%) with $\tau < 6 \cdot 10^{-6}$ sec, while for histograms IV and V the number of neglected divisions with $\tau < 8 \cdot 10^{-6}$ sec and $\tau < 10 \cdot 10^{-6}$ sec was, respectively, 143 (45%) and 177 (56%).

As the number of divisions rejected becomes greater, the histogram undergoes more and more noticeable changes. Of these, we must distinguish between regular changes and purely random fluctuations in the distribution of the intervals resulting from the reduction in the total number of readings used. If we leave the latter group out of consideration, we need consider only the regular changes. These include the displacement of the maximum and the increase in the statistical spread of the intervals, which is the same as increasing the dispersion of the distribution curve W(t). These regular changes may be described in a quantitative sense in terms of the mean values of the intervals t_m and $\bar{t}$ found after each selection, as well as in terms of the calculated probabilities of spot division in the range $W(4 < t < 16)$. The results of these calculations are plotted in the form of curves in Fig. 118. With increase in the number of intentionally missed observations, δn, plotted in the positive direction of the axis of abscissas, t_m and $\bar{t}$ increase monotonically, first slowly, and then more rapidly. At the same time, the probability of spot division in the time interval under study $W(4 < t < 16)$ decreases monotonically. Thus, the failures to record divisions corresponding to rapidly decaying spots only serve to increase the mean values of the intervals, and to increase the dispersion of the distribution. It is perfectly obvious that distortions of this type in the distribution $W(t)$ will be observed independently of whether the lost points are the result of intentionally ignored short-lived spots, which do not satisfy the criterion for spot division being used, or whether they come from insufficient resolving power in the method of observation. In either case the effect of the lost points may be found with the aid of the curves in Fig. 118.

The results of statistical selection of the observations show with a great degree of clarity that the predominant grouping of intervals in the range of values corresponding to the maximum observed in the histogram of Fig. 114 cannot possibly be due to insufficient resolving power in the method of observation, or to using too coarse a criterion for division. Actually, as has already been pointed out, the resolving power of this method of observation ensures recording all the divisions which satisfy the criterion for division chosen. As for the criterion itself, it is, from the physical nature of cathode-spot division, the most sensitive criterion of division that can be thought of in practice. Further, by using the results of artificial statistical selection, shown in Fig. 118, it may easily be shown in a completely rigorous fashion that even if we take all the possible branchings, including those which do not satisfy the division criterion, there will be no substantial change in the nature of the distribution of the intervals between succesive spot divisions. To do this, we need only extrapolate the curves in the figure in the negative direction of the axis of abscissas. The values of δn laid off in this direction are naturally to be interpreted as the number of uncounted branchings in the image of the cathode spot which do not satisfy the division criterion. It follows from the data of Table XIV that the number of uncounted branches under the conditions corresponding to the original histogram (Fig. 117) was about 80, or about 21% of the total possible number of branches. Extrapolating the curves in the negative direction of the axis of abscissas to the above values produces a 15-20% decrease in the mean values of the interavals and an insignificant increase in spot-division probability in the range 4-16 μsec. It may be seen from this, first, that including the branches which do not satisfy the division criterion has only a slight effect on the interval distribution W(t). Second, the result of including these readings will be to increase the frequency of spot division somewhat more, while the statistical spread in the values of the intervals around the most probable value t_m will simply be reduced.

The results of this study may be formulated briefly in the following way:

1. The predominant grouping of the intervals between successive cathode-spot divisions in a narrow range of values of the order of 10^{-5} sec shown by the histograms represents the true distribution of the intervals.
2. This selectivity in the values of the intervals is sufficient ground for stating that spot division follows a regular law.
3. The fact that division is found to follow a law, together with the high frequency of the process, indicates that cathode-spot division arises from a cause that is always in operation.

The conclusion that spot division follows a regular law is based on quite simple considerations. Having a maximum in a distribution curve always indicates that there is a causal relation between the predominant values of these quantities. As far as the statistical spread in values around this predominant value is concerned, it is always due to random secondary processes. Thus, in the Maxwellian distribution of velocities, having a most probable value of the velocity reflects the law relating the distribution with the mean energy of the gas, while the statistical spread in velocities around this value is simply the result of exchange processes which introduce the element of randomness. If the principal problem of the investigation is to establish this causal relationship, the dominant feature of the distribution must be considered to be the grouping of velocities around a most probable value, and in this sense it would be correct to call it a regular distribution. Independently of what words are used to describe the present distribution, the existence of a maximum indicates unconditionally that there is a definite regular process which fixes the observed grouping of values, while the spread in values is simply a measure of the degree of distortion of the causal relation lying at the basis of the phenomenon, brought about by random secondary factors. If we apply these ideas to the interval distribution W(t), the grouping around the most probable value t_m must be related to the existence of a continually acting cause of division. A careful analysis of the results of artificial selection of observations shows, on the one hand, that the statistical spread in the intervals comes from the random nature of the distribution of the individual spots in duration τ. Thus the spread is in no way an indication that the process of cathode-spot division is itself of a random nature.

The regular nature of cathode-spot division on a homogeneous mercury cathode is emphasized by the fact that a tendency toward division is observed under these conditions (with no magnetic field) at practically all currents at which the arc is capable of prolonged existence. The method of statistical selection points to the most natural way of interpreting the changes which the distribution undergoes when the current is reduced. We are struck by the fact that the changes in distribution of the intervals when the current is reduced (Fig. 114) are of exactly the same sort as those which occur when we increase the number of uncounted divisions in the series of histograms in Fig. 117. This similiarity suggests the idea that the reduction in spot-division frequency and the increase in spread of the intervals at low currents are caused by the general reduction in spot lifetime observed when the current is decreased, and by the associated rapidly increasing losses from readings where the duration becomes lower than the critical value. These losses may be evaluated from Eq. (82) with the simplifying assumption that the lifetime of each branch decreases in proportion to the current. Then, by using the results of statistical selection shown in Fig. 118, it is also easy to find what changes in distribution of the intervals are produced by the lost readings. The results of these calculations are shown by the dashed curves in Fig. 115 together with the experimental data. The small differences in form between the experimental and the theoretical curves may be attributed to the lack of accurate data on the variation of spot lifetime with the current. More important is the fact that the quantitative changes in interval distribution resulting from the calculation made for the range correspond to those in reality. This agreement confirms the correctness of the assumption made concerning the role played by the loss of some of the readings in producing the changes observed. Thus the apparent reduction in intensity of cathode-spot division at low currents is no proof that the cause of division ceases to act at low currents. This effect must be ascribed to a reduction in lifetime of the dividing spots.

d. The Effect Which the Criterion for Division Has on the Results of the Observations. In any attempt to investigate cathode-spot division, a certain element of arbitrariness is introduced into the problem by the criterion used for spot division. A legitimate question arises in this connection: what effect can changing the criterion for division have on the results of the observations and the way in which they are interpreted? It should be quite obvious from the results of the investigation given above that the effect which the choice of division criterion has on the results of the observations is, in the final count, determined by the number of branches recorded. Any change in the criterion which leads to recording a larger number of autonomous spots will bring about the following changes in the results of the observations:

1. displacement of the maximum in the interval distribution W(t) toward small times;
2. reduction in the mean values of the intervals $\bar{t}$ between successive divisions, and increase in the spot-division frequency;
3. reduction in the statistical spread of the intervals around the most probable value;
4. reduction in the values of current at which apparent stoppage of the spot-division process occurs.

The above changes show that if a more sensitive criterion of division is taken, such as using a method with higher resolving power, division will appear to be a more and more regular process. But this means that no matter how the criterion is chosen, there really is a constantly acting cause of the division. There are reasons for assuming that the criterion for spot division used in the present work may be regarded as the most sensitive imaginable physical criterion, and also that the resolving power of the method of observations used corresponds to this criterion. Hence it follows that the data obtained in the work most accurately reflect the nature of the cathode-spot division process being studied. If a coarser criterion is chosen, or a method with smaller resolving power is used, division will more and more take on the appearance of a random process. In all probability the idea existing until recently that cathode-spot division is a random phenomenon arose as a result of using exceedingly coarse methods of observation.

Thus, division of the cathode spot on a homogeneous mercury cathode must from all indications be regarded as a process following a regular law, which occurs as the result of a constantly acting cause. In the low-pressure arc, this cause is the irregularity of the inherent magnetic field of the arc at the spot. It was shown above that spot division as well as the motion in an external magnetic field obey the maximum field principle. From this point of view, spot division may be regarded as simply one of the basic forms of continuous re-formation, controlled by the distribution of the total magnetic field in the cathode region of the arc. At the basis of the process are the same phenomena and laws which control the motion of the spot over the cathode as a whole in an external field. The only difference between the two forms of spot re-formation is that, in the one case, the spot as a whole interacts with the external field, while in the other case the individual parts of the spot interact through magnetic fields. Thus, spot re-formation of division type may in principle be reduced to re-formation of the type which occurs during ordered motion in an external magnetic field.

C. CHAOTIC MOTION OF THE CATHODE SPOT ON A HOMOGENEOUS LIQUID CATHODE

53. Formulation of the Problem of a Possible Relation Between Chaotic Motion and Division of the Spot

In proceeding to the last of the problems proposed in this book—the investigation of the long known, but so far unexplained phenomenon of chaotic motion of the spot over the surface of the cathode—we shall begin with a number of general comments concerning the possible nature of the phenomenon.

In considering the nature of the chaotic motion of various objects, there is a very widespread opinion to the effect that such motion can only be produced by random phenomena or random causes. An idea of this sort does not give a completely accurate picture of the nature of chaotic motion. The very fact that the motion is continual shows that it is in all cases produced by causes that are subject to law. The only thing that is random is the direction of action on the object, and this is the principal characteristic of this form of motion. If we apply this idea to the cathode spot, we can easily see that one of the possible causes of the disordered motion over the metal may be provided by the regular division of the spot, with subsequent mutual repulsion between the autonomous spots, as has been shown to exist in the previous section. Actually, the unstable autonomous spots formed in every division will move apart in opposite directions as a result of mutual repulsion, until one of them disappears after an interval of time of some length τ_i. As a result of each such division and subsequent repulsion of the autonomous spots formed, the more stable, main spot remaining on the cathode will be displaced some distance s_i from its original position. The direction of each such displacement depends on the orientation of the chain of cells at the instant division occurs. If there is no external magnetic field, all directions of orientation of the chain are equally probable. Hence it follows that if there are no external factors in the form of a magnetic field acting on the spot, the directions of the displacements s_i resulting from successive divisions will be distributed according to a purely random law. Thus, systematic division of the cathode spot, with all directions of division equally probable, and the individual parts repelling one another with the secondary spot subsequently decaying, is a situation which necessarily leads to random motion of the spot over the cathode. The question is simply whether or not spot division followed by mutual repulsion of the individual parts is the dominant cause of the chaotic motion actually observed. This question arises in view of the fact that there may be, and beyond doubt actually are, other effects that are capable of producing the same kind of spot motion. According to the above cathode-spot model, one of the sources of the random motion may be the motion of the center of emission inside the metal evaporation region which occurs as the result of uncontrollable disturbances in symmetry of the charge distribution in the ionization region. Continuous motion of the spot over the cathode may also be caused by hydrodynamic and gasdynamic effects from the violent boiling of metal under the spot, as has already been pointed out by a number of authors [2, 35, 69]. With so many possibilities, the true causes of the chaotic motion of the spot can only be found through a quantitative study of the question, as we have done below.

The quantitative relation between cathode-spot division and the random motion that it produces may easily be found by simplifying the problem with the single assumption that no new division occurs until after one of the autonomous spots has decayed. Under these conditions, the interaction between the spots is of the simplest kind, since there cannot be more than two spots on the cathode at any instant of time. This last condition will be maintained for small discharge currents up to about 10-15 A. If we make this assumption and are given the random distribution of the directions of the successive displacements, then the mean square distance $\overline{\rho^2}$ which the spot moves from an arbitrary initial point as a result of n divisions is given by the sum of the squares of all the elementary displacements:

$$\overline{\rho^2} = \sum_{i=1}^{n} s_i^2 . \tag{85}$$

If we know the mean value of the square of the elementary spot displacement resulting from division, $\overline{s^2}$, and the mean division frequency f, the sum may be replaced by the product $(\overline{fs^2})t$, and (85) may be written in the more convenient form

$$\overline{\rho^2} = (\overline{fs^2})\, t, \tag{86}$$

where t is the total observation time.

The problem formulated here of finding the role played by division in the chaotic motion of the spot reduces to finding whether or not Eq. (86) is satisfied. For this purpose, the values of the coefficient of t, found from experiments on spot division at different currents, must be compared with the values of the ratio $\overline{\rho^2}/t$, found under the same conditions by direct observation of the chaotic motion of the spot. It must be emphasized that in the comparison we are not making any assumptions concerning the distribution of the values of s_i which can limit the generality of the conclusions. If the comparison shows that the values of the ratio $\overline{\rho^2}/t$ are many times greater than the coefficient $(\overline{fs^2})$, this means that division does not play an essential role in the chaotic motion of the spot over the cathode. On the other hand, it may be assumed that a relation between the two phenomena has been proved to exist if the values being compared are of the same order of magnitude.

54. Experimental Proof of the Dominant Role Played by Division in the Chaotic Motion of the Spot on a Homogeneous Liquid Cathode

In order to make an experimental check of Eq. (86), the cathode spot was photographed with an ordinary camera on isochromatic film through a rotating red filter in the form of a disk with radial slits uniformly spaced around the circumference. The slotted-disk arrangement made it possible to get a series of discrete images of the cathode spot on the film showing the instantaneous positions of the spot on the cathode at equal time intervals. The time intervals could be varied between wide limits by changing the number of slits or the velocity of rotation of the disk. In addition to the bright discrete images of the cathode spot on the film, using a perforated disk that was semitransparent for the spot radiation

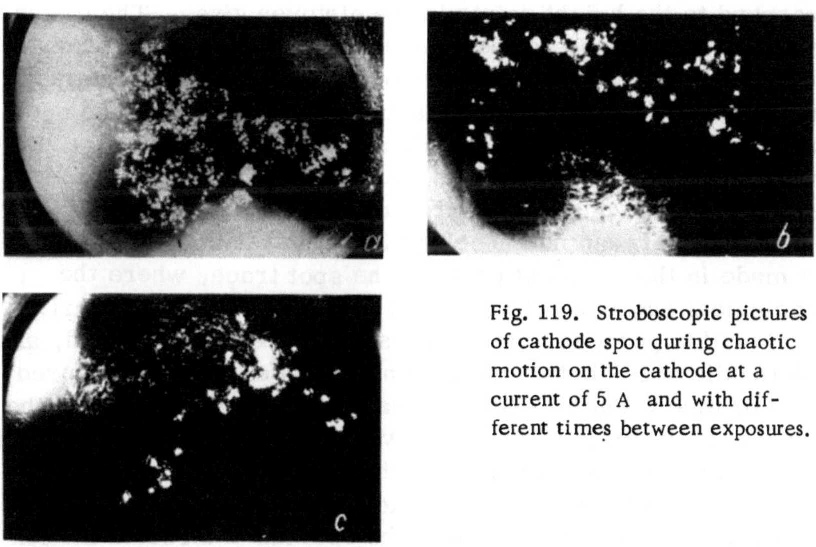

Fig. 119. Stroboscopic pictures of cathode spot during chaotic motion on the cathode at a current of 5 A and with different times between exposures.

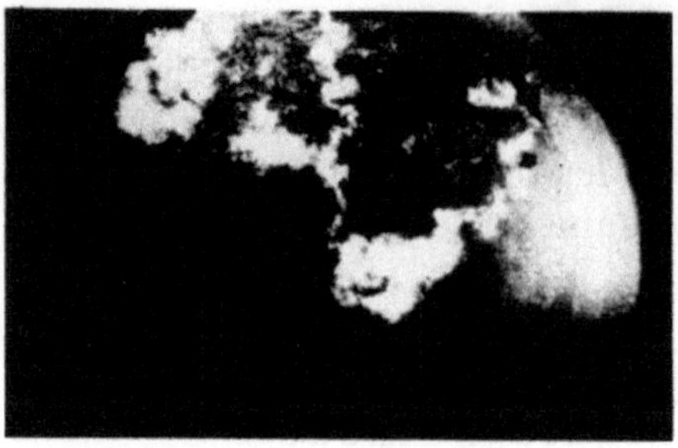

Fig. 120. Cathode-spot trace with a clearly defined directed motion component.

also gave a gray continuous trace left by the spot during its chaotic motion over the cathode. This made it possible to trace out the chronological sequence of the discrete images of the spot.

A series of typical pictures, taken with the above-described stroboscopic apparatus at a current of about 5 A and almost a second of total exposure, are shown in Fig. 119. The time intervals between discrete exposures in the pictures increase in the order a, b, c. The cathode spot leaves a rather confused trace on the film, which can only be followed from the continous gray image, which does not show in the reproductions given here. But, even with this, the problem can only be solved for quite small total exposures, and by no means in all parts of the pictures. The discrete exposures correspond to the bright points in the pictures given. The distances between them increase as the time interval between exposures is made greater. Getting the data from the pictures is a matter of measuring the distances between successive positions of the spot on the cathode for a number of values of the time intervals t. Determining the mean values of the square of the spot deflection, $\overline{\rho^2}$, for each value of t chosen required making about 100 readings of the distances between neighboring discrete spot images, which were made in the simplest parts of the spot trace, where the chronological sequence of the discrete images could be clearly seen. For this purpose, the pictures were greatly magnified, after which the distances between the points could easily be measured with an ordinary scale. Then the readings were squared, and the squares were averaged. The mean values derived from these measurements will be somewhat too high, for the simple reason that in the least complex parts of the trace, where the measurements were made, the random motion of the spot has a directed motion superimposed on it. It was precisely because of the directed motion that these parts of the trace looked simpler. Departures of this sort from purely random motion are clearly visible on the picture (Fig. 120) made without using the slotted disc. The picture shows rather extensive portions of the path of the spot where there is a clearly defined directed component of the motion. It is easily shown that this is the result of temporary distortions of the discharge path coming from periodic changes in the position of the anode spot. Each time the anode spot shifted to a new part of the anode, the cathode spot followed it along the cathode, at the same time performing its random motion. In this case we are encountering essentially the same phenomenon of spot deflection as a result of disturbing the symmetry of the magnetic field of the arc which turned out to be the reason for the change in direction in an inclined field (see Sec. 47c).

The results of measuring the mean square spot displacement, $\overline{\rho^2}$, for different observations times t are given in Fig. 121. To fit the data of all the measurements on one graph, and at the same time to reduce the effect of statistical spread of the points, the square roots are plotted instead of the actual values of $\overline{\rho^2}$ and t. For each of the three currents, 5, 10, and 15 A, the points are grouped with a certain amount of spread around a sloping line passing through the origin. This shows that there is a linear relation of the

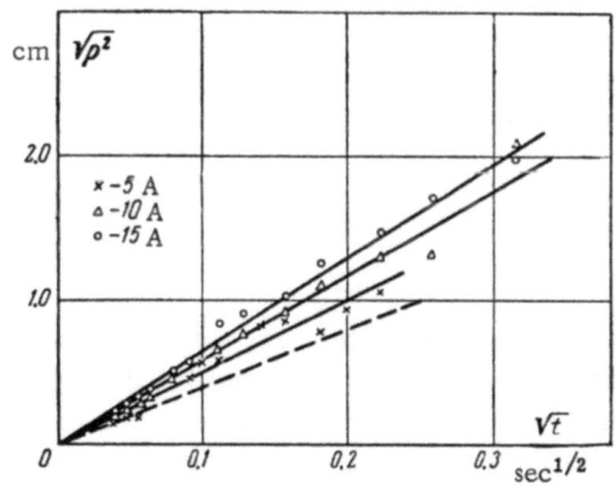

Fig. 121. Results of a quantitative study of the chaotic motion of the cathode spot on mercury at currents of 5, 10, and 15 A. The dotted line shows Schmidt's data.

type (86) between the quantities $\overline{\rho^2}$ and t, and in addition that the motion is predominantly random. As a matter of fact, if this relation is obeyed, in spite of the departures from random motion encountered, it can only mean that for these observation times the random component of the spot motion is predominant. The slope of the lines increases with increase in current, which shows that there is an increase in the rate of motion. The results given are in good agreement, both qualitatively and quantitatively, with the data of Schmidt [69], obtained by another method of observation for a discharge current of about 5 A, and shown for comparison by the dashed line. The small differences between the data of this author and the results of our measurements must obviously be atributed to the sources of error already mentioned. From the data of Fig. 121, it is a simple matter to find the total mean square spot displacements per unit time, $\overline{\rho^2}/t$, for each current. These values are numerically equal to the square of the slope of the corresponding line.

It must be recalled that the final problem of the present study is to check Eq. (86) by comparing our values of $\overline{\rho^2}/t$ with the values obtained by multiplying the spot-division frequency f by the mean square of the elementary displacement $\overline{s^2}$. We have data on the division frequency f at different currents. They were already given in Table XVII. Thus, in order to compare the quantities that we are interested in, there remains only the determination of $\overline{s^2}$. It was possible to solve this problem by measuring the maximum distance l_1 between each pair of spots on the photographic scanning pictures. The value of the final spot displacement s_i resulting from each such division will be $l_i/2$. Then the mean square of the elementary spot displacement may be found by the simple averaging operation:

$$\overline{s^2} = \frac{1}{n} \sum_{i=1}^{n} \left(\frac{l_i}{2} \right)^2 .$$

The values of s^2 found in this way are given along with the other quantities under discussion in Table XVIII. They are of the order of 10^{-4} to 10^{-3} cm^2. The extremely small size of each of the elementary spot displacements is, however, compensated for by the high values of the division frequency. This explains the relatively high velocity of the random motion produced by division of

TABLE XVIII

I, A	f, sec^{-1}	$\overline{s^2}$, cm^2	$f\overline{s^2}$, cm^2/sec	$\overline{\rho^2}/t$, cm^2/sec
5	$3.2 \cdot 10^4$	$3.3 \cdot 10^{-4}$	10.6	24.9
10	$5.7 \cdot 10^4$	$5.3 \cdot 10^{-4}$	30.2	34.3
15	$6.0 \cdot 10^4$	$7.5 \cdot 10^{-4}$	45.0	42.0

the spot. If we compare the values of $\overline{fs^2}$ and $\overline{\rho^2}/t$ given in the last two columns of the table, we see first that they are of the same order of magnitude. The fact that the numbers in the first of the columns under comparison are as a whole somewhat less than the numbers in the second column may be attributed to failures to record spot divisions. This is confirmed in particular by the fact that, as the current is increased, which reduces the importance of uncounted divisions, the difference between the two sets of numbers drops to a few percent. The practically complete agreement, within the limits of possible error, between the quantities being compared leads to the unequivocal conclusions that the reason for the chaotic motion observed is the division of the cathode spot.

This conclusion does not eliminate the possibility that there are other causes for the random motion in addition to spot division. In the mirror scanning pictures of the spot image, it is sometimes possible to observe unexpected displacements which have no visible connection with the division processes. The conclusion formulated above must be understood simply as a statement to the effect that division is the dominant factor in the random motion of the cathode spot under these experimental conditions. We will recall that the studies described are for the normal conditions in an arc with a homogeneous mercury cathode and an equilibrium mercury vapor pressure of about 1.2 μ Hg. It is not impossible that if the experimental conditions are greatly changed some other cause of the motion will predominate, such as gasdynamic effects from the violent boiling of the mercury in the cathode spot. Some preliminary predictions can be made for experimental conditions of this sort. It follows from the data in the last table that the random motion associated with spot division is slowed up as the current is reduced. The principal reason for this is that as the current is reduced there is a large reduction in the mean square elementary displacement of the spot in a single division, $\overline{s^2}$. This reduction is a result of shortening the duration of the combined existence of each pair of spots and of weakening the interaction between them. It can be imagined that, if the current is low enough, the spot motion will occur principally as a result of gasdynamic or hydrodynamic effects. In contrast to this, the reason for the random motion of the spot on a solid cathode may be the melting of the metal under the spot. The role of spot division in causing motion over the cathode will also become smaller when extraneous gases are introduced into the discharge space and the gas density in the medium is increased. There will be some critical gas density at which mutual repulsion of the spots no longer occurs. Under these conditions, spot division can no longer be the dominant cause of the motion. It should be noted finally that the effect of spot division may be partially paralyzed by a magnetic field tangent to the cathode. Such a field always orients the spot in such a way that division occurs in a direction normal to the direction of the ordered motion. Under these conditions, the random displacements of the spot can have only one degree of freedom, so that they take on the special form of transverse deflections of the spot from the direction of the trajectory.

In summing up this chapter, we shall limit ourselves to pointing out the fundamental results. One of the fundamental results must be considered to be that a detailed investigation has shown that there is an intimate connection between the three basic forms of spot motion on a homogeneous cathode, namely, directed motion of the spot in a magnetic field, division of the spot, and random motion over the cathode. It has also been shown that at the basis of all these forms of motion there is essentially one and the same process of continual cathode spot re-formation, the direction of which in the low-pressure arc depends entirely on the distribution of magnetic field intensity in the vicinity of the cathode spot. This shows up clearly in the fact that the maximum field principle proved to apply to all the forms of spot motion under discussion. However the direction of spot re-formation is correctly described by this principle only under the obligatory condition that account is taken of the inherent magnetic field at the spot. The need to include the inherent field of the arc shows the important role it plays in the behavior of the cathode spot, and along with this in maintaining the arc cycle.

CONCLUSION

In the above we have given a detailed report of the results of a variegated investigation of the mercury arc including a large number of diversified experiments, concentrated around questions relating to the stability of the arc cycle. Beginning with a statistical study of spontaneous extinction, and the effect of various external and internal factors on the stability of the arc, we then passed to the oscillation processes in the arc and, finally, made an analysis of the structure of the cathode spot and of the continuous changes which it undergoes, including directed motion in a magnetic field, division, and random motion over the cathode. On closer consideration, all these apparently unrelated phenomena proved to be merely different links in the same chain of internal instability phenomena in the mercury arc cathode. We have observed them under all experimental conditions, including those under which an arc of this type should have its maximum stability, as, for example, in a discharge with a boiling cathode. The changes in arc behavior observed in this way were purely quantitative in character. It must be concluded from this that there are profound causes lying at the basis of the phenomena observed which lead to the actual mechanism of the cold-arc discharge, so that in this case we have every reason to speak of the internal instability of the arc discharge. It may be seen from the study made on the transient phenomena in the cathode region of the mercury arc that this form of discharge is not a definite state of equilibrim between the processes in the arc cycle, but is rather a state in which all the processes are continually changing in intensity, with accompanying large departures from equilibrium. During these departures from equilibrium, large changes occur in all the quantities which are considered to be characteristic of the arc cycle, such as the cathode drop, the current density at the cathode, and the ratio of ion to electron current. A closed arc cycle can only be maintained by alternately enhancing some of the processes at the expense of reducing the intensity of others. At low currents, the instability shows up in its most severe form, namely, as spontaneous extinction of the discharge, which means in essence that under these conditions the disturbance in equilibrium goes so far that the cycle can no longer be restored by the causal relations existing among the various processes in the cycle. Restoring the arc when it gets into one of these critical states is a matter of probability; i.e., the critical state can end either by the discharge being restored or by extinction, depending on the chance occurrence of conditions. That there is a definite probability of restoration of the arc when it is in one of these critical states explains why there is a random distribution of the time intervals between extinctions. The probability of restoring the arc increases with increase in current, since the conditions existing in the arc become simpler, with the result that the mean arc duration increases exponentially. If the restoration probability becomes quite large, spontaneous extinction of the discharge becomes of second order of importance as a sign of instability in comparison with other indications, such as voltage oscillations at the arc electrodes as well as division of the cathode spot and the way it moves continuously over the metal.

The problems in arc stability formulated and solved in this work will help to give more accurate information on the cold arc. In summing up the work done as a whole, we divide, for the sake of clarity, the new information obtained into three categories, depending on the general nature of the information. To the first category must be assigned the new phenomena and effects observed, such as the true internal instability of the arc, the large increase in stability of the cathode spot when the cathode is frozen or is boiling, the exceedingly large stabilizing effect exerted on the spot by a magnetic field, the periodic appearance at small currents of the transitional form of the arc, and the shifting neon luminescence in a combined discharge, consisting of hemispherical regions showing the instability of the cathode drop. To the same category must be assigned the data obtained in the work on the elementary cathode-spot cells and the continuous decay and re-formation phenomena in the spot, on the high frequency and regular nature of spot division, and on the interaction between the various parts of the spot.

The second category includes the explanations given in the work for well-known effects and relationships which had never been fully pursued. Without re-counting these results of the work in detail, we shall point out just a few of them. These include the data taken in the work to explain the random nature of arc extinction and the exponential increase in arc duration with increase in current, interpreting the way in which various external and internal factors affect this relationship, and the mechanism given in the work by which the emission current is controlled, as well as the solution of the problem of the reasons for the retrograde motion and division of the spot.

To the third category, which somewhat broadens our ideas on this form of discharge, must be assigned the causal relationship established between the various phenomena in the arc cycle, which has so far managed to elude physicists. The possibility found in the work of establishing the causal relation between these phenomena is, in the author's opinion, the most essential feature of the approach to the arc problem used here, which is based on a study of the stability of the arc cycle. We must stop to consider this aspect of the question in somewhat more detail. As may be seen from the report given, two basic lines of approach were opened up for investigating the relation between the phenomena. On the one hand it turned out that, as manifestations of internal instability of the arc, there was an intimate relation between spontaneous extinction, the various types of oscillation processes in the arc, and the tendency of the cathode spot to be in continual motion over the metal. On the other hand, from the data on the instability of the spot and the way the arc is affected by a magnetic field, it proved to be possible to establish a relation between the different forms of spot motion on a homogeneous liquid cathode, consisting of directed motion in a magnetic field, division, and random motion over the cathode. These types of motion were only different forms of one and the same continuous cathode-spot re-formation process, resulting from the instability of the spot, and controlled by the distribution of the total magnetic field intensity in the vicinity of the spot. As was shown by a detailed study of the behavior of the spot under varying experimental conditions, the direction of re-formation under arbitrary conditions is correctly described by the maximum field principle formulated in the work, including the inherent magnetic field of the arc. Thus, the basic forms of cathode-spot motion have a simple explanation if we take account of the role played by the inherent field of the arc, which in this case consists in introducing asymmetry or inhomogeneity into the resultant field distribution in the vicinity of the spot.

The present work is only one of several parts of the investigation of the cold arc, so that it would be premature to use these data to arrive at any conclusions concerning the mechanism of the cold-cathode arc. We simply have to limit ourselves here to pointing out that the material obtained in the work is generally compatible with the cold-electron theory of the arc discharge. Based on the ideas and laws established, the work must be continued in future in the direction of investigating the stability and behavior of the cathode spot on other metals. However, in view of the exceedingly broad extent of such investigations, it is beyond the powers of any one investigator to carry them out, and the work must be done by collective efforts.

The author expresses the hope that the results of the present work will to some degree stimulate further investigation of the arc and solution of a number of technical problems.

APPENDICES

1. METHOD OF TESTING MASS-PRODUCED MERCURY TUBES

The rejective testing of tubes for backfires under nominal operating conditions as practiced at the present time involves large expenditures of time and energy. With the increase in production of tubes, the question of tests of this sort is becoming a serious problem. The best solution of the problem would seem to be to develop new methods of testing tubes which do not require so much time or such large expenditures of energy. One such possibility has been revealed by the results of the study of mercury arc stability at low currents, as presented in Chapter 2. Actually, as was found in Part A of this chapter, even small changes in the internal conditions of the discharge, such as the surface state of the cathode, the composition and pressure of the gaseous medium, etc., produce appreciable changes in stability. These changes are revealed by the fact that the curves giving the mean arc duration ϑ as a function of current undergo various specific changes with a corresponding change in the conditions. Thus, it is possible to tell from the nature of the change in the curves what it is that is producing the change. By measuring the mean duration of the arc in tubes at small currents, such defects as contamination of the cathode by sputtered metal and graphite, or loss of vacuum, may be detected. Thus, measuring the mean duration of the arc at low currents can form the basis of a method for the rejective testing of mass-produced tubes. It is convenient for this purpose to use a low-current operating arc maintained by a d-c source.

The complete method of measuring the mean arc duration described in Chapter 2 may be used for the tests. The method may be greatly simplified by not measuring accurate mean values for the discharge current. Then the problem reduces to measuring the duration of the arc in the tubes under test with fixed external circuit conditions, and comparing the results of the measurements with some established standard. The method not only envisages the possibility of simple rejective tests, but also of making more complex diagnostic tests on the tubes. For tests of the first type, the problem reduces to observing all possible deviations from the normal conditions in the tubes, which requires about two minutes per tube. Samples showing more than the allowable deviation can then be given further diagnostic tests to find the cause of the difficulty. Here a curve should be taken showing arc duration as a function of current in the 0.1-2 A range. Taking such a characteristic requires about 20 min. Practical use of these tube tests should lead to considerable economies in energy and time. It is desirable that the tests give information in time on leaks in the tube seals, which are the principal cause of early failure. Work done so far on applying the tests to mass production of tubes of various types has shown that it is to be recommended as an auxiliary method for preliminary rejection of tubes.

2. MAGNETIC ANCHORING OF THE CATHODE SPOT AND PROTECTION OF THE METAL SHELL OF MERCURY TUBES

The laws found to govern the motion of the cathode spot on mercury can be used as the basis of new methods for anchoring the spot and keeping it away from the metal shell of mercury tubes having an uninsulated cathode.

It was shown above that in a low-pressure mercury cathode arc the spot exhibits a tendency to move in a magnetic field in the direction of the greatest increase in the resultant field at the edge of the spot, as found by including the inherent field of the arc. By making use of this property of the spot, it may be forced to stay for an unlimited length of time in a definite part of the cathode, far from the walls of the metal shell of the tube: where the spot is held it will execute an ordered or random motion. The principle used in constructing a cathode in which the spot is held magnetically in a closed annular region with a radial field may be seen from Figs. 90 and 91 already given. The first figure shows a section of a cathode with a concentric-pole-face electromagnet having a narrow annular gap located below the surface of the mercury. The second figure shows the direction of the magnetic lines of force above the poles. The cathode spot executes rapid motion around a circular trajectory in the annular zone of the cathode with a radial field located over the gap (see Fig. 96).

In order to check the principle of magnetic anchoring of the cathode spot at large currents under the normal operating conditions for a mercury tube with a metal shell, a large-size metal cathode was designed and built suitable for tests using the metal shell of a type RMNV-200 tube. The test showed that under certain conditions the field retained its anchoring action up to currents of about 200 A (maximum value 600 A). The anchoring process may be thought of as consisting of two stages, namely: the field suppresses the random motion of the spot over the cathode by orienting the chain of cells in the spot along the lines of force and then opposing the mutual ejection of the autonomous spots formed in divisions from the radial field zone. The second process is of no great importance at low currents, and the anchoring action of the field is simply to suppress the random motion of the spot, for which a field of not more than about 300 Oe is sufficient. At large currents, on the other hand, the spot becomes free as a result of mutual repulsion of the autonomous spots, which causes one of them to move out of the radial field zone. The field intensity required to counteract this type of disturbance with the arrangement used may be estimated roughly from the empirical formula

$$H = 4.5 I_m,$$

where I_m is the maximum value of the current.

For currents of about 600 A, H must be more than 2500 Oe. In view of the difficulty of producing any stronger field on the mercury surface, this value of current must be regarded as the limit for the anchoring device used.

With magnetic anchoring, the spot always moves on the free surface of the mercury without touching any extraneous metals. This not only maintains pure mercury discharge conditions, but at the same time eliminates the possibility of the anchoring mechanism being destroyed by the cathode spot. On the other hand this method of anchoring has some substantial defects. These include the complexity of the cathode construction and the need to expend additional energy. The energy losses come not so much from maintaining the magnetic field, which can be completely avoided by using a permanent magnet, as from the increase in operating voltage of the arc when it is in a transverse field. The arc characteristics under these conditions are discussed in [188]. It is sufficient to point out here that the increase in voltage on an arc with correctly chosen operating conditions is about 2 V. Accordingly the additional power loss is about 200 W per 100 A. Further, increasing the voltage on the arc increases the probability of forming new spots outside the strong magnetic field zone, which leads to sudden loss of anchoring. Most often, the new spot is formed right on the amalgamated shell, particularly if there are films on the mercury. To eliminate this phenomenon, the shell of the tube must be shielded from the plasma.

In view of the drawbacks to magnetic spot anchoring, it is better to use some other methods of protecting the shell of the tube, where the spot is kept out of the dangerous outside part of the cathode by the inherent magnetic field of the arc without using any external field. One of the theoretically possible ways of making use of the laws of motion of the cathode spot to protect the shell of the tubes is based on the fact that, when the spot gets close to a barrier or wall made of a ferromagnetic metal, the symmetry in the distribution of the field around the edge of the spot is disturbed. Here the field between the wall and the spot is greatly weakened, while it is strengthened on the other side, as shown schematically in Fig. 122. According to the maximum field principle, the result of this disturbance in symmetry will be the repulsion of the spot from the wall. Accordingly, in order to protect the shell from the cathode spot it is in principle only necessary to make the shell out

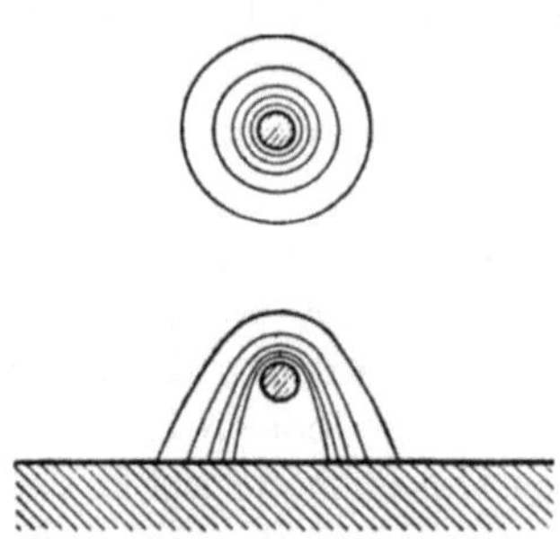

Fig. 122. Disturbance in symmetry of the magnetic field of the arc when the cathode spot is close to a ferromagnetic wall.

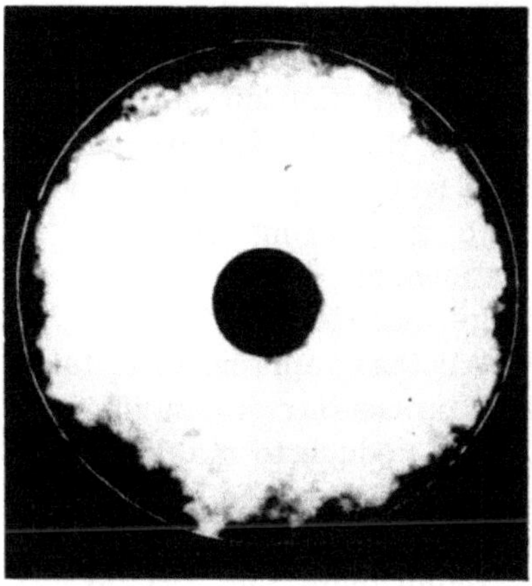

Fig. 123. Picture illustrating the repulsion of the cathode spot from a steel barrier.

of ordinary magnetic steel or to place a steel barrier, consisting of a ring projecting out of the mercury, in front of it. That the spot is repelled by a barrier of this sort is shown by the picture in Fig. 123. The picture was taken at an arc current of about 20 A with a 1-min exposure. In the picture, the inside surface of the barrier illuminated by the discharge appears as a narrow luminous circular border. The dark circle in the middle marks the position of the hollow igniter, while the surrounding illumination is the surface of the mercury cathode with the spot moving over it. There is dark space left all along the barrier between it and the illuminated part, which shows that the cathode spot is repelled by the barrier. Because of the repulsion, the spot does not even get through the narrow openings in the barrier, which is made up of four separate sectors, although the inner and outer mercury reservoirs communicate through the openings. The repulsive action is only disturbed in those parts of the barrier which are wet by the mercury, which is due to the distortion of the form of the mercury meniscus at these points. Since the action of the barrier depends on the form of the meniscus, this method of protecting the shell is less useful and obliges us to give preference to another method.

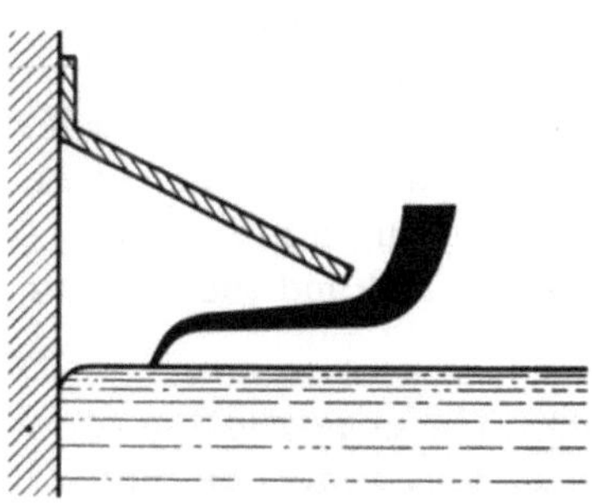

Fig. 124. Illustrating the action of a protecting ring on the tube shell.

The second method of protecting the tube shell from the cathode spot is clear from Fig. 124. The figure shows a vertical section of part of a cathode with a metal barrier fastened on the bottom, provided with an additional canopy or protecting ring.

The ring may be fastened directly to the wall of the tube, so that it becomes unnecessary to have a barrier. If the cathode spot runs under the protecting ring it pulls a string of current along with it, but the bend disturbs the symmetry in the distribution of the arc magnetic field around the spot in such a way that the spot is pushed out from under the ring. The action of the field only increases with increase in current, making this method of protecting the shell more reliable. At currents of 30 A and above the effect of the inherent field on the spot is already so large that the barrier can absolutely not be overcome by the spot. All that remains is to take measures to prevent the spot from falling on the barrier during a pause in the main current. These measures reduce to making the ring wide enough. The required width of the ring in the radial direction may be calculated from the data on the rate of random motion of the spot, which may easily be found from Fig. 121. It is easily found from the graph that during a pause in the main current, which usually lasts about 0.01 sec, the spot can cover in its random motion a distance in a straight line which is on the average not greater than 1 cm. Taking the width of the ring with some margin to be 10 cm, we can be sure that it will work successfully at all currents, as has been confirmed experimentally.

In making practical use of this method of protecting the tube shell, the ring must be set in an inclined position to keep mercury from accumulating on it.

3. MEASURING THE AMOUNT OF MERCURY ATOMIZED AND EVAPORATED BY THE CATHODE SPOT AND THE RECOIL FORCES EXPERIENCED BY THE CATHODE

The values given in the literature for the amount of mercury atomized and evaporated by the cathode spot, and for the recoil forces experienced by the cathode, are all quite different. In order to check and get better values for these data, the author has made some measurements of these quantities by weighing a mercury cathode directly in the discharge tube. The method and the results of the measurements are given below.

a. Determination of Mercury Losses. The construction of the tube used in the experiments is shown in Fig. 125. The mercury cathode is suspended on the spring balance 8 in the small cup 1, welded together from thin sheet molybdenum. The cathode and

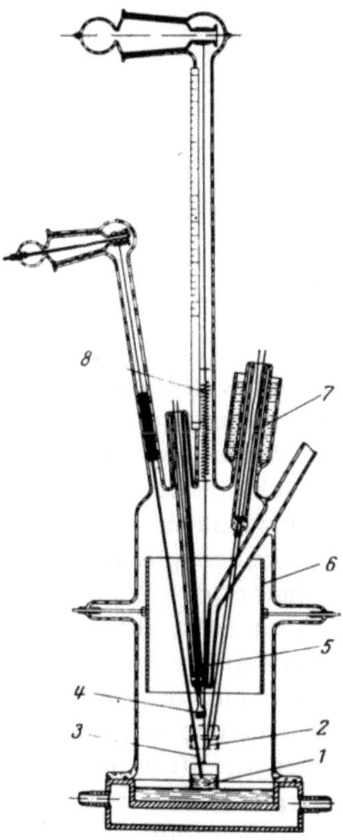

Fig. 125. Construction of tube used to measure mercury losses from the cathode spot.

the spring could be pulled up by means of a flexible metal thread wound around the shaft of the ground glass plug in the upper part of the tube. During the weighing, the plug was rotated to the point where the indicator on the lower end of the spring coincided with the mark on the wall of the tube which served as the starting reading. From the position of a second indicator on the upper end of the spring, the weight of the cup containing the mercury was read on a scale fastened outside the wall of the tube. The apparatus was calibrated ahead of time by using known weights. Then the tube was evacuated and the cup was filled with mercury by distillation. After the first mercury weighing, the cup was lowered until it touched the rest of the mercury filling the lower part of the tube, which was connected to the glass anode bulb with a flat ground joint. Electrical and thermal contact with the mercury in the shell was provided through the bottom, which consisted of sheet molybdenum 0.1 mm thick. At any instant of time t_1, a cathode spot could be excited on the mercury in the cup by means of the contact needle 3, which was lowered on a thread from a second opening in the upper part of the tube. In order to prevent the spot from jumping out of the cup, the walls of the cup were protected on the inside by a thin porcelain cylinder. The heat developed in the cell was conducted out through the tin metal bottom to the mercury outside, and from it through the metal wall of the tube to the water circulating in the space below the mercury reservoir.

After the required amount of exposure at a given discharge

TABLE XIX

Current, A	Water temperature, °C	Δt, sec	Δm, mg	T_0, °C	Δm_s, mg	$\frac{\Delta m}{\Delta t \cdot i}, \frac{mg}{A \cdot sec}$
10	27	69	5 000	84	1 380	5.2
10	28	175	11 000	95	4 370	3.8
10	28	170	9 000	88	3 400	3.3
5	23	182	4 000	68	728	3.6
5	29	180	3 500	57	270	3.6
14.5	24	72	7 000	92	2 160	4.6
14.6	22	176	16 000	93	5 280	4.1
18.5	24	60	65 000	103	2 100	4.0
					Mean	~4.0

current, the cup was pulled out of the mercury, which automatically stopped the arc. The exposure time was long enough so that the greater part of the exposure occurred at a constant mercury temperature in the cell, which we shall call T_0. The instant of removal t_2 was determined with a second meter, after which the cup was quickly drawn up to make contact with the thermocouple 2, which gave the mercury temperature T_1, and the instant of measuring the temperature, t_3, was read. During the measurements, the temperature of the two upper junctions of the thermocouple was maintained at a constant level by the water in the vessel 7. surrounding the junctions. After measuring the temperature, the cup was weighed again and the instant t_4 of weighing was recorded. After this, the cathode was again drawn up to the thermocouple and the temperature of the cooling mercury was measured after definite intervals of time at the instants t_5, t_6, etc.

Two goals were being pursued in the system developed for making the measurements. First, the system made it possible to find the total cathode weight loss resulting from the action of the

TABLE XX

Current, A	Water temperature, °C	Δt, sec	Δm, mg	T_0, °C	Δm_s, mg	$\frac{\Delta m}{\Delta t \cdot i}, \frac{mg}{A \cdot sec}$
2	22	2 400	2 800	~40	800	0.42
1	22	1 920	3 000	44	1 150	0.96
4	24	1 800	20 000	72	16 200	0.53
2	26	1 680	6 500	59	3 020	1.03
1	16	1 080	600	34	216	0.36
2	15	3 600	3 500	38	1 080	0.33
					Mean	~0.6

cathode spot. Second, it made it possible to find the temperature T_0 of the mercury, which was necessary for making a correction to the amount of mercury Δm_S evaporated from the surface of the mercury between the first and second weighings. The temperature T_0 was found by extrapolating the mercury cooling curves to the instant t_2 at which the cup was taken out. In view of the rather slow cooling of the mercury when it is suspended in a vacuum, the extrapolation is highly reliable.

Calculating Δm_S for each value found for the temperature T_0 required making auxiliary measurements of the rate of evaporation of mercury from the total surface of the mercury mirror in the cup at the same temperatures. For this purpose, the mercury in the cup was brought into contact with the thermocouple and heated to the required temperature T_0 by the radiation from the incandescent tungsten filament 4. When the temperature had become steady, the cup was weighed and left at this temperature for a long time (of the order of 1 hr). A second weighing at the end of this period of time gave the amount of mercury evaporated, which was used to determine the correction Δm_S. It was found in making these auxiliary measurements that the evaporation rate of the mercury depends on the level in the cup, decreasing rapidly as the level drops. In view of this condition, the auxiliary measurements of the rate of temperature evaporation of the mercury were made with the same weights of mercury as the main experiments on finding the mercury losses in the arc discharge.

In setting up the measurements, no attempt was made at a separate determination of the mercury actually evaporated in the cathode spot and that thrown out by the spot in the form of drops of all possible sizes. The amount of mercury evaporated by the spot is apparently so small that the correction for mercury evaporated from the whole surface will be many times greater than the quantity measured. The measurements do not deserve any confidence under these conditions. It is thus better to speak simply of the total mercury losses from the spot, where the predominant role is played by the mercury atomized by the spot. In determining the losses, means were provided for measuring either with a freely moving or with an anchored spot. In the latter case, a nickel anchor was mounted in the center of the cup. Otherwise the method of making measurements with an anchored spot did not differ from that described.

The results of measurements made with moving and anchored spots are given in Tables XIX and XX. The first column of each table gives the discharge current, while the second column gives the temperature of the cooling water. The third column gives the exposure time, and the fourth the measured weight losses of the mercury. The fifth column gives the temperatures T_0 as found by extrapolating to the instant the cup was pulled out. The sixth column gives the correction for mercury evaporation over the whole cathode surface during the discharge. The last column gives the mercury lost from the cathode spot itself referred to 1 cal. The mercury losses per calorie are independent of the current and for a freely moving spot are on the average 4 mg/cal, while for an anchored spot they are about 0.6 mg/cal. In all probability, the losses in both cases are due principally to atomization of the mercury. The reduction in losses in an arc with an anchored spot must be attributed to the reduction in dimensions of the mercury droplets atomized under these conditions.

b. "Weighing" the Cathode Spot. In addition to measuring the mercury losses from the cathode spot, some experiments were made to find the force which the cathode is subjected to while the arc is operating. For this purpose, a miniature cup 1 was made, designed to hold only a small quantity of mercury, so that a more sensitive spring balance could be used in the weighing. In "weighing" the cathode spot, the cell containing the mercury was drawn up far enough so that the bottom was not in contact with the surface of the mercury in the reservoir of the tube. Contact with the mercury in the reservoir was made with a thin molybdenum wire welded to the bottom of the cup and in contact with the mercury in the reservoir. When a cathode spot was excited, the cup dropped several millimeters from the effect of the arc, thus elongating the spring. The elongation of the spring gave the force acting on the cathode during operation of the arc. When the discharge was stopped, the cup again rose, to a level somewhat higher than before. The reduction in weight of the cup was the result of atomization by the cathode spot of a certain amount of mercury during the experiment. The recoil forces found in the measurements increased in proportion to the current and were about 56 dynes/A for an arc with either a moving or an anchored spot.

In comparing the results of the present measurements with those of other authors, we can note the following. The figures

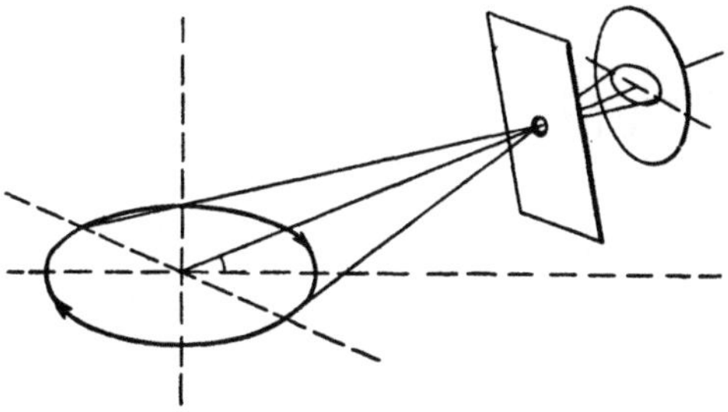

Fig. 126. Diagram of location of circular spot trajectory, diaphragm, and smoked glass.

found here for the total mercury losses from a freely moving cathode spot are approximately a factor of 2 less than those given by Güntherschulze [7]. Within the limits of possible error, the mercury losses per calorie were independent of the current, while according to Issendorf they increase with the current [29]. These differences may be attributed to the large amount of mercury evaporation from the whole cathode surface in these papers. The results of the measurements on an anchored spot are nearly the same as those which were published by Tonks [37], but they must be regarded as the total mercury losses from the spot due to atomization and evaporation and not to evaporation alone, as was assumed by Tonks. As far as the recoil forces on the cathode are concerned, they are practically the same as Kobel's data [35].

4. GENERATION OF MERCURY DROPS BY THE CATHODE SPOT

It is well known that in its random motion, the spot atomizes the mercury to give fine drops which fly off at a high velocity in all directions. This may easily be seen, for example, if a piece of smoked glass is put into the discharge. The incident drops are reflected from the glass, leaving traces in the soot, which give an idea of the number and size of the drops [41]. From the point of view of the mechanism by which the mercury drops are formed in the arc, a study of atomization of the mercury by the cathode spot when it is executing rapid ordered motion in a magnetic field is of special interest. Observing drops of mercury under these conditions means that they do not come from random motion of the spot over the cathode, with associated agitation of the mercury. Further, with ordered motion, it is a simple matter to control both the direction of motion and the velocity of the spot.

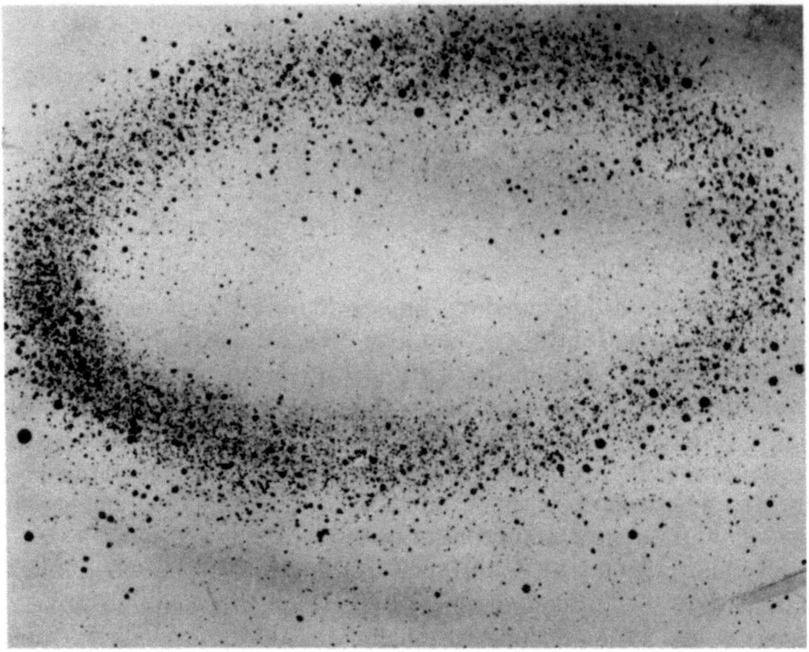

Fig. 127. Ellipsoidal grouping of mercury drop traces.

In the experiments on ordered motion, it was convenient to use the tube with the cathode (see Fig. 90) where the spot moved around a perfectly circular trajectory in a radial field. In order to observe the mercury drops atomized by the spot, a round piece of glass with a thin layer of soot on one side was mounted on a ground joint in an inclined side arm of the tube, the soot-covered side facing the cathode. A diaphragm with a 1-mm opening was placed in front of the glass and could be closed as desired with a shutter moved from outside by a magnet. The relative location of the circular trajectory of the spot, the diaphragm, and the smoked glass is shown schematically in Fig. 126. As the cathode spot moved around the circle, some of the mercury drops flying off passed through the opening in the diaphragm and struck the smoked

---

Fig. 128. Photomicrographs of sections of ellipse showing traces of drops emitted in the direction of motion (a) and in the opposite direction (b).

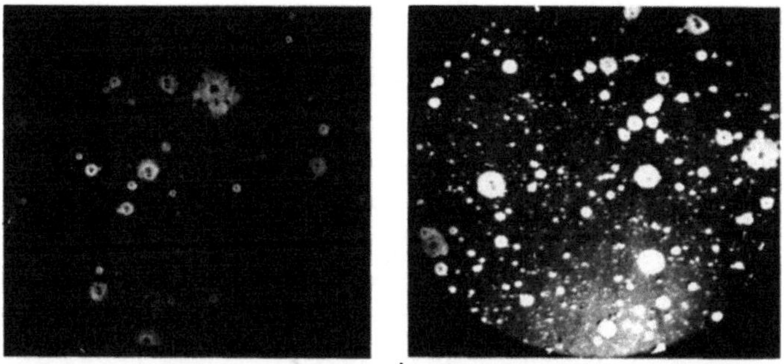

6 m/sec

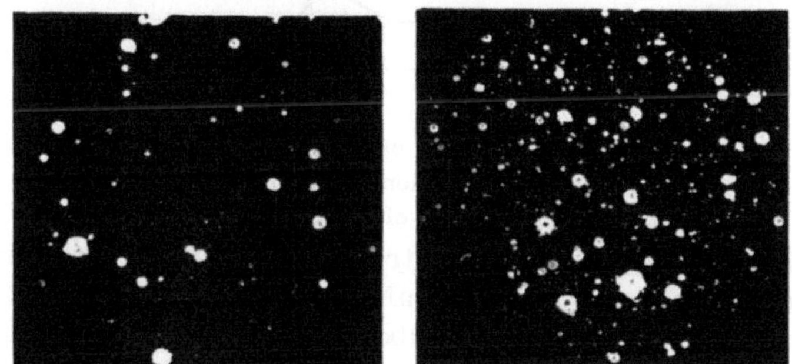

12.3 m/sec

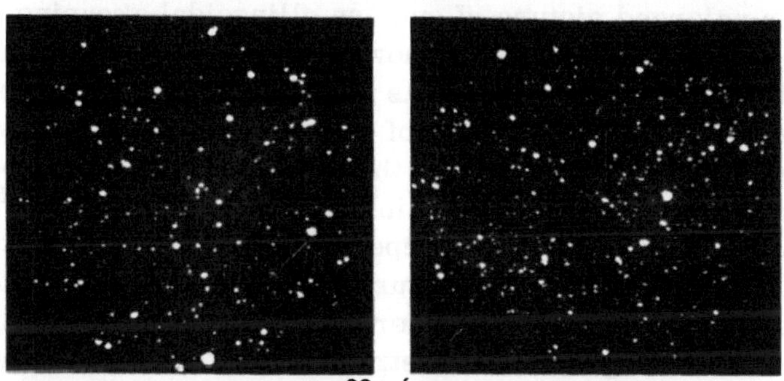

29 m/sec

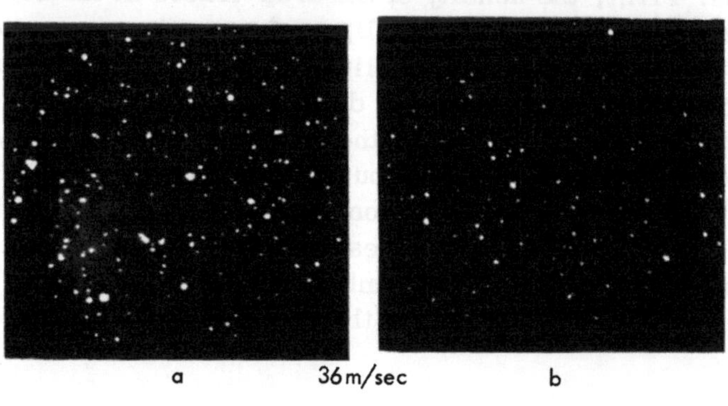

a 36 m/sec b

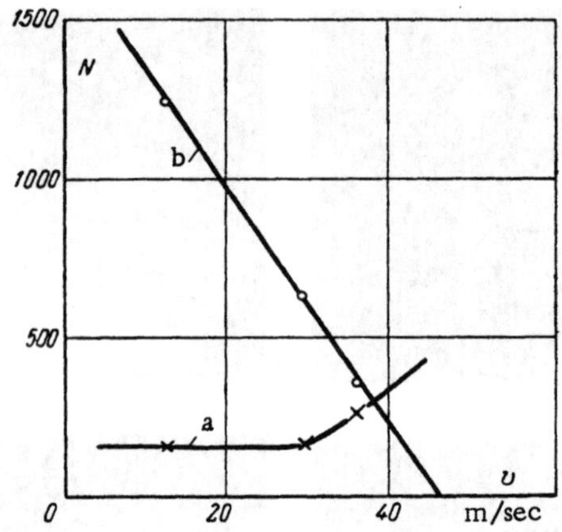

Fig. 129. Relative number of drops thrown off by the cathode spot in the direction of motion (a) and backward (b) as a function of velocity.

glass, thus leaving traces. The traces of the drops were arranged with a certain amount of spread along the ellipse forming the projected image of the circle described by the spot on the cathode. This is direct proof that the drops are emitted by the cathode spot itself. An enlarged picture of such an ellipsoidal grouping of drop traces on the smoked glass is shown in Fig. 127. In these experiments, the discharge current was held at 8.5 A. By opening the diaphragm for different lengths of time, different densities of the drop traces could be obtained in the ellipse. The usual exposure time was 30 min, during which time the spot completed 10^5-10^6 revolutions around the circle, depending on the rate of motion.

In examining enlarged pictures of the pieces of smoked glass which had been exposed to the mercury drops, the following laws were observed to govern the distribution of the number and size of the drops. First, the density of the drop traces in the soot was unequally distributed around the ellipse. It was always greatest at one end of the long axis of the ellipse and smallest at the other end. This shows that the cathode spot does not atomize the same number of drops along, and opposite to, the direction of motion. By comparing the irregularity in distribution of the traces around the ellipse with the direction of rotation of the spot, it was easily seen that the spot sends out the greatest number of drops to the rear, while only a relatively small number of drops is atomized in the direction of motion. Second, as the velocity is increased, the number and dimensions of the drops atomized by the spot decrease monotonically. The difference in the number of drops coming off in different directions and the way in which the number and dimensions of the drops vary with the rate of motion of the cathode spot are clearly visible in the two series of photomicrographs of different parts of the ellipse, as shown in Fig. 128. The two series correspond to the points of intersection of the ellipse with its long axis, and they are arranged in the order of increasing spot velocity. The right-hand series of pictures shows enlarged traces of the drops thrown to the rear by the spot. In the left-hand series of pictures, the traces are formed by drops emitted in the direction of motion of the spot. The number of drops on the pictures in the right-hand series is much greater than the number in the left-hand series. The dimensions of the drops decrease from top to bottom. By recording all the drop traces on the pictures and measuring their dimensions, it was possible to find the relation between the number of drops and the velocity, as well as how they were distributed in dimensions. The distribution of the drops in dimensions could in all cases be represented by a falling curve; i.e., as the diameter increased, the number of drops decreased monotonically. The relation between the total number of drop traces on the pictures (in the field of vision of the microscope) and the rate of motion of the spot is shown in Fig. 129. The upper curve shows the change in the number of drops thrown to the rear by the spot, and the lower curve, for drops emitted in the direction of motion. As the velocity of the spot is increased, the number of drops thrown to the rear decreases rapidly. In contrast to this, the number of drops emitted in the direction of motion remains practically constant. At velocities of about 40 m/sec the number of drops is about the same in both directions.

The difference observed in the number of drops emitted in the two directions in all probability comes from the surface contour of the mercury in the spot, which is seen to be asymmetric. It may be assumed that when the spot moves over the cathode a hill is formed in front and a valley behind. As the velocity is increased, the distortion in the surface becomes less, which explains why the number of drops in both directions becomes the same. The reduction in drop dimensions as the velocity increases agrees well with the idea that the reason for the atomization of the mercury is the explosive boiling of the superheated liquid, since the depth of heating will be greatly reduced at high velocities. The decrease in drop size at high spot velocities may be used in practice to eliminate the backfires produced by mercury drops in the tubes. A study of drop backfires occurring when the cathode spot is executing ordered motion has shown that they become less probable as the velocity of the spot is increased, which is a direct result of the reduction in dimensions and in the number of drops produced.

LITERATURE CITED

1. Kaptsov, N.A., Electrical Phenomena in Gases and Vacuum [in Russian], Gostekhizdat, 1947.
2. Loeb, L., Fundamental Processes of Electrical Discharges in Gases [Russian translation], Gostekhizdat, 1950.
3. Stark, J., Ann. Phys. 18, p. 213, 1905.
4. Smith, C.G., Phys. Rev. 69, p.96, 1946.
5. Güntherschulze, A., Ztschr. Phys. 11, p.71, 1922.
6. Stark, J., Ztschr. Phys. 5, p.750, 1904.
7. Güntherschulze, A., Ztschr. Phys. 11, p.74, 1922.
8. Tonks, L., Physics 6, p.294, 1935.
9. Cobine, J.D., and Gallagher, C.J., Phys. Rev. 74, p.1524, 1948.
10. Froome, K.D., Nature (Lond.) 167, p.446, 1946.
11. Froome, K.D., Proc. Phys. Soc. (Lond.) 60, p.424, 1948.
12. Froome, K.D., Proc. Phys. Soc. (Lond.) 62, p.805, 1949.
13. Froome, K.D., Proc. Phys. Soc. (Lond.) 63, p.377, 1950.
14. Froome, K.D., J. Scient. Instr. 25, p.371, 1948.
15. Bertele, Brit. J. Appl. Phys. 3, p.358, 1952.
16. Froome, K.D., Brit. J. Appl. Phys. 4, p.91, 1953.
17. Smith, C.G., Brit. J. Appl. Phys. 4, p.252, 1953.
18. Hernqvist, K.G., Phys. Rev. 109, p.636, 1958.
19. Killian, T.J., Phys. Rev. 31, p.1122, 1928.
20. Nottingham, W.B., J. Frankl. Inst. 208, p.299, 1929.
21. Lamar, E.S., and Compton, K. T., Phys. Rev. 37, p.1069, 1931.
22. Compton, K.T., Phys. Rev. 37, p.1077, 1931.
23. Granovskii, V.L. and Bykhovskaya, L.N., Dokl. Akad. Nauk SSSR 49, No.5, 1945; Zhurn. Eksp. i Teor. Fiz. 16, p.823, 1946.
24. Beams, J.W., Phys. Rev. 44, p.803, 1933.
25. Lübcke, E., Ztschr. techn. Phys. 12, p.598, 1929.

26. Tanberg, R., and Berkey, W.E., Phys. Rev. 38, p. 296, 1931.
27. Ishida, Hiyama, Inst. Phys. Chem. Res. (Tokyo) 36, p.390, 1939.
28. Finkelburg, W., Phys. Rev. 45, p.341, 1934.
29. Issendorf, J., Phys. Ztschr. 29, p.857, 1928.
30. Compton, K.T., and Voorhis, C.C., Proc. Nat. Acad. Science 13, p.336, 1927.
31. Oliphant, M.L.E., and Moon, P.B., Proc. Roy. Soc. A, 127, p.388, 1930.
32. Shekhter, Sh.Sh., Zhurn. Eksp. i. Teor. Fiz. 7, p.751, 1937.
33. Granovskii, V.L., Electric Current in a Gas [in Russian], Gostekhizdat, 1952.
34. Chaudhri, R.M., Nature 177, p.1226,1956.
35. Kobel, E., Phys. Rev. 36, p.1636, 1930.
36. Tanberg, R., Phys. Rev. 35, p. 1080, 1930.
37. Tonks, L., Phys. Rev. 54, p.634, 1938.
38. Issendorf, J., Schenkel, M., and Seeliger, R., Wiss. Veröffentl. Siemens-Konzern 9, p.73, 1930.
39. Askinazi, A.E., Gurevich, M.A., and Sena, L.A., Elektrichestvo No. 9, p.37, 1947.
40. Golubchin, G.N., and Sena, L.A., Elektrichestvo No.6, p.60, 1956.
41. Udris, Ya.Ya., Trudy VÉI, No.63, Investigation in the Field of Electric Discharges in Gases [in Russian], Gosénergoizdat, 1958.
42. Duffield, Burnham, Davis, Phil. Trans. Roy. Soc. A, 220, p.109, 1920; Sellerio, Phil. Mag. 44, p.765, 1922.
43. Tonks, L., Phys. Rev. 46, p.279, 1934.
44. Slepian, J., and Mason, R.C., Phys. Rev. 37, p.779, 1931.
45. Haynes, I.R., Phys. Rev. 73, p.891, 1948.
46. Schuster, A., and Hemsalech, G., Proc. Roy. Soc. 64, p.331, 1899.
47. Mohler, J.F., Astrophys.J. 15, p.125, 1902.
48. Lawrence, E.O., and Dunnington, F.G., Phys. Rev. 35, p.396, 1930.
49. Robertson, R.M., Phys. Rev. 53, p.578, 1938.
50. Easton, E.C., Lucas, F.B., and Creedy, F., El. Eng. 53, p. 1454, 1934.

51. Somerville, J. M., and Blevin, W. R., Phys. Rev. 76, p. 982, 1949.
52. Žižka, E., Czechosl. J. Phys. 7, p. 306, 1957.
53. Hermoch, V., Czechosl. J. Phys. 9, p. 221, 1959.
54. Lander, J. J., and Germer, L. H., J. Appl. Phys. 19, p. 910, 1948.
55. Germer, L. H., and Haworth, F. E., Phys. Rev. 73, p. 1121, 1948.
56. Germer, L. H., and Haworth, F. E., J. Appl. Phys. 20, p. 1085, 1949.
57. Germer, L. H., J. Appl. Phys. 22, p. 955, 1951.
58. Germer, L. H., J. Appl. Phys. 22, p. 1133, 1951.
59. Germer, L. H., and Smith, J. L., J. Appl. Phys 23, p. 553, 1952.
60. Lazarenko, B. R., and Lazarenko, N. I., Electrical Erosion of Metals [in Russian], Gosénergoizdat (TsBTI NKÉP, No. 1, 1944; No. 2, 1946).
61. Gusev, V. N., Anode-Mechanical Treatment of Metals [in Russian], Moscow, 1952.
62. Zaimovskii, A. S., and Usov, V. V., Metals and Alloys in Electrical Engineering [in Russian], Gosénergoizdat, 1949, p. 485.
63. Somerville, J. M., Blevin, W. R., and Flether, N. H., Proc. Phys. Soc. B, 65, p. 963, 1952.
64. Hermoch, V., Czechosl. J. Phys. 4, p. 28, 1954.
65. Hermoch, V., Zitka, B., and Sobra, K., Czechosl. J. Phys. 4, p. 486, 1954.
66. Zingerman, A. S., Zhurn. Tekhn. Fiz. 26, p. 5, p. 2008, 1956.
67. Libin, I. Sh., Radiotekhnika i Elektronika No. 4, p. 1026, 1959.
68. Mandel'shtam, S. L., and Raiskii, S. M., Izvestiya Akad. Nauk SSSR, Physics Series, 13, p. 549, 1949.
69. Schmidt, E., Ann. Phys. 4, p. 246, 1949.
70. Steer, G., Ztschr. ang. Phys. 5, p. 116, 1953.
71. Kloss, A., Elektrotechn. Obzor 48, p. 634, 1959.
72. Stark, J., Phys. Ztschr. 4, p. 440, 1903.
73. Weintraub, E., Phil. Mag. 7, p. 95, 1904.
74. Minorsky, N., J. de Phys. et Rad. 9, p. 127, 1928.
75. Tanberg, R., Nature 124, p. 371, 1929.

76. Smith, C.G., Phys. Rev. 62, p.48, 1942.
77. Smith, C.G., Phys. Rev. 64, p.40, 1943.
78. Smith, C.G., Trans. AIEE 63, p.883, 1944.
79. Gallagher, C.J., and Cobine, J.D., Phys. Rev. 71, p.481, 1947.
80. Longini, R.L., Phys. Rev. 71, p.642, 1947.
81. Himler, H., and Cohn, G.J., El. Eng. 67, p.1148, 1948.
82. Smith, C.G., Phys. Rev. 73, p.543, 1948.
83. Gallagher, C.J., and Cobine, J.D., El. Eng. 68, p.469, 1949.
84. Gallagher, C.J., J. Appl. Phys. 21, p.768, 1950.
85. Rothstein, J., Phys. Rev. 78, p.331, 1950.
86. Jamamura, S., J. Appl. Phys. 21, p.193, 1950.
87. Smith, C.G., Phys. Rev. 82, p.570, 1951.
88. Smith, C.G., Phys. Rev. 83, p.194; 84, p.1075, 1951.
89. St. John, R.M., and Winans, J.G., Phys. Rev. 94, pp. 5, 1094, 1954.
90. Miller, C.G., and Sanders, N.L., Phys. Rev. 93, p.654, 1954.
91. Robson, A.E., and Engel, A., Phys. Rev. 93, p.1121, 1954.
92. Ware, A.A., Proc. Phys. Soc. A, 67, p.869, 1954.
93. St. John, R.M. and Winans, J.G., Phys. Rev. 98, p.1664, 1955.
94. Robson, A.E., and Engel, A., Phys. Rev. 104, p.15, 1956.
95. Smith, C.G., J. Appl. Phys. 28, p. 1328, 1957.
96. Eidinger, A., and Reider, W., Arch. Electrotechn. 43, p.94, 1957.
97. Ecker, G., Müller, K.G., Ztschr. Phys. 151, p.577, 1958.
98. Zei, D., and Winans, J.G., J. Appl. Phys. 30, p.1813, 1959.
99. Bertele, H., Proc. Inst. El. Eng. 101, 11, p. 493, 1954.
100. Tucker, R., Direct Current 2, p. 40, 1954.
101. Sena, L.A., Zhurn. Eksp. i Teor. Fiz. 15, p.427, 1945.
102. Boyle, W.S., Kisliuk, P., and Germer, L.H., J. Appl. Phys. 26, p.720, 1955.
103. Fowler, R.H., and Nordheim, L.H., Proc. Roy. Soc. A, 119, p.173, 1928.
104. Paschen, F., Ann. Phys. u. Chem. 37, p. 69, 1889.
105. Howell, A.N., Trans. Am. Inst. El. Eng. 58, p.193, 1939.

106. Boulloud, A., Ann. Phys. 8, p.909, 1953.
107. Young, D.R., J. Appl. Phys. 21, p.222, 1950.
108. Boyle, W.S., and Kisliuk, P., Phys. Rev. 97, p.255, 1955.
109. Mackey, G.M.J., and Ferguson, C.V., J. Frankl. Inst. 181, p.209, 1916.
110. Seeliger, R., Phys. Ztschr. 27, p.22, 1926.
111. Wehrli, M., Ztschr. Phys. 44, p.301, 1927.
112. Schrum, G.M., and Wiest, H.I., Trans. Am. Inst. El. Eng. 50, p.650, 1931.
113. Druyvesteyn, M.J., Ztschr. Phys. 73, p.727, 1932.
114. Becken, O., and Seeliger, R., Ann. Phys. 24, p.608, 1935.
115. Engel, A., and Steenbeck, M., Physics and Technology of Electrical Discharge in Gases [Russian translation], ONTI NKTP, Vol.II, 1936.
116. Mackeown, S.S., El. Eng. 51, p.386, 1932.
117. Boyle, W.S., and Haworth, F.E., Phys. Rev. 101, p.935, 1956.
118. Bauer, A., Ztschr. Phys. 138, p.35, 1954; Bauer, A., Ann. Phys. 18, p.387, 1956.
119. Murphy, E.L., and Good, R.H., Phys. Rev. 102, p.1464, 1956.
120. Thouret, W., Weisel, W., and Günther, P., Ztschr. Phys. 130, p.621, 1951.
121. Maxfield, F.A., Hegbar, H.R., and Eaton, I.R., Trans. Am. Inst. El. Eng. 59, p.816, 1059, 1940.
122. Boldyr', Yu. D., and Greben', K. L., Zhurn. Tekhn. Fiz. 10, p. 1140, 1940.
123. Sena, L.A., Panov, I.P., and Fridlyand, R.M., Izvestiya NIIPT No.1, p.39, 1957.
124. Sena, L.A., and Fridlyand, R.M., Zhurn. Tekhn. Fiz. 29, p.3, 1959.
125. Slepian, J., and Ludwig, L., Trans. AIEE 52, p. 693, 1933.
126. Gage, I., Gen. El. Rev. 38, p.464, 1935; Mierdel, G., Wiss. Veröffentl. Siemens-Werken 15, p.35, 1936.
127. Shlyaposhnikov, B.M., Zhurn. Tekhn. Fiz. 16, p.1031, 1946.
128. Warmoltz, N., Philipps Research Reports 9, p. 104, 1947; 2, p.426, 1947; 6, p.388, 1951.
129. Burstyn, W., Ztschr. Hochfrequenztechn. 23, p.8, 1924.
130. Mierdel, G., Ztschr. techn. Phys. 17, p.452, 1936.
131. Engelbrecht, O., Arch. Electrotechn. 36, p.515, 1942.

132. Mantrov, V.M., Trudy VÉI No.63, Gosenérgoizdat, 1958.
133. Hernqvist, K.G., J. Appl. Phys. 27, p. 1226, 1956.
134. Sena, L.A., Izvestiya NIIPT Coll. 2, p.32, 1957.
135. Suits, C.G., and Hocker, J.P., Phys. Rev. 53, p.670, 1938.
136. Sena, L.A., and Fridlyand, R.M., Izvestiya NIIPT Coll. 2, p.20, 1957.
137. Lufcy, C.W., and Copeland, P.L., J. Appl. Phys. 16, p. 740, 1944.
138. Copeland, P.L., and Sparing, W.H., J. Appl. Phys. 16, p. 302, 1945.
139. Kaufman, W., Ann. Phys. 2, p.158, 1900.
140. Mitkevich, V. F., Izvestiya S. Peterburskogo Politekhnicheskogo Instituta 4, No. 5, 1905.
141. Stark, J., Ann. Phys. 12, p. 692, 1903.
142. Compton, K.T., Phys. Rev. 21, p. 266, 1923.
143. Ramberg, W., Ann. Phys. 12, p.319, 1932.
144. Stolt, H., Ztschr Phys. 26, p.95, 1924; 31, p.240, 1925.
145. Slepian, J., Phys. Rev. 27, p.407, 1926.
146. Langmuir, I., Gen. El. Rev. 26, p.731, 1923.
147. Mackeown, S.S., Phys. Rev. 34, p.611, 1929.
148. Slepian, J., J. Frankl. Inst. 201, p.79, 1926.
149. Weisel, W., Rompe, R., and Schön, M., Ztschr. Phys. 115, p.179, 1940.
150. Druyvesteyn, M.J., Nature 137, p.580, 1930.
151. Birch, F., Phys. Rev. 40, p.1054; 41, p.641, 1932.
152. Robson, A.E., and Engel, A., Nature 175, p.646, 1955; Robson, A.E., Radiotekhnika i Elektronika 4, p. 1295, 1959.
153. Wasserrab, T., Ztschr. Phys. 130, p.311, 1951.
154. Lee, T.H., J. Appl. Phys. 28, p. 920, 1957.
155. Dyke, W.P., and Trolan, J.K., Phys. Rev. 89, p.799, 1953.
156. Engel, A., and Robson, A.E., J. Appl. Phys. 29, p.734, 1958.
157. Lee, T.H., J. Appl. Phys. 29, p.734, 1958.
158. Cassie, A.M., Nature 181, p.476, 1958.
159. Klyarfel'd, B.N., and Sobolev, V., Zhurn. Tekhn. Fiz. 17, p. 319, 1947.
160. Reece, M.P., Nature 181, p.475, 1958.
161. Ives, H.E., J. Frankl. Inst. 198, p.437, 1924.

162. Rompe, R., and Weisel, W., Ann. Phys. 1, p.350, 1947.
163. Kesaev, I.G., Dokl. Akad. Nauk SSSR 122, p.371, 1958.
164. Kesaev, I.G., Dokl. Akad. Nauk SSSR 124, p.563, 1959.
165. Kesaev, I.G., Zhurn. Tekhn. Fiz. 29, p.1462, 1959.
166. Kesaev, I.G., Zhurn. Tekhn. Fiz. 30, 6, p.674, 1960.
167. Kesaev, I.G., and Levshenkova, L.A., Zhurn. Tekhn. Fiz. 30, p.815, 1960.
168. Collection of papers edited by A.R. Shulman, Characteristic Electron Energy Losses in Solids [Russian translation], Izd. Inostr. Lit., Moscow, 1959.
169. Benjamin, M., and Jenkins, R.O., Proc. Roy. Soc. A, 176, p.262, 1940.
170. Andreev, I.S., Zhurn. Tekhn. Fiz. 22, p.1428, 1952.
171. Elinson, M.I., and Vasil'ev, G.F., Zhurn. Tekhn. Fiz. 26, p.1669, 1956.
172. Elinson, M.I., Gor'kov, V.A., and Vasil'ev, G.F., Radiotekhnika i Elektronika 3, p.438, 1958.
173. Smirnov, B.G., and Shuppe, G.N., Zhurn. Tekhn. Fiz. 22, p.973, 1952.
174. Dyke, W.P., Trolan, J.K., Dolan, W.W., and Grundhauser, E.J., J. Appl. Phys. 25, p.106, 1954.
175. Smith, G.F., Phys. Rev. 94, p.295, 1954.
176. Müller, E.W., J. Appl. Phys. 26, p.732, 1955.
177. Elinson, M.I., and Vasil'ev, G.F., Cold Emission of Electrons [in Russian], Gos. izd. fiz.-mat. lit., 1958.
178. Drechsler, M., and Müller, E.W., Ztschr. Phys. 134, p.208, 1953.
179. Townsend, J.S., Electrons in Gases, Hutchinson, London, 1947.
180. Bostick, W.H., and Levin, M.A., Phys. Rev. 97, p.13, 1955.
181. Simon, A., Phys. Rev. 98, p.317, 1955.
182. Kruskal, M., and Schwarzschild, M., Proc. Roy. Soc. A, 233, p.348, 1954.
183. Shafranov, V.D., Atomnaya Energiya 5, p.38, 1956.
184. Kesaev, I.G., Dokl. Akad. Nauk SSSR 112, p.619, 1957.
185. Kesaev, I.G., Dokl. Akad. Nauk SSSR 113, p.71, 1957.
186. Kesaev, I.G., Radiotekhnika i Elektronika 4, p.1289, 1959.
187. Kesaev, I.G., Radiotekhnika i Elektronika 5, p.986, 1960.
188. Kesaev, I.G., and Pashkova, V.V., Zhurn. Tekhn. Fiz. 29, p.287, 1959.

MIX
Papier aus verantwortungsvollen Quellen
Paper from responsible sources
FSC® C105338

If you have any concerns about our products,
you can contact us on
ProductSafety@springernature.com

In case Publisher is established outside the EU,
the EU authorized representative is:
Springer Nature Customer Service Center GmbH
Europaplatz 3, 69115 Heidelberg, Germany

Printed by Libri Plureos GmbH
in Hamburg, Germany